DATA MINING FOR
BUSINESS ANALYTICS

DATA MINING FOR BUSINESS ANALYTICS

Concepts, Techniques, and
Applications with JMP Pro®

GALIT SHMUELI
PETER C. BRUCE
MIA L. STEPHENS
NITIN R. PATEL

WILEY

Published by John Wiley & Sons, Inc., Hoboken, New Jersey.
Published simultaneously in Canada.

For general information on our other products and services or for technical support, please contact our Customer Care Department within the United States at (800) 762-2974, outside the United States at (317) 572-3993 or fax (317) 572-4002.

Wiley also publishes its books in a variety of electronic formats. Some content that appears in print may not be available in electronic formats. For more information about Wiley products, visit our web site at www.wiley.com.

Library of Congress Cataloging-in-Publication Data:

Names: Shmueli, Galit, 1971- author. | Bruce, Peter C., 1953- author. |
    Stephens, Mia L., author. | Patel, Nitin R. (Nitin Ratilal), author.
Title: Data mining for business analytics : concepts, techniques, and
    applications with JMP Pro / Galit Shmueli, Peter C. Bruce, Mia L. Stephens,
    Nitin R. Patel.
Description: Hoboken, New Jersey : John Wiley & Sons, 2017. | Includes index.
Identifiers: LCCN 2015048305| ISBN 9781118877432 (cloth) | ISBN 9781118877524
    (epub)
Subjects: LCSH: Business mathematics–Computer programs. | Business–Data
    processing. | JMP (Computer file) | Data mining.
Classification: LCC HF5691 .S43245 2016 | DDC 006.3/12–dc23 LC record available at http://lccn.loc.
    gov/2015048305

10 9 8 7 6 5 4 3 2 1

To our families

Boaz and Noa
Liz, Lisa, and Allison
Michael, Madi, Olivia, and in memory of E.C. Jr.
Tehmi, Arjun, and in memory of Aneesh

CONTENTS

FOREWORD   xvii

PREFACE   xix

ACKNOWLEDGMENTS   xxi

PART I PRELIMINARIES

1   Introduction   3

    1.1   What Is Business Analytics? 3
        Who Uses Predictive Analytics? 4
    1.2   What Is Data Mining? 5
    1.3   Data Mining and Related Terms 5
    1.4   Big Data 6
    1.5   Data Science 7
    1.6   Why Are There So Many Different Methods? 7
    1.7   Terminology and Notation 8
    1.8   Roadmap to This Book 10
        Order of Topics 11
        Using JMP Pro, Statistical Discovery Software from SAS 11

2   Overview of the Data Mining Process   14

    2.1   Introduction 14
    2.2   Core Ideas in Data Mining 15
        Classification 15
        Prediction 15
        Association Rules and Recommendation Systems 15

Predictive Analytics 16
Data Reduction and Dimension Reduction 16
Data Exploration and Visualization 16
Supervised and Unsupervised Learning 16
2.3   The Steps in Data Mining 17
2.4   Preliminary Steps 19
Organization of Datasets 19
Sampling from a Database 19
Oversampling Rare Events in Classification Tasks 19
Preprocessing and Cleaning the Data 20
Changing Modeling Types in JMP 20
Standardizing Data in JMP 25
2.5   Predictive Power and Overfitting 25
Creation and Use of Data Partitions 25
Partitioning Data for Crossvalidation in JMP Pro 27
Overfitting 27
2.6   Building a Predictive Model with JMP Pro 29
Predicting Home Values in a Boston Neighborhood 29
Modeling Process 30
Setting the Random Seed in JMP 34
2.7   Using JMP Pro for Data Mining 38
2.8   Automating Data Mining Solutions 40
Data Mining Software Tools: the State of the Market by Herb Edelstein 41
Problems 44

PART II DATA EXPLORATION AND DIMENSION REDUCTION

3   Data Visualization   51

3.1   Uses of Data Visualization 51
3.2   Data Examples 52
Example 1: Boston Housing Data 53
Example 2: Ridership on Amtrak Trains 53
3.3   Basic Charts: Bar Charts, Line Graphs, and Scatterplots 54
Using The JMP Graph Builder 54
Distribution Plots: Boxplots and Histograms 56
Tools for Data Visualization in JMP 59
Heatmaps (Color Maps and Cell Plots): Visualizing Correlations and Missing Values 59
3.4   Multidimensional Visualization 61
Adding Variables: Color, Size, Shape, Multiple Panels, and Animation 62
Manipulations: Rescaling, Aggregation and Hierarchies, Zooming, Filtering 65
Reference: Trend Lines and Labels 68
Adding Trendlines in the Graph Builder 69
Scaling Up: Large Datasets 70

Multivariate Plot: Parallel Coordinates Plot 71
Interactive Visualization 72
3.5   Specialized Visualizations 73
Visualizing Networked Data 74
Visualizing Hierarchical Data: More on Treemaps 75
Visualizing Geographical Data: Maps 76
3.6   Summary of Major Visualizations and Operations, According to Data Mining Goal 77
Prediction 77
Classification 78
Time Series Forecasting 78
Unsupervised Learning 79
Problems 79

4   Dimension Reduction   **81**

4.1   Introduction 81
4.2   Curse of Dimensionality 82
4.3   Practical Considerations 82
Example 1: House Prices in Boston 82
4.4   Data Summaries 83
Summary Statistics 83
Tabulating Data (Pivot Tables) 85
4.5   Correlation Analysis 87
4.6   Reducing the Number of Categories in Categorical Variables 87
4.7   Converting a Categorical Variable to a Continuous Variable 90
4.8   Principal Components Analysis 90
Example 2: Breakfast Cereals 91
Principal Components 95
Normalizing the Data 97
Using Principal Components for Classification and Prediction 100
4.9   Dimension Reduction Using Regression Models 100
4.10   Dimension Reduction Using Classification and Regression Trees 100
Problems 101

PART III PERFORMANCE EVALUATION

5   Evaluating Predictive Performance   **105**

5.1   Introduction 105
5.2   Evaluating Predictive Performance 106
Benchmark: The Average 106
Prediction Accuracy Measures 107
Comparing Training and Validation Performance 108
5.3   Judging Classifier Performance 109
Benchmark: The Naive Rule 109
Class Separation 109
The Classification Matrix 109

Using the Validation Data 111
Accuracy Measures 111
Propensities and Cutoff for Classification 112
Cutoff Values for Triage 112
Changing the Cutoff Values for a Confussion Matrix in JMP 114
Performance in Unequal Importance of Classes 115
False-Positive and False-Negative Rates 116
Asymmetric Misclassification Costs 116
Asymmetric Misclassification Costs in JMP 119
Generalization to More Than Two Classes 120
5.4   Judging Ranking Performance 120
Lift Curves 120
Beyond Two Classes 122
Lift Curves Incorporating Costs and Benefits 122
5.5   Oversampling 123
Oversampling the Training Set 126
Stratified Sampling and Oversampling in JMP 126
Evaluating Model Performance Using a Nonoversampled
Validation Set 126
Evaluating Model Performance If Only Oversampled Validation
Set Exists 127
Applying Sampling Weights in JMP 128
Problems 129

PART IV PREDICTION AND CLASSIFICATION METHODS

6   Multiple Linear Regression   **133**

6.1   Introduction 133
6.2   Explanatory versus Predictive Modeling 134
6.3   Estimating the Regression Equation and Prediction 135
Example: Predicting the Price of Used Toyota Corolla
Automobiles 136
Coding of Categorical Variables in Regression 138
Additional Options for Regression Models in JMP 140
6.4   Variable Selection in Linear Regression 141
Reducing the Number of Predictors 141
How to Reduce the Number of Predictors 142
Manual Variable Selection 142
Automated Variable Selection 142
Coding of Categorical Variables in Stepwise Regression 143
Working with the All Possible Models Output 145
When Using a Stopping Algorithm in JMP 147
Other Regression Procedures in JMP Pro—Generalized
Regression 149
Problems 150

7   *k*-Nearest Neighbors (*k*-NN)   **155**

7.1   The *k*-NN Classifier (Categorical Outcome) 155
         Determining Neighbors 155
         Classification Rule 156
         Example: Riding Mowers 156
         Choosing *k* 157
         k Nearest Neighbors in JMP Pro 158
            The Cutoff Value for Classification 159
         k-NN Predictions and Prediction Formulas in JMP Pro 161
            k-NN with More Than Two Classes 161
7.2   *k*-NN for a Numerical Response 161
         Pandora 161
7.3   Advantages and Shortcomings of *k*-NN Algorithms 163
       Problems 164

8   The Naive Bayes Classifier   **167**

8.1   Introduction 167
         Naive Bayes Method 167
         Cutoff Probability Method 168
         Conditional Probability 168
         Example 1: Predicting Fraudulent Financial Reporting 168
8.2   Applying the Full (Exact) Bayesian Classifier 169
         Using the "Assign to the Most Probable Class" Method 169
         Using the Cutoff Probability Method 169
         Practical Difficulty with the Complete (Exact) Bayes Procedure 170
         Solution: Naive Bayes 170
         Example 2: Predicting Fraudulent Financial Reports, Two
         Predictors 172
       Using the JMP Naive Bayes Add-in 174
         Example 3: Predicting Delayed Flights 174
8.3   Advantages and Shortcomings of the Naive Bayes Classifier 179
         Spam Filtering 179
       Problems 180

9   Classification and Regression Trees   **183**

9.1   Introduction 183
9.2   Classification Trees 184
         Recursive Partitioning 184
         Example 1: Riding Mowers 185
         Categorical Predictors 186
9.3   Growing a Tree 187
         Growing a Tree Example 187
         Classifying a New Observation 188
       Fitting Classification Trees in JMP Pro 191
         Growing a Tree with CART 192

9.4   Evaluating the Performance of a Classification Tree 192
            Example 2: Acceptance of Personal Loan 192
9.5   Avoiding Overfitting 193
            Stopping Tree Growth: CHAID 194
            Growing a Full Tree and Pruning It Back 194
            How JMP Limits Tree Size 196
9.6   Classification Rules from Trees 196
9.7   Classification Trees for More Than Two Classes 198
9.8   Regression Trees 199
            Prediction 199
            Evaluating Performance 200
9.9   Advantages and Weaknesses of a Tree 200
9.10   Improving Prediction: Multiple Trees 204
            Fitting Ensemble Tree Models in JMP Pro 206
9.11   CART and Measures of Impurity 207
        Problems 207

10 Logistic Regression   211

10.1   Introduction 211
            Logistic Regression and Consumer Choice Theory 212
10.2   The Logistic Regression Model 213
            Example: Acceptance of Personal Loan (Universal Bank) 214
            Indicator (Dummy) Variables in JMP 216
            Model with a Single Predictor 216
            Fitting One Predictor Logistic Models in JMP 218
            Estimating the Logistic Model from Data: Multiple Predictors 218
            Fitting Logistic Models in JMP with More Than One Predictor 221
10.3   Evaluating Classification Performance 221
            Variable Selection 222
10.4   Example of Complete Analysis: Predicting Delayed Flights 223
            Data Preprocessing 225
            Model Fitting, Estimation and Interpretation—A Simple Model 226
            Model Fitting, Estimation and Interpretation—The Full Model 227
            Model Performance 229
            Variable Selection 230
            Regrouping and Recoding Variables in JMP 232
10.5   Appendixes: Logistic Regression for Profiling 234
            Appendix A: Why Linear Regression Is Problematic for a
            Categorical Response 234
            Appendix B: Evaluating Explanatory Power 236
            Appendix C: Logistic Regression for More Than Two Classes 238
            Nominal Classes 238
        Problems 241

11 Neural Nets   245

11.1   Introduction 245
11.2   Concept and Structure of a Neural Network 246

11.3   Fitting a Network to Data 246
                Example 1: Tiny Dataset 246
                Computing Output of Nodes 248
                Preprocessing the Data 251
            Activation Functions and Data Processing Features in JMP Pro 251
                Training the Model 251
                Fitting a Neural Network in JMP Pro 254
                Using the Output for Prediction and Classification 256
                Example 2: Classifying Accident Severity 258
                Avoiding overfitting 259
11.4   User Input in JMP Pro 260
                Unsupervised Feature Extraction and Deep Learning 263
11.5   Exploring the Relationship between Predictors and Response 264
                Understanding Neural Models in JMP Pro 264
11.6   Advantages and Weaknesses of Neural Networks 264
            Problems 265

12 Discriminant Analysis   **268**

12.1   Introduction 268
                Example 1: Riding Mowers 269
                Example 2: Personal Loan Acceptance (Universal Bank) 269
12.2   Distance of an Observation from a Class 270
12.3   From Distances to Propensities and Classifications 272
                Linear Discriminant Analysis in JMP 275
12.4   Classification Performance of Discriminant Analysis 275
12.5   Prior Probabilities 277
12.6   Classifying More Than Two Classes 278
                Example 3: Medical Dispatch to Accident Scenes 278
                Using Categorical Predictors in Discriminant Analysis in JMP 279
12.7   Advantages and Weaknesses 280
            Problems 282

13 Combining Methods: Ensembles and Uplift Modeling   **285**

13.1   Ensembles 285
                Why Ensembles Can Improve Predictive Power 286
            The Wisdom of Crowds 287
                Simple Averaging 287
                Bagging 288
                Boosting 288
                Creating Ensemble Models in JMP Pro 289
                Advantages and Weaknesses of Ensembles 289
13.2   Uplift (Persuasion) Modeling 290
                A-B Testing 290
                Uplift 290
                Gathering the Data 291
                A Simple Model 292
                Modeling Individual Uplift 293

Using the Results of an Uplift Model 294
Creating Uplift Models in JMP Pro 294
Using the Uplift Platform in JMP Pro 295
13.3   Summary 295
Problems 297

PART V MINING RELATIONSHIPS AMONG RECORDS

14 Cluster Analysis   301

14.1   Introduction 301
Example: Public Utilities 302
14.2   Measuring Distance between Two Observations 305
Euclidean Distance 305
Normalizing Numerical Measurements 305
Other Distance Measures for Numerical Data 306
Distance Measures for Categorical Data 308
Distance Measures for Mixed Data 308
14.3   Measuring Distance between Two Clusters 309
Minimum Distance 309
Maximum Distance 309
Average Distance 309
Centroid Distance 309
14.4   Hierarchical (Agglomerative) Clustering 311
Hierarchical Clustering in JMP and JMP Pro 311
Hierarchical Agglomerative Clustering Algorithm 312
Single Linkage 312
Complete Linkage 313
Average Linkage 313
Centroid Linkage 313
Ward's Method 314
Dendrograms: Displaying Clustering Process and Results 314
Validating Clusters 316
Two-Way Clustering 318
Limitations of Hierarchical Clustering 319
14.5   Nonhierarchical Clustering: The k-Means Algorithm 320
k-Means Clustering Algorithm 321
Initial Partition into K Clusters 322
K-Means Clustering in JMP 322
Problems 329

PART VI FORECASTING TIME SERIES

15 Handling Time Series   335

15.1   Introduction 335
15.2   Descriptive versus Predictive Modeling 336
15.3   Popular Forecasting Methods in Business 337
Combining Methods 337

15.4   Time Series Components 337
           Example: Ridership on Amtrak Trains 337
15.5   Data Partitioning and Performance Evaluation 341
           Benchmark Performance: Naive Forecasts 342
           Generating Future Forecasts 342
       Partitioning Time Series Data in JMP and Validating
       Time Series Models 342
       Problems 343

16 Regression-Based Forecasting   **346**

16.1   A Model with Trend 346
           Linear Trend 346
       Fitting a Model with Linear Trend in JMP 348
       Creating Actual versus Predicted Plots and Residual Plots in JMP 350
           Exponential Trend 350
       Computing Forecast Errors for Exponential Trend Models 352
           Polynomial Trend 352
       Fitting a Polynomial Trend in JMP 353
16.2   A Model with Seasonality 353
16.3   A Model with Trend and Seasonality 356
16.4   Autocorrelation and ARIMA Models 356
           Computing Autocorrelation 356
           Improving Forecasts by Integrating Autocorrelation Information 360
       Fitting AR (Autoregression) Models in the JMP Time Series
       Platform 361
           Fitting AR Models to Residuals 361
           Evaluating Predictability 363
       Summary: Fitting Regression-Based Time Series Models in JMP 365
       Problems 366

17 Smoothing Methods   **377**

17.1   Introduction 377
17.2   Moving Average 378
           Centered Moving Average for Visualization 378
           Trailing Moving Average for Forecasting 379
       Computing a Trailing Moving Average Forecast in JMP 380
           Choosing Window Width (w) 382
17.3   Simple Exponential Smoothing 382
           Choosing Smoothing Parameter α 383
       Fitting Simple Exponential Smoothing Models in JMP 384
       Creating Plots for Actual versus Forecasted Series and Residuals Series
       Using the Graph Builder 386
           Relation between Moving Average and Simple Exponential
           Smoothing 386
17.4   Advanced Exponential Smoothing 387
           Series with a Trend 387
           Series with a Trend and Seasonality 388
       Problems 390

PART VII CASES

18 Cases 402

    18.1   Charles Book Club 401
           The Book Industry 401
           Database Marketing at Charles 402
           Data Mining Techniques 403
           Assignment 405
    18.2   German Credit 409
           Background 409
           Data 409
           Assignment 409
    18.3   Tayko Software Cataloger 410
           Background 410
           The Mailing Experiment 413
           Data 413
           Assignment 413
    18.4   Political Persuasion 415
           Background 415
           Predictive Analytics Arrives in US Politics 415
           Political Targeting 416
           Uplift 416
           Data 417
           Assignment 417
    18.5   Taxi Cancellations 419
           Business Situation 419
           Assignment 419
    18.6   Segmenting Consumers of Bath Soap 420
           Business Situation 420
           Key Problems 421
           Data 421
           Measuring Brand Loyalty 421
           Assignment 421
    18.7   Direct-Mail Fundraising 423
           Background 423
           Data 424
           Assignment 425
    18.8   Predicting Bankruptcy 425
           Predicting Corporate Bankruptcy 426
           Assignment 428
    18.9   Time Series Case: Forecasting Public Transportation Demand 428
           Background 428
           Problem Description 428
           Available Data 428
           Assignment Goal 429
           Assignment 429
           Tips and Suggested Steps 429

References 431

Data Files Used in the Book 433

Index 435

FOREWORD

No matter what your chosen profession or place of work, your future will almost certainly be saturated with data. The modern world is defined by the bits of data pulsing from billions of keyboards and trillions of card swipes—emanating from every manner of electronic device and system—transmitted instantaneously around the globe. The sheer amount of data is measured in volumes difficult to comprehend. But it's not about how much data you have; it's what you do with it, and how quickly, that counts most. Grappling with this messy world of data and putting it to good use will be key to productive and well-functioning organizations and successful managerial careers, not just in the obvious places circling Silicon Valley such as Google and Facebook but in insurance companies, banks, auto manufacturers, airlines, hospitals, and indeed nearly everywhere.

That's where *Data Mining for Business Analytics: Concepts, Techniques and Applications with JMP Pro®* can help. Professor Shmueli and her coauthors provide a very useful guide for students of business to learn the important concepts and methods for navigating complex datasets. Born out of the authors' years of experience teaching the subject, the book has evolved from earlier editions to keep pace with the changing landscape of business analytics in graduate and undergraduate education. Most important, new with this edition is the integration of JMP Pro®, a statistical tool from SAS Institute, which is provided as the vehicle for working with data in problem sets. Learning analytics is ultimately about doing things to and with data to generate insights. Mastering one's dexterity with powerful statistical tools is a necessary and critical step in the learning process.

If you've set your sights on leading in a digital world, this book is a great place to start preparing yourself for the future.

MICHAEL RAPPA
Institute for Advanced Analytics
North Carolina State University

PREFACE

The textbook *Data Mining for Business Intelligence* first appeared in early 2007. Since then, it has been used by numerous practitioners and in many courses, ranging from dedicated data mining classes to more general business analytics courses (including our own experience teaching this material both online and in person for more than 10 years). Following feedback from instructors teaching MBA, undergraduate, and executive courses, and from students, the second edition saw revisions to some of the existing chapters and included two new topics: data visualization and time series forecasting.

This book is the first edition to fully integrate JMP Pro®[1] rather than the Microsoft Office Excel add-in, XLMiner. JMP Pro® is a desktop statistical package from SAS Institute that runs natively on Mac and Windows machines. All examples, special topics boxes, instructions, and exercises presented in this book are based on JMP 12 Pro, the professional version of JMP, which has a rich array of built- in tools for interactive data visualization, analysis, and modeling.[2]

There are other important changes in this edition. The first noticeable change is the title: other than the addition of JMP Pro®, we now use *Business Analytics* in place of *Business Intelligence*. This update reflects the change in terminology since the second edition: BI today refers mainly to reporting and data visualization (what is happening now), while BA has taken over the advanced analytics, which include predictive analytics and data mining. In this new edition we therefore also updated these terms in the book, using them as is currently common.

We added a new chapter, *Combining Methods: Ensembles and Uplift Modeling* (Chapter 13). This chapter, which is the last in Part IV on Prediction and Classification Methods, introduces two important approaches. The first—ensembles—is the combination of multiple models for improving predictive power. Ensembles have routinely proved their

---

[1] JMP Pro®, Version 12. SAS Institute Inc., Cary, NC 27513. See Chapter 1 for information on how to get JMP Pro®.

[2] Relevant new features in JMP Pro 13 are noted in the chapters.

usefulness in practical applications and in data mining contests. The second topic—uplift modeling—introduces an improved approach for measuring the impact of an intervention or treatment. Similar to other chapters, this new chapter includes real-world examples and end-of-chapter problems.

Other changes include the addition of two new cases based on real data (one on political persuasion and uplift modeling, and another on taxi cancellations), and the removal of one chapter, Association Rules (association rules is a feature not available in JMP 12 Pro, but will be a new feature in JMP 13 Pro).

Since the second edition's appearance, the landscape of courses using the textbook has greatly expanded: whereas initially the book was used mainly in semester-long elective MBA-level courses, it is now used in a variety of courses in Business Analytics degree and certificate programs, ranging from undergraduate programs, to post-graduate and executive education programs. Courses in such programs also vary in their duration and coverage. In many cases, our book is used across multiple courses. The book is designed to continue supporting the general Predictive Analytics or Data Mining course as well as supporting a set of courses in dedicated business analytics programs.

A general "Business Analytics," "Predictive Analytics," or Data Mining course, common in MBA and undergraduate programs as a one-semester elective, would cover Parts I–III, and choose a subset of methods from Parts IV and V. Instructors can choose to use cases as team assignments, class discussions, or projects. For a two-semester course, Part VI might be considered. For a set of courses in a dedicated Business Analytics program, here are a few courses that have been using the second edition of *Data Mining for Business Intelligence*:

Predictive Analytics: Supervised Learning   In a dedicated Business Analytics program, the topic of Predictive Analytics is typically instructed across a set of courses. The first course would cover Parts I–IV and instructors typically choose a subset of methods from Part IV according to the course length. We recommend including the new Chapter 13 in such a course.

Predictive Analytics: Unsupervised Learning   This course introduces data exploration and visualization, dimension reduction, mining relationships, and clustering (Parts III and V). If this course follows the Predictive Analytics: Supervised Learning course, then it is useful to examine examples and approaches that integrate unsupervised and supervised learning.

Forecasting Analytics   A dedicated course on time series forecasting would rely on Part VI.

In all courses, we strongly recommend including a project component, where data are either collected by students according to their interest or provided by the instructor (e.g., from the many data mining competition datasets available). From our experience and the experience of other instructors, such projects enhance learning and provide students with an excellent opportunity to understand the strengths of data mining and the challenges that arise in working with data and solving real business problems.

ACKNOWLEDGMENTS

The authors thank the many people who assisted us in improving the first edition and improving it further in the second edition, and now in this JMP edition. Anthony Babinec, who has been using drafts of this book for years in his Data Mining courses at Statistics.com, provided us with detailed and expert corrections. The Statistics.com team has also provided valuable checking, trouble-shooting, and critiquing: Kuber Deokar, Instructional Operations Supervisor, and Shweta Jadhav and Dhanashree Vishwasrao, Assistant Teachers. We also thank the many students who have used and commented on earlier editions of this text at Statistics.com.

Similarly, Dan Toy and John Elder IV greeted our project with enthusiasm and provided detailed and useful comments on earlier drafts. Boaz Shmueli and Raquelle Azran gave detailed editorial comments and suggestions on the first two editions; Noa Shmueli provided edits on the new edition; Bruce McCullough and Adam Hughes did the same for the first edition. Ravi Bapna, who used an early draft in a Data Mining course at the Indian School of Business, provided invaluable comments and helpful suggestions. Useful comments and feedback have also come from the many instructors, too numerous to mention, who have used the book in their classes.

From the Smith School of Business at the University of Maryland, colleagues Shrivardhan Lele, Wolfgang Jank, and Paul Zantek provided practical advice and comments. We thank Robert Windle, and MBA students Timothy Roach, Pablo Macouzet, and Nathan Birckhead for invaluable datasets. We also thank MBA students Rob Whitener and Daniel Curtis for the heatmap and map charts. And we thank the many MBA students for fruitful discussions and interesting data mining projects that have helped shape and improve the book.

This book would not have seen the light of day without the nurturing support of the faculty at the Sloan School of Management at MIT. Our special thanks to Dimitris Bertsimas, James Orlin, Robert Freund, Roy Welsch, Gordon Kaufmann, and Gabriel Bitran. As teaching assistants for the data mining course at Sloan, Adam Mersereau gave detailed comments on the notes and cases that were the genesis of this book, Romy Shioda helped with the preparation of several cases and exercises used here, and Mahesh Kumar helped with the material on clustering. We are grateful to the MBA students at Sloan for stimulating discussions in the class that led to refinement of the notes.

Chris Albright, Gregory Piatetsky-Shapiro, Wayne Winston, and Uday Karmarkar gave us helpful advice. Anand Bodapati provided both data and advice. Suresh Ankolekar and Mayank Shah helped develop several cases and provided valuable pedagogical comments. Vinni Bhandari helped write the Charles Book Club case.

We would like to thank Marvin Zelen, L. J. Wei, and Cyrus Mehta at Harvard, as well as Anil Gore at Pune University, for thought-provoking discussions on the relationship between statistics and data mining. Our thanks to Richard Larson of the Engineering Systems Division, MIT, for sparking many stimulating ideas on the role of data mining in modeling complex systems. They helped us develop a balanced philosophical perspective on the emerging field of data mining.

Our thanks to Ajay Sathe, and his Cytel colleagues who helped launch this project: Suresh Ankolekar, Poonam Baviskar, Kuber Deokar, Rupali Desai, YogeshGajjar, Ajit Ghanekar, Ayan Khare, Bharat Lande, Dipankar Mukhopadhyay, S. V.Sabnis, Usha Sathe, Anurag Srivastava, V. Subramaniam, Ramesh Raman, and Sanhita Yeolkar.

Steve Quigley at Wiley showed confidence in this book from the beginning, helped us navigate through the publishing process with great speed, and together with Curt Hinrichs's encouragement and support helped make this JMP Pro® edition possible. Jon Gurstelle, Allison McGinniss, Sari Friedman, and Katrina Maceda at Wiley, and Shikha Pahuja from Thomson Digital, were all helpful and responsive as we finalized this new JMP Pro® edition.

We also thank Catherine Plaisant at the University of Maryland's Human–Computer Interaction Lab, who helped out in a major way by contributing exercises and illustrations to the data visualization chapter, Marietta Tretter at Texas A&M for her helpful comments and thoughts on the time series chapters, and Stephen Few and Ben Shneiderman for feedback and suggestions on the data visualization chapter and overall design tips.

Gregory Piatetsky-Shapiro, founder of KDNuggets.com, has been generous with his time and counsel over the many years of this project. Ken Strasma, founder of the microtargeting firm HaystaqDNA and director of targeting for the 2004 Kerry campaign and the 2008 Obama campaign, provided the scenario and data for the section on uplift modeling.

Finally, we'd like to thank the reviewers of this first JMP Pro® edition for their feedback and suggestions, and members of the JMP Documentation, Education and Development teams, for their support, patience, and responsiveness to our endless questions and requests. We thank L. Allison Jones-Farmer, Maria Weese, Ian Cox, Di Michelson, Marie Gaudard, Curt Hinrichs, Rob Carver, Jim Grayson, Brady Brady, Jian Cao, Chris Gotwalt, and Fang Chen. Most important, we thank John Sall, whose inovatation, inspiration, and continued dedication to providing accessible and user-friendly desktop statistical software made JMP, and this book, possible.

PART I

PRELIMINARIES

1

INTRODUCTION

1.1 WHAT IS BUSINESS ANALYTICS?

Business analytics is the practice and art of bringing quantitative data to bear on decision-making. The term means different things to different organizations. Consider the role of analytics in helping newspapers survive the transition to a digital world.

One tabloid newspaper with a working-class readership in Britain had launched a web version of the paper, and did tests on its home page to determine which images produced more hits: cats, dogs, or monkeys. This simple application, for this company, was considered analytics. By contrast, the *Washington Post* has a highly influential audience that is of interest to big defense contractors: it is perhaps the only newspaper where you routinely see advertisements for aircraft carriers. In the digital environment, the *Post* can track readers by time of day, location, and user subscription information. In this fashion the display of the aircraft carrier advertisement in the online paper may be focused on a very small group of individuals—say, the members of the House and Senate Armed Services Committees who will be voting on the Pentagon's budget.

Business analytics, or more generically, *analytics*, includes a range of data analysis methods. Many powerful applications involve little more than counting, rule checking, and basic arithmetic. For some organizations, this is what is meant by analytics.

The next level of business analytics, now termed *business intelligence*, refers to the use of data visualization and reporting for becoming aware and understanding "what happened and what is happening." This is done by use of charts, tables, and dashboards to display, examine, and explore data. Business intelligence, which earlier consisted mainly of generating static reports, has evolved into more user-friendly and effective tools and practices, such as creating interactive dashboards that allow the user not only to access real-time data, but also to directly interact with it. Effective dashboards are those that tie directly to company data, and give managers a tool to see quickly what might not readily be apparent in a large complex database. One such tool for industrial operations managers displays customer orders in one two-dimensional display using color and bubble size as

Data Mining for Business Analytics: Concepts, Techniques, and Applications with JMP Pro®, First Edition.
Galit Shmueli, Peter C. Bruce, Mia L. Stephens, and Nitin R. Patel.
© 2017 John Wiley & Sons, Inc. Published 2017 by John Wiley & Sons, Inc.

added variables. The resulting 2 by 2 matrix shows customer name, type of product, size of order, and length of time to produce.

Business analytics includes more sophisticated data analysis methods, such as statistical models and data mining algorithms used for exploring data, quantifying and explaining relationships between measurements, and predicting new records. Methods like regression models are used to describe and quantify "on average" relationships (e.g., between advertising and sales), to predict new records (e.g., whether a new patient will react positively to a medication), and to forecast future values (e.g., next week's web traffic).

WHO USES PREDICTIVE ANALYTICS?

The widespread adoption of predictive analytics, coupled with the accelerating availability of data, has increased organizations' capabilities throughout the economy. A few examples:

Credit scoring: One long-established use of predictive modeling techniques for business prediction is credit scoring. A credit score is not some arbitrary judgement of creditworthiness; it is based mainly on a predictive model that uses prior data to predict repayment behavior.

Future purchases: A more recent (and controversial) example is Target's use of predictive modeling to classify sales prospects as "pregnant" or "not-pregnant." Those classified as pregnant could then be sent sales promotions at an early stage of pregnancy, giving Target a head start on a significant purchase stream.

Tax evasion: The US Internal Revenue Service found it was 25 times more likely to find tax evasion when enforcement activity was based on predictive models, allowing agents to focus on the most likely tax cheats (Siegel, 2013).

The business analytics toolkit also includes statistical experiments, the most common of which is known to marketers as A-B testing. These are often used for pricing decisions:

- Orbitz, the travel site, has found that it could price hotel options higher for Mac users than Windows users.
- Staples online store found it could charge more for staplers if a customer lived far from a Staples store.

Beware the organizational setting where analytics is a solution in search of a problem: a manager, knowing that business analytics and data mining are hot areas, decides that her organization must deploy them too, to capture that hidden value that must be lurking somewhere. Successful use of analytics and data mining requires both an understanding of the business context where value is to be captured, and an understanding of exactly what the data mining methods do.

1.2   WHAT IS DATA MINING?

In this book *data mining* refers to business analytics methods that go beyond counts, descriptive techniques, reporting, and methods based on business rules. While we do introduce data visualization, which is commonly the first step into more advanced analytics, the book focuses mostly on the more advanced data analytics tools. Specifically, data mining *includes statistical and machine learning methods that inform decision-making*, often in automated fashion. Prediction is typically an important component, often at the individual level. Rather than "what is the relationship between advertising and sales," we might be interested in "what specific advertisement, or recommended product, should be shown to a given online shopper at this moment?" Or we might be interested in clustering customers into different "personas" that receive different marketing treatment, then assigning each new prospect to one of these personas.

The era of big data has accelerated the use of data mining. Data mining methods, with their power and automaticity, have the ability to cope with huge amounts of data and extract value.

1.3   DATA MINING AND RELATED TERMS

The field of analytics is growing rapidly, both in terms of the breadth of applications, and in terms of the number of organizations using advanced analytics. As a result, there is considerable overlap and inconsistency in terms of definitions.

The term *data mining* itself means different things to different people. To the general public, it may have a general, somewhat hazy and pejorative meaning of digging through vast stores of (often personal) data in search of something interesting. One major consulting firm has a "data mining department," but its responsibilities are in the area of studying and graphing past data in search of general trends. And, to confuse matters, their more advanced predictive models are the responsibility of an "advanced analytics department." Other terms that organizations use are *predictive analytics*, *predictive modeling*, and *machine learning*.

Data mining stands at the confluence of the fields of statistics and machine learning (also known as *artificial intelligence*). A variety of techniques for exploring data and building models have been around for a long time in the world of statistics: linear regression, logistic regression, discriminant analysis, and principal components analysis, for example. But the core tenets of classical statistics—computing is difficult and data are scarce—do not apply in data mining applications where both data and computing power are plentiful.

This is what gives rise to Daryl Pregibon's description of data mining as "statistics at scale and speed" (Pregibon, 1999). Another major difference between the fields of statistics and machine learning is the focus in statistics on inference from a sample to the population regarding an "average effect"—for example, "a $1 price increase will reduce average demand by 2 boxes." In contrast, the focus in machine learning is on predicting individual records— "the predicted demand for person i given a $1 price increase is 1 box, while for person j it is 3 boxes." The emphasis that classical statistics places on inference (determining whether a pattern or interesting result might have happened by chance in our sample) is missing in data mining.

In comparison to statistics, data mining deals with large datasets in an open-ended fashion, making it impossible to put the strict limits around the question being addressed that inference would require. As a result the general approach to data mining is vulnerable to the danger of *overfitting*, where a model is fit so closely to the available sample of data that it describes not merely structural characteristics of the data, but random peculiarities as well. In engineering terms, the model is fitting the noise, not just the signal.

In this book, we use the term *machine learning* to refer to algorithms that learn directly from data, especially local data, often in layered or iterative fashion. In contrast, we use *statistical models* to refer to methods that apply global structure to the data. A simple example is a linear regression model (statistical) versus a k-nearest neighbors algorithm (machine learning). A given record would be treated by linear regression in accord with an overall linear equation that applies to *all* the records. In k-nearest neighbors, that record would be classified in accord with the values of a small number of nearby record.

However, many practitioners, particularly those from the IT and computer science communities, use the term *machine learning* to refer to all the methods discussed in this book.

1.4 BIG DATA

Data mining and big data go hand in hand. *Big data* is a relative term—data today are big by reference to the past, and to the methods and devices available to deal with them. The challenge big data presents is often characterized by the four V's - volume, velocity, variety, and veracity. *Volume* refers to the amount of data. *Velocity* refers to the flow rate—the speed at which it is being generated and changed. *Variety* refers to the different types of data being generated (currency, dates, numbers, text, etc.). *Veracity* refers to the fact that data is being generated by organic distributed processes (e.g., millions of people signing up for services or free downloads) and not subject to the controls or quality checks that apply to data collected for a study.

Most large organizations face both the challenge and the opportunity of big data because most routine data processes now generate data that can be stored and, possibly, analyzed. The scale can be visualized by comparing the data in a traditional statistical analysis on the large size (e.g., 15 variables and 5000 records) to the Walmart database. If you consider the traditional statistical study to be the size of a period at the end of a sentence, then the Walmart database is the size of a football field. And that probably does not include other data associated with Walmart—social media data, for example, which comes in the form of unstructured text.

If the analytical challenge is substantial, so can be the reward:

- OKCupid, the dating site, uses statistical models with their data to predict what forms of message content are most likely to produce a response.
- Telenor, a Norwegian mobile phone service company, was able to reduce subscriber turnover 37% by using models to predict which customers were most likely to leave, and then lavishing attention on them.
- Allstate, the insurance company, tripled the accuracy of predicting injury liability in auto claims by incorporating more information about vehicle type.

The examples above are from Eric Siegel's *Predictive Analytics* (2013, Wiley).

Some extremely valuable tasks were not even feasible before the era of big data. Consider web searches, the technology on which Google was built. In early days, a search for "Ricky Ricardo Little Red Riding Hood" would have yielded various links to the "I Love Lucy" show, other links to Ricardo's career as a band leader, and links to the children's story of Little Red Riding Hood. Only once the Google database had accumulated sufficient data (including records of what users clicked on) would the search yield, in the top position, links to the specific *I Love Lucy* episode in which Ricky enacts, in a comic mixture of Spanish and English, Little Red Riding Hood for his infant son.

1.5   DATA SCIENCE

The ubiquity, size, value, and importance of big data has given rise to a new profession: the *data scientist*. *Data science* is a mix of skills in the areas of statistics, machine learning, math, programming, business, and IT. The term itself is thus broader than the other concepts we discussed above, and it is a rare individual who combines deep skills in all the constituent areas. Harlan Harris, in *Analyzing the Analyzers* (with Sean Murphy and Marck Vaisman, O'Reilly 2013) describes the skill sets of most data scientists as resembling a "T"—deep in one area (the vertical bar of the T), and shallower in other areas (the top of the T).

At a large data science conference session (Strata-Hadoop World, October 2014) most attendees felt that programming was an essential skill, though there was a sizable minority who felt otherwise. And, although big data is the motivating power behind the growth of data science, most data scientists do not actually spend most of their time working with terabyte-size or larger data.

Data of the terabyte or larger size would be involved at the deployment stage of a model. There are manifold challenges at that stage, most of them IT and programming issues related to data handling and tying together different components of a system. Much work must precede that phase. It is that earlier piloting and prototyping phase on which this book focuses—developing the statistical and machine learning models that will eventually be plugged into a deployed system. What methods do you use with what sorts of data and problems? How do the methods work? What are their requirements, their strengths, their weaknesses? How do you assess their performances?

1.6   WHY ARE THERE SO MANY DIFFERENT METHODS?

As can be seen in this book or any other resource on data mining, there are many different methods for prediction and classification. You might ask yourself why they coexist, and whether some are better than others. The answer is that each method has advantages and disadvantages. The usefulness of a method can depend on factors such as the size of the dataset, the types of patterns that exist in the data, whether the data meet some underlying assumptions of the method, how noisy the data are, and the particular goal of the analysis. A small illustration is shown in Figure 1.1, where the goal is to find a combination of *household income level* and *household lot size* that separate buyers (solid circles) from nonbuyers (hollow circles) of riding mowers. The first method (left panel) looks only for horizontal lines to separate buyers from nonbuyers, whereas the second method (right panel) looks for a single diagonal line.

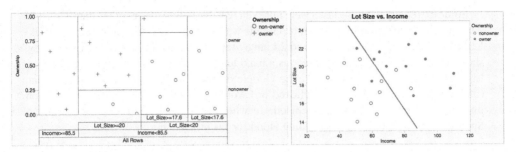

FIGURE 1.1   Two methods for separating buyers from nonbuyers.

Different methods can lead to different results, and their performances can vary. It is therefore customary in data mining to apply several different methods and select the one that is most useful for the goal at hand.

1.7   TERMINOLOGY AND NOTATION

Because of the hybrid origins of data mining, its practitioners often use multiple terms to refer to the same thing. For example, in the machine learning (artificial intelligence) field, the variable being predicted is the output variable or target variable. To a statistician, it is the dependent variable or the response. Here is a summary of terms used:

Algorithm   Refers to a specific procedure used to implement a particular data mining technique: classification tree, discriminant analysis, and the like.

Attribute   See **Predictor**.

Case   See **Observation**.

Confidence   Has a broad meaning in statistics (*confidence interval*), concerning the degree of error in an estimate that results from selecting one sample as opposed to another.

Dependent Variable   See **Response**.

Estimation   See **Prediction**.

Feature   See **Predictor**.

Holdout Sample   Is a sample of data not used in fitting a model, used to assess the performance of that model; this book uses the term *validation set* or, if one is used in the problem, *test set* instead of *holdout sample*.

Input Variable   See **Predictor**.

Model   Refers to an algorithm as applied to a dataset, complete with its settings (many of the algorithms have parameters that the user can adjust).

Observation   Is the unit of analysis on which the measurements are taken (a customer, a transaction, etc.); also called *instance*, *sample*, *example*, *case*, *record*, *pattern*, or *row*. (Each row typically represents a record; each column, a variable. Note that the use of the term "sample" here is different from its usual meaning in statistics, where it refers to all the data sampled from a larger data source, not simply one record.)

Outcome Variable   See **Response**.

Output Variable   See **Response**.

$P(A|B)$ Is the conditional probability of event A occurring given that event B has occurred. Read as "the probability that A will occur given that B has occurred."

Pattern   Is a set of measurements on an observation (e.g., the height, weight, and age of a person).

Prediction   Means the prediction of the value of a continuous output variable; also called *estimation.*

Predictor   Usually denoted by X, is also called a *feature, input variable, independent variable*, or from a database perspective, a *field*.

Record   See **Observation**.

Response   Usually denoted by Y, is the variable being predicted in supervised learning; also called *dependent variable, output variable, target variable*, or *outcome variable.*

Score   Refers to a predicted value or class. *Scoring new data* means to use a model developed with training data to predict output values in new data.

Success Class   Is the class of interest in a binary outcome (e.g., *purchasers* in the outcome *purchase/no purchase*).

Supervised Learning   Refers to the process of providing an algorithm (logistic regression, regression tree, etc.) with records in which an output variable of interest is known and the algorithm "learns" how to predict this value with new records where the output is unknown.

Target   See **Response**.

Test Data   (or **test set**) Refers to that portion of the data used only at the end of the model building and selection process to assess how well the final model might perform on additional data.

Training Data   (or **training set**) Refers to that portion of data used to fit a model.

Unsupervised Learning   Refers to analysis in which one attempts to learn something about the data other than predicting an output value of interest (e.g., whether it falls into clusters).

Validation Data   (or **validation set**) Refers to that portion of the data used to assess how well the model fits, to adjust some models, and to select the best model from among those that have been tried.

Variable   Is any measurement on the records, including both the input (X) variables and the output (Y) variable.

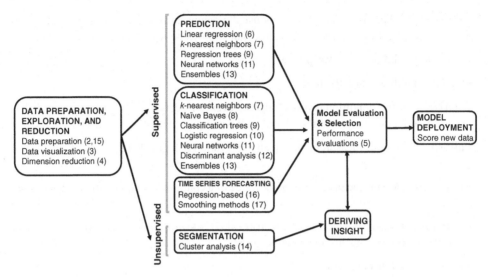

FIGURE 1.2 Data mining from a process perspective. Numbers in Parentheses indicate chapter numbers.

1.8 ROADMAP TO THIS BOOK

The book covers many of the widely used predictive and classification methods as well as other data mining tools. Figure 1.2 outlines data mining from a process perspective and where the topics in this book fit in. Chapter numbers are indicated beside the topic. Table 1.1 provides a different perspective: it organizes data mining procedures according to the type and structure of the data.

TABLE 1.1 Organization of data mining methods in this book, according to the nature of the data

	Supervised		Unsupervised
	Continuous Response	Categorical Response	No Response
Continuous Predictors	Linear regression (6) Neural nets (11) k-nearest neighbors (7) Ensembles (13)	Logistic regression (10) Neural nets (11) Discriminant analysis (12) k-nearest neighbors (7) Ensembles (13)	Principal components (4) Cluster analysis (14)
Categorical Predictors	Linear regression (6) Neural nets (11) Regression trees (9) Ensembles (13)	Neural nets (11) Classification trees (9) Logistic regression (10) Naive Bayes (8) Ensembles (13)	

Note: Numbers in parentheses indicate chapter number.

Order of Topics

The book is divided into five parts: Part I (Chapters 1 and 2) gives a general overview of data mining and its components. Part II (Chapters 3 and 4) focuses on the early stages of data exploration and dimension reduction.

Part III (Chapter 5) discusses performance evaluation. Although it contains a single chapter, we discuss a variety of topics, from predictive performance metrics to misclassification costs. The principles covered in this part are crucial for the proper evaluation and comparison of supervised learning methods.

Part IV includes eight chapters (Chapters 6 through 13), covering a variety of popular supervised learning methods (for classification and/or prediction). Within this part, the topics are generally organized according to the level of sophistication of the algorithms, their popularity, and ease of understanding. The final chapter introduces ensembles and combinations of methods.

Part V presents one very popular method for unsupervised mining of relationships, cluster analysis (Chapter 14).

Part VI includes three chapters (Chapters 15 through 17), with the focus on forecasting time series. The first chapter covers general issues related to handling and understanding time series. The next two chapters present two popular forecasting approaches: regression-based forecasting and smoothing methods.

Finally, Part VII includes a set of cases.

Although the topics in the book can be covered in the order of the chapters, each chapter stands alone. We advise, however, to read Parts I through III before proceeding to chapters in parts IV and V. Similarly, Chapter 15 should precede other chapters in Part VI.

USING JMP PRO, STATISTICAL DISCOVERY SOFTWARE FROM SAS

To facilitate the learning experience, this book uses JMP Pro (Figure 1.3), a desktop statistical package from SAS for Mac OS and Windows operating systems (see jmp.com/system for complete system requirements).

JMP comes in two primary flavors: JMP (the standard version) and JMP Pro (the professional version). The standard version of JMP is dynamic and interactive, and it offers a variety of built in tools for graphing and analyzing data. JMP has extensive tools for data visualization and data preparation, along with statistical and data mining techniques for classification, prediction, and forecasting. It offers a variety of supervised data mining tools, including neural nets, classification and regression trees, logistic regression, linear regression, and discriminant analysis. It also offers unsupervised algorithms, such as principal components analysis, k-means clustering, and hierarchical clustering.

JMP Pro has all of the functionality of JMP, but adds advanced tools for predictive modeling, including k-nearest neighbor, uplift modeling, advanced trees, additional options for creating neural networks, and ensemble models. It also provides a number of modeling utilities for preparing data for modeling, and includes built-in tools for model validation, model comparison, and model selection.

For these reasons we use JMP Pro throughout this book. JMP Pro is required for the built-in tools for model cross-validation and comparison that aren't available in the standard version of JMP. However, the standard version of JMP can be used for creating many of the graphs, summaries, analyses, and models presented in this book.

While we provide JMP instructions throughout this book, there are many resources available for new users. For tips on getting started with JMP go to jmp.com/gettingstarted. An in-depth introduction to JMP, *Discovering JMP*, is available online or in the JMP Help files (under *Help > Books* in JMP). Additional resources on specific topics, along with short videos, can be found in the JMP Learning Library at jmp.com/learn. For additional details on the features available only in JMP Pro, see jmp.com/pro.

JMP Pro is available through department or campus licenses at most colleges and universities and through site licenses in many organizations—see your software IT administrator for availability and download information. If you do not have access to a site license of JMP Pro, would like more information, or would like to request an evaluation of JMP Pro for classroom use, write to academic@jmp.com. If you do not qualify for an academic license, a trial of JMP Pro can also be requested at jmp.com/trial.

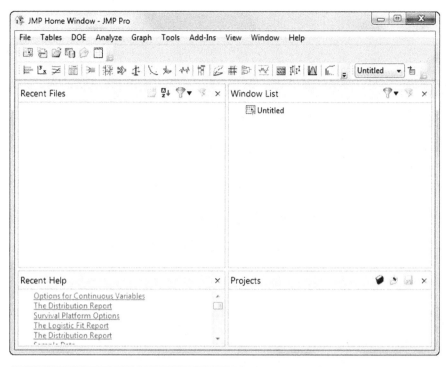

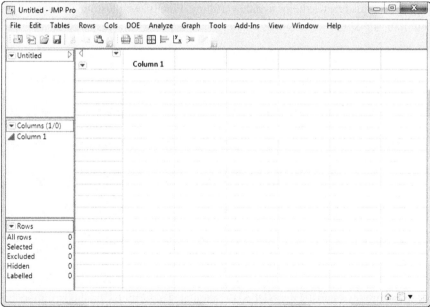

FIGURE 1.3   JMP Home Window (top) and Data Table (on a Windows PC).

2

OVERVIEW OF THE DATA MINING PROCESS

In this chapter we give an overview of the steps involved in data mining, starting from a clear goal definition and ending with model deployment. The general steps are shown schematically in Figure 2.1. We also discuss issues related to data collection, cleaning, and preprocessing. We explain the notion of data partitioning, where methods are trained on a set of training data and then model performance is evaluated on a separate set of validation data, and how this practice helps avoid overfitting. Finally, we illustrate the steps of model building by applying them to data.

2.1   INTRODUCTION

In Chapter 1 we saw some very general definitions of data mining. In this chapter we introduce the variety of methods sometimes referred to as *data mining*. The core of this book focuses on what has come to be called *predictive analytics*, the tasks of classification and prediction as well as pattern discovery, that have become key elements of a *business analytics* function in most large firms. These terms are described and illustrated below.

Not covered in this book to any great extent are two simpler database methods that are sometimes considered to be data mining techniques: (1) OLAP (online analytical processing) and (2) SQL (structured query language). OLAP and SQL searches on databases are descriptive in nature ("find all credit card customers in a certain zip code with annual charges > \$20,000, who own their own home and who pay the entire amount of their monthly bill at least 95% of the time") and are based on business rules set by the user. They do not involve statistical modeling or automated algorithmic methods, although SQL queries are often used to obtain the data for use in data mining with other tools. (Note that, in JMP 12, the *Query Builder* was added under *File > Database* for selecting and importing data from an SQL database without writing SQL statements.)

Data Mining for Business Analytics: Concepts, Techniques, and Applications with JMP Pro®, First Edition.
Galit Shmueli, Peter C. Bruce, Mia L. Stephens, and Nitin R. Patel.
© 2017 John Wiley & Sons, Inc. Published 2017 by John Wiley & Sons, Inc.

| Define purpose | Obtain data | Explore & clean data | Determine DM task | Choose DM methods | Apply methods & select final model | Evaluate performance | Deploy |

FIGURE 2.1   Schematic of the data modeling process.

2.2 CORE IDEAS IN DATA MINING

Classification

Classification is perhaps the most basic form of data analysis. The recipient of an offer can respond or not respond. An applicant for a loan can repay on time, repay late, or declare bankruptcy. A credit card transaction can be normal or fraudulent. A packet of data traveling on a network can be benign or threatening. A bus in a fleet can be available for service or unavailable. The victim of an illness can be recovered, still be ill, or be deceased.

A common task in data mining is to examine data where the classification is unknown or will occur in the future, with the goal of predicting what that classification is or will be. Similar data where the classification is known are used to develop rules, which are then applied to the data with the unknown classification.

Prediction

Prediction is similar to classification, except that we are trying to predict the value of a numerical variable (e.g., amount of purchase) rather than a class (e.g., purchaser or nonpurchaser). Of course, in classification we are trying to predict a class, but the term *prediction* in this book refers to the prediction of the value of a continuous variable. (Sometimes in the data mining literature, the terms *estimation* and *regression* are used to refer to the prediction of the value of a continuous variable, and *prediction* may be used for both continuous and categorical data.)

Association Rules and Recommendation Systems

Large databases of customer transactions lend themselves naturally to the analysis of associations among items purchased, or "what goes with what." *Association rules*, or *affinity analysis*, is designed to find such general associations patterns among items in large databases. The rules can then be used in a variety of ways. For example, grocery stores can use such information for product placement. They can use the rules for weekly promotional offers or for bundling products. Association rules derived from a hospital database on patients' symptoms during consecutive hospitalizations can help find "which symptom is followed by what other symptom" and help predict future symptoms for returning patients.

Online recommendation systems, such as those used on Amazon.comand Netflix.com, use *Collaborative Filtering*, a method that uses individual users' preferences and tastes given their historic purchase, rating, browsing, or any other measurable behavior indicative of preference, as well as other users' histories. In contrast to association rules that generate rules general to an entire population, collaborative filtering generates "what goes with what" at the individual user level. Hence collaborative filtering is used in many recommendation systems that aim to deliver personalized recommendations to users with a wide range of preferences. (Note that these topics are not covered in this book.)

Predictive Analytics

Classification, prediction, and, to some extent, association rules and collaborative filtering constitute the analytical methods employed in *predictive analytics*. However, the term predictive analytics is sometimes used to also include data pattern identification methods such as clustering.

Data Reduction and Dimension Reduction

The performance of data mining algorithms is often improved when the number of variables is limited, and when large numbers of records can be grouped into homogeneous groups. For example, rather than dealing with thousands of product types, an analyst might wish to group them into a smaller number of groups and build separate models for each group. Or a marketer might want to classify customers into different "personas," and must first group customers into homogeneous groups to define the personas. This process of consolidating a large number of records (or cases) into a smaller set is termed *data reduction*. Methods for reducing the number of cases are often called *clustering*.

Reducing the number of variables is typically called *dimension reduction*. Dimension reduction is a common initial step before deploying supervised learning methods, intended to improve predictive power, manageability, and interpretability.

Data Exploration and Visualization

One of the earliest stages of engaging with a dataset is exploring it. Exploration is aimed at understanding the global landscape of the data and for detecting unusual values. Exploration is used for data cleaning and manipulation as well as for visual discovery and "hypothesis generation".

Methods for exploring data include looking at various data aggregations and summaries, both numerically and graphically. This includes looking at each variable separately as well as looking at relationships between variables. The purpose is to discover patterns and exceptions. Exploration by creating charts and dashboards is called *data visualization* or *visual analytics*. For numerical variables, we use histograms and boxplots to learn about the distribution of their values, to detect outliers (extreme observations), and to find other information that is relevant to the analysis task. Similarly, for categorical variables, we use bar charts. We can also look at scatterplots of pairs of numerical variables to learn about possible relationships, the type of relationship, and, again, to detect outliers. Visualization can be greatly enhanced by adding features such as color, zooming, interactive navigation, and data filtering.

Supervised and Unsupervised Learning

A fundamental distinction among data mining techniques is between supervised and unsupervised methods. *Supervised learning algorithms* are those used in classification and prediction. We must have data available in which the value of the outcome of interest (e.g., purchase or no purchase) is known. These *training data* are the data from which the classification or prediction algorithm "learns," or is "trained," about the relationship between predictor variables and the outcome variable. Once the algorithm has learned from the training data, it is then applied to another sample of data (the *validation data*) where the outcome is known, to see how well it does in comparison to other models. If many different models are being tried out, it is prudent to save a third sample, which also includes known outcomes (the *test data*) to use with the model finally selected to predict how well it will do. The model can then be used to classify or predict the outcome of interest in new cases where the outcome is unknown.

Simple linear regression is an example of a supervised learning algorithm (although rarely called that in the introductory statistics course where you probably first encountered it). The Y variable is the (known) outcome variable and the X variable is a predictor variable. A regression line is drawn to minimize the sum of squared deviations between the actual Y values and the values predicted by this line. The regression line can now be used to predict Y values for new values of X for which we do not know the Y value.

Unsupervised learning algorithms are those used where there is no outcome variable to predict or classify. Hence, there is no "learning" from cases where such an outcome variable is known. Association rules, dimension reduction methods, and clustering techniques are all unsupervised learning methods.

Supervised and unsupervised methods are sometimes used in conjunction. For example, unsupervised clustering methods are used to separate loan applicants into several risk-level groups. Then supervised algorithms are applied separately to each risk-level group for predicting loan default propensity.

2.3 THE STEPS IN DATA MINING

This book focuses on understanding and using data mining algorithms (steps 4 to 7 below). However, some of the most serious errors in analytics projects result from a poor understanding of the problem—an understanding that must be developed before we get into the details of algorithms to be used. Here is a list of steps to be taken in a typical data mining effort:

1. *Develop an understanding of the purpose of the data mining project.* What is the problem? How will the stakeholder use the results? Who will be affected by the results? Will the analysis be a one-shot effort or an ongoing procedure?

2. *Obtain the dataset to be used in the analysis.* This often involves random sampling from a large database to capture records to be used in an analysis. It may also involve pulling together data from different databases or sources. The databases could be internal (e.g., past purchases made by customers) or external (credit ratings). While data mining deals with very large databases, usually the analysis to be done requires only thousands or tens of thousands of records.

3. *Explore, clean, and preprocess the data.* This step involves verifying that the data are in reasonable condition. How should missing data be handled? Are the values in a reasonable range, given what you would expect for each variable? Are there obvious outliers? The data are reviewed graphically: for example, a matrix of scatterplots showing the relationship of each variable with every other variable. We also need to ensure consistency in the definitions of fields, units of measurement, time periods, and so on. In this step, new variables are also typically created from existing ones. For example, "duration" can be computed from start and end dates.

4. *Reduce the data dimension, if necessary.* Dimension reduction can involve operations such as eliminating unneeded variables, transforming variables (e.g., turning "money spent" into "spent > \$100" vs. "spent ≤ \$100"), and creating new variables (e.g., a variable that records whether at least one of several products was purchased). Make sure that you know what each variable means and whether it is sensible to include it in the model.

5. *Determine the data mining task* (classification, prediction, clustering, etc.). This involves translating the general question or problem of step 1 into a more specific statistical question.

6. *Partition the data* (for supervised tasks). If the task is supervised (classification or prediction), partition the dataset into three parts: training, validation, and test datasets.

7. *Choose the data mining techniques to be used* (regression, neural nets, hierarchical clustering, etc.).

8. *Use algorithms to perform the task.* This is typically an iterative process—trying multiple variants, and often using multiple variants of the same algorithm (choosing different variables or settings within the algorithm). Where appropriate, feedback from the algorithm's performance on validation data is used to refine the settings.

9. *Interpret the results of the algorithms.* This involves making a choice as to the best algorithm to deploy, and where possible, testing the final choice on the test data to get an idea as to how well it will perform. (Recall that each algorithm may also be tested on the validation data for tuning purposes; in this way the validation data become a part of the fitting process and are likely to underestimate the error in the deployment of the model that is finally chosen.)

10. *Deploy the model.* This step involves integrating the model into operational systems and running it on real records to produce decisions or actions. For example, the model might be applied to a purchased list of possible customers, and the action might be "include in the mailing if the predicted amount of purchase is > \$10." A key step here is "scoring" the new records, or using the chosen model to predict the target value for each new record.

The foregoing steps encompass the steps in SEMMA, a methodology developed by the parent company of JMP, SAS Institute:

Sample   Take a sample from the dataset; partition into training, validation, and test datasets.

Explore   Examine the dataset statistically and graphically.

Modify   Transform the variables and impute missing values.

Model   Fit predictive models (e.g., regression tree, collaborative filtering).

Assess   Compare models using a validation dataset.

IBM SPSS Modeler (previously SPSS-Clementine) has a similar methodology, termed CRISP-DM (CRoss-Industry Standard Process for Data Mining).

2.4 PRELIMINARY STEPS

Organization of Datasets

Datasets are nearly always constructed and displayed so that variables are in columns and records are in rows. In the example shown in Section 2.6 (home value in West Roxbury, Boston, in 2014), 14 variables are recorded for over 5000 homes. The JMP data table (West Roxbury Housing.jmp) is organized so that each row represents a home—the first home's assessed value was $344,200, its tax was $4430, its size was 9965 ft^2, it was built in 1880, and so on. In supervised learning situations, one of these variables will be the outcome variable, typically listed at the end or the beginning (in this case it is TOTAL VALUE, in the first column).

Sampling from a Database

Quite often, we want to perform our data mining analysis on less than the total number of records that are available. Data mining algorithms will have varying limitations on what they can handle in terms of the numbers of records and variables, limitations that may be specific to computing power and capacity as well as software limitations. Even within those limits, many algorithms will execute faster with smaller samples.

Accurate models can often be built with as few as several hundred or thousand records (as discussed later). Hence we may want to sample a subset of records for model building.

Oversampling Rare Events in Classification Tasks

If the event we are interested in classifying is rare, such as customers purchasing a product in response to a mailing, or fraudulent credit card transactions, sampling a random subset of records may yield so few events (e.g., purchases) that we have little information on them. We would end up with lots of data on nonpurchasers and non fraudulent transactions but little on which to base a model that distinguishes purchasers from nonpurchasers or fraudulent from nonfraudulent. In such cases we would want our sampling procedure to overweight the rare class (purchasers or frauds) relative to the majority class (nonpurchasers, nonfrauds) so that our sample would end up with a healthy complement of purchasers or frauds.

Assuring an adequate number of responder or "success" cases to train the model is just part of the picture. A more important factor is the costs of *misclassification*. That is, the cost of incorrectly classifying outcomes. Whenever the response rate is extremely low, we are likely to attach more importance to identifying a responder than to identifying a nonresponder. In direct-response advertising (whether by traditional mail, email, or web advertising), we may encounter only one or two responders for every hundred records—the value of finding such a customer far outweighs the costs of reaching him or her. In trying to identify fraudulent transactions, or customers unlikely to repay debt, the costs of failing to find the fraud or the nonpaying customer are likely to exceed the cost of more detailed review of a legitimate transaction or customer.

More generally, we want to train our model with the asymmetric costs in mind so that the algorithm will catch the more valuable responders, probably at the cost of "catching" and misclassifying more nonresponders as responders than would be the case if we assume equal costs. This subject is discussed in detail in Chapter 5.

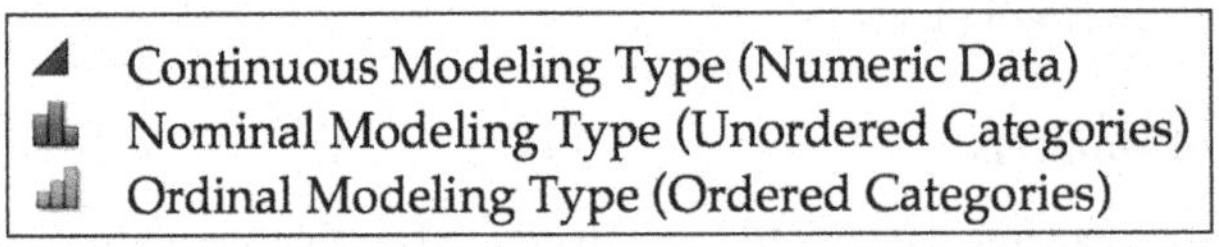

◢ Continuous Modeling Type (Numeric Data)
▆ Nominal Modeling Type (Unordered Categories)
▄ Ordinal Modeling Type (Ordered Categories)

FIGURE 2.2   Three JMP modeling types.

Preprocessing and Cleaning the Data

Types of Variables   There are several ways of classifying variables. Variables can be numerical or text (character/string). They can be continuous (able to assume any real numerical value, usually in a given range), integer (taking only integer values), or categorical (assuming one of a limited number of values). Categorical variables can be either coded as numerical (1, 2, 3) or text (payments current, payments not current, bankrupt). Categorical variables can also be unordered (called *nominal variables*) with categories such as North America, Europe, and Asia; or they can be ordered (called *ordinal variables*) with categories such as high value, low value, and nil value.

In JMP, *modeling types* are applied to distinguish the type of variable, and are used in many platforms to determine the appropriate graphs and analyses. Colored icons are used to indicate the modeling types applied (see Figure 2.2). Continuous variables are coded with the *continuous* modeling type (represented by blue triangles), variables with unordered categories are given the *nominal* modeling type (represented by red bars), and variables that represent ordered categories are given the *ordinal* modeling type (represented by green bars).

CHANGING MODELING TYPES IN JMP

Modeling types in JMP should be specified prior to analysis. To change a modeling type, right-click on a column in the *Columns Panel* of the data table and select *Column Info*. Then, select the correct Modeling Type. Note that for continuous data you may also need to change the *Data Type* to *Numeric*.

Figure 2.3 shows the *Columns Panel*, along with the general layout of JMP data tables. The dataset is *Companies*, which is available from JMP Sample Data Library (under the *Help* menu). This figure was borrowed from the one-page guide, *JMP Data Tables*, found at jmp.com/learn.

Working with Continuous Variables   Continuous variables can be handled by most data mining routines. In JMP, all supervised routines can take continuous predictor variables. The machine learning roots of data mining grew out of problems with categorical outcomes; the roots of statistics lie in the analysis of continuous variables. Sometimes it is desirable to convert continuous variables to categorical variables. This is often done in the case of outcome variables, where the numerical variable is mapped to a decision (e.g., credit scores above a certain level mean "grant credit," a medical test result above a certain level means "start treatment"). This type of conversion can be easily done in JMP.

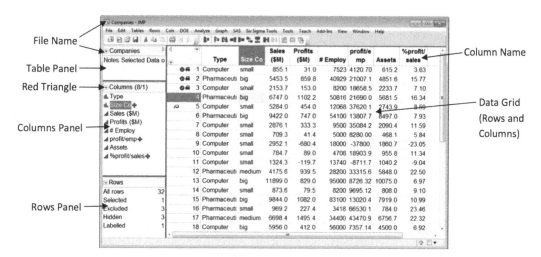

FIGURE 2.3   Example dataset, showing the columns panel (on the left) and the general layout of JMP data tables.

Handling Categorical Variables   Categorical variables can also be handled by most routines but often require special handling. If the categorical variable is ordered (age group, degree of creditworthiness, etc.), we can sometimes code the categories numerically (1, 2, 3, ...) and treat the variable as if it were a continuous variable. The smaller the number of categories, and the less they represent equal increments of value, the more problematic this approach becomes, but it often works well enough.

Nominal categorical variables, however, often cannot be used as is. In many cases they must be decomposed into a series of binary variables, called *indicator variables* or *dummy variables*. For example, a single categorical variable that can have possible values of "student," "unemployed," "employed," or "retired" could be split into four separate indicator variables:

> *Student*—Yes/No
> *Unemployed*—Yes/No
> *Employed*—Yes/No
> *Retired*—Yes/No

In many cases only three of the indicator variables need to be used; if the values of three are known, the fourth is also known. For example, given that these four values are the only possible ones, we can know that if a person is neither student, unemployed, nor employed, he or she must be retired. In some routines (e.g., linear regression and logistic regression) you should not use all four indicator variables—the redundant information will cause the algorithm to fail. JMP has a utility to convert categorical variables to binary indicator variables (*Make Indicator Columns*, from the *Cols > Utilities* menu). But, for the most part, JMP automatically codes categorical variables behind the scenes if needed, thereby not requiring the user to manually create indicator variables.

Variable Selection   More is not necessarily better when it comes to selecting variables for a model. Other things being equal, parsimony, or compactness, is a desirable feature in a model. For one thing, the more variables we include, the greater the number of records we will need to assess relationships among the variables. Fifteen records may suffice to give us a rough idea of the relationship between Y and a single predictor variable X. If we now want information about the relationship between Y and 15 predictor variables $X_1 \ldots X_{15}$, then 15 records will not be enough (each estimated relationship would have an average of only one record's worth of information, making the estimate very unreliable). In addition, models based on many variables are often less robust, as they require the collection of more variables in the future, are subject to more data quality and availability issues, and require more data cleaning and preprocessing.

How Many Variables and How Much Data?   Statisticians give us procedures to learn with some precision how many records we would need to achieve a given degree of reliability with a given dataset and a given model. These are called "power calculations" and are intended to assure that an average population effect will be estimated with sufficient precision from a sample. Data miners' needs are usually different, since the focus is not on identifying an average effect but rather predicting individual records. This purpose typically requires larger samples than those used for statistical inference. A good rule of thumb is to have 10 records for every predictor variable. Another, used by Delmaster and Hancock (2001, p. 68) for classification procedures, is to have at least $6 \times m \times p$ records, where m is the number of outcome classes and p is the number of variables.

In general, compactness or parsimony is a desirable feature in a data mining model. Even when we start with a small number of variables, we often end up with many more after creating new variables (e.g., converting a categorical variable into a set of dummy variables). Data visualization and dimension reduction methods help reduce the number of variables so that redundancies are avoided.

Even when we have an ample supply of data, there are good reasons to pay close attention to the variables that are included in a model. Someone with domain knowledge (i.e., knowledge of the business process and the data) should be consulted, as knowledge of what the variables represent is typically critical for building a good model and avoiding errors.

For example, suppose that we're trying to predict the total purchase amount spent by customers, and we have a few predictor columns that are coded $X_1, X_2, X_3, \ldots$, where we don't know what those codes mean. We might find that X_1 is an excellent predictor of the total amount spent. However, if we discover that X_1 is the amount spent on shipping, calculated as a percentage of the purchase amount, then obviously a model that uses shipping amount cannot be used to predict purchase amount, since the shipping amount is not known until the transaction is completed. Another example is if we are trying to predict loan default at the time a customer applies for a loan. If our dataset includes only information on approved loan applications, we will not have information about what distinguishes defaulters from nondefaulters among denied applicants. A model based on approved loans alone can therefore not be used to predict defaulting behavior at the time of loan application, but rather only once a loan is approved.

Outliers   The more data we are dealing with, the greater the chance of encountering erroneous values resulting from measurement error, data-entry error, or the like. If the erroneous value is in the same range as the rest of the data, it may be harmless. If it is well outside the range of the rest of the data (e.g., a misplaced decimal), it might have a substantial effect on some of the data mining procedures we plan to use.

Values that lie far away from the bulk of the data are called *outliers*. The term *far away* is deliberately left vague because what is or is not called an outlier is basically an arbitrary decision. Analysts use rules of thumb such as "anything over 3 standard deviations away from the mean is an outlier," but no statistical rule can tell us whether such an outlier is the result of an error. In this statistical sense, an outlier is not necessarily an invalid data point, it is just a distant data point.

The purpose of identifying outliers is usually to call attention to values that need further review. We might come up with an explanation looking at the data—in the case of a misplaced decimal, this is likely. We might have no explanation, but know that the value is wrong—a temperature of 178° F for a sick person. Or, we might conclude that the value is within the realm of possibility and leave it alone. All these are judgments best made by someone with *domain knowledge*, knowledge of the particular application being considered: direct mail, mortgage finance, and so on, as opposed to technical knowledge of statistical or data mining procedures. Statistical procedures can do little beyond identifying the record as something that needs review.

If manual review is feasible, some outliers may be identified and corrected. In any case, if the number of records with outliers is very small, they might be treated as missing data. How do we inspect for outliers? One technique is to plot all of the variables and review the graphs for very large or very small values in each graph. Another option is to examine the minimum and maximum values of each column using the JMP *Columns Viewer* (from the *Cols* menu). For a more automated approach that considers each record as a unit, automated procedures could be used to identify outliers across many variables. Several procedures are available in the JMP Pro *Explore Outliers* column modeling utility.

Missing Values   Typically some records will contain missing values. In JMP, missing values for continuous variables are denoted with a "." in the cell. For categorical (nominal or ordinal) variables, the cell is empty. In some modeling routines, such as linear regression, records with missing values are omitted. If the number of records with missing values is small, this might not be a problem. However, if we have a large number of variables, a small proportion of missing values can affect a lot of records. Even with only 30 variables, if only 5% of the values are missing (spread randomly and independently among cases and variables), almost 80% of the records would have to be omitted from the analysis. (The chance that a given record would escape having a missing value is $0.95^{30} = 0.215$.)

An alternative to omitting records with missing values is to replace the missing value with an imputed value, based on the other values for that variable across all records. For example, if among 30 variables, household income is missing for a particular record, we might substitute the mean household income across all records. Doing so does not, of course, add any information about how household income affects the outcome variable. It merely allows us to proceed with the analysis and not lose the information contained in this record for the other 29 variables. Note that using such a technique will understate the variability in a dataset. However, we can assess variability and the performance of our data mining technique, using the validation data, and therefore this need not present a major problem. More sophisticated procedures do exist—for example, using regression, based on other variables, to fill in the missing values. These methods have been elaborated mainly for analysis of medical and scientific studies, where each patient or subject record comes at great expense. In data mining, where data are typically plentiful, simpler methods usually suffice.

Some datasets contain variables that have a very large number of missing values. In other words, a measurement is missing for a large number of records. In that case, dropping records with missing values will lead to a large loss of data. Imputing the missing values might also be useless, as the imputations are based on a small number of existing records. An alternative is to examine the importance of the predictor. If it is not very crucial, it can be dropped. If it is important, perhaps a proxy variable with fewer missing values can be used instead. When such a predictor is deemed central, the best solution is to invest in obtaining the missing data.

Significant time may be required to deal with missing data, as not all situations are susceptible to automated solutions. In a messy dataset, for example, a "0" might mean two things: (1) the value is missing or (2) the value is actually zero. In the credit industry, a "0" in the "past due" variable might mean a customer who is fully paid up, or a customer with no credit history at all—two very different situations. Likewise, in many situations a "999" is used to denote missing values. Human judgment may be required for individual cases or to determine a special rule to deal with the situation. *Note:* It is important to distinguish between missing values and zeros. JMP will include zeros in calculations and analyses but will omit missing values from analyses. If a variable contains codes for missing value, the *Missing Value Code* column property for the variable can be used to indicate this.

In JMP Pro, the *Explore Missing Values* column modeling utility (from the *Cols* menu) can be used to both explore and impute missing values (in JMP Pro only), and the *Missing Data Pattern* utility, from the *Tables* menu, can be used to explore whether there are any patterns among the missing values. In JMP modeling platforms, *Informative Missing* coding is available to address missing values.

Normalizing (Standardizing) and Rescaling Data   Some algorithms require that the data be normalized before the algorithm can be implemented effectively. To normalize a variable, we subtract the mean from each value and then divide by the standard deviation. This operation is also sometimes called *standardizing*. In effect, we are expressing each value as the "number of standard deviations away from the mean," also called a *z-score*.

Normalizing is one way to bring all variables to the same scale. Another popular approach is rescaling each variable to a [0,1] scale. This is done by subtracting the minimum value and then dividing by the range. Subtracting the minimum shifts the variable origin to zero. Dividing by the range shrinks or expands the data to the range [0,1].

To consider why normalizing or scaling to [0,1] might be necessary, consider the case of clustering. Clustering typically involves calculating a distance measure that reflects how far each record is from a cluster center or from other records. With multiple variables, different units will be used: days, dollars, counts, and so on. If the dollars are in the thousands and everything else is in the tens, the dollar variable will come to dominate the distance measure. Moreover, changing units from (say) days to hours or months can alter the outcome completely.

Data mining software, including JMP, typically have an option to normalize the data in those algorithms where it may be required. It is generally an option rather than an automatic feature of such algorithms, since there are situations where we want each variable to contribute to the distance measure in proportion to its original scale. Variables in JMP can also be standardized using the *Formula Editor*, or by using the dynamic transformation feature from the data table or from any analysis column selection list.

> **STANDARDIZING DATA IN JMP**
>
> Standardizing data prior to analysis or modeling in JMP is typically not required—it is generally done automatically or is provided as a default option in JMP analysis platforms (where needed). However, standardization can easily be done using the dynamic transformation feature in JMP. From an analysis dialog, right-click on a variable in the column selection list and select *Distributional > Standardize*. From the data table, right-click on the column header for the continuous variable and select *New Formula Column > Transform > Standardize*.

2.5   PREDICTIVE POWER AND OVERFITTING

In supervised learning, a key question presents itself: How well will our prediction or classification model perform when we apply it to new data? We are particularly interested in comparing the performance of various models so that we can choose the one we think will do the best when it is implemented in practice. A key concept is to make sure that our chosen model generalizes beyond the dataset that we have at hand. To ensure generalization, we use the concept of *data partitioning* and try to avoid *overfitting*. These two important concepts are described next.

Creation and Use of Data Partitions

At first glance, we might think it best to choose the model that did the best job of classifying or predicting the outcome variable of interest with the data at hand. However, when we use the same data both to develop the model and to assess its performance, we introduce an "optimism" bias. This is because when we choose the model that works best with the data, this model's superior performance comes from two sources:

- A superior model
- Chance aspects of the data that happen to match the chosen model better than they match other models

The latter is a particularly serious problem with techniques (e.g., trees and neural nets) that do not impose linear or other structure on the data, and thus end up overfitting it.

To address the overfitting problem, we simply divide (partition) our data and develop our model using only one of the partitions. After we have a model, we try it out on another partition and see how it performs, which we can measure in several ways. In a classification model, we can count the proportion of held-back records that were misclassified. In a prediction model, we can measure the residuals (prediction errors) between the predicted values and the actual values. This evaluation approach in effect mimics the deployment scenario where our model is applied to data that it hasn't "seen." This model-building and evaluation process is generically referred to as *model crossvalidation*.

We typically deal with two or three partitions: a training set, a validation set, and sometimes an additional test set. Partitioning the data into training, validation, and test sets is done either randomly according to predetermined proportions or by specifying which records go into which partition according to some relevant variable (e.g., in timeseries forecasting, the data are partitioned according to their chronological order). In most cases the partitioning should be done randomly to avoid getting a biased partition. It is also possible (although cumbersome) to divide the data into more than three partitions by successive partitioning (e.g., divide the initial data into three partitions, then take one of those partitions and partition it further).

Training Partition   The training partition, typically the largest partition, contains the data used to build the various models we are examining. The same training partition is generally used to develop multiple models.

Validation Partition   The validation partition (sometimes called the *test partition*) is used to assess the predictive performance of each model so that you can compare models and choose the best one. In some algorithms (e.g., classification and regression trees, k-nearest neighbors), the validation partition may be used in automated fashion to tune and improve the model.

Test Partition   The test partition (sometimes called the *holdout* or *evaluation partition*) is used to assess the performance of the chosen model with new data.

Why have both a validation and a test partition? When we use the validation data to assess multiple models and then choose the model that performs best with the validation data, we again encounter another (lesser) facet of the overfitting problem—chance aspects of the validation data that happen to match the chosen model better than they match other models. In other words, by using the validation data to choose one of several models, the performance of the chosen model on the validation data will be overly optimistic.

The random features of the validation data that enhance the apparent performance of the chosen model will probably not be present in new data to which the model is applied. Therefore we might have overestimated the accuracy of our model. The more models we test, the more likely it is that one of them will be particularly effective in modeling the noise in the validation data. Applying the model to the test data, which it has not seen before, will provide an unbiased estimate of how well the model will perform with new data. Figure 2.4 shows the three data partitions and their use in the data mining process. When we are concerned mainly with finding the best model and less with exactly how well it will do, we might use only training and validation partitions.

Note that with many algorithms, such as classification and regression trees, the model is built using the training data, but the validation set is used to find the best model on the training data. So the validation data are an integral part of fitting the initial model in JMP. For a truly independent assessment of the model, the third test partition should be used.

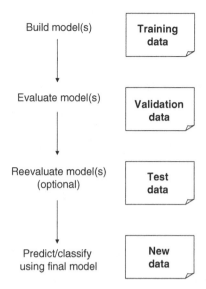

FIGURE 2.4   Three data partitions and their role in the data mining process.

PARTITIONING DATA FOR CROSSVALIDATION IN JMP PRO

The JMP Pro column modeling utility *Make Validation Column* has options for partitioning a dataset. This utility creates a new variable, *Validation*, that contains the values "Training" (stored within JMP as "0"), "Validation" (1) and "Test" (2), according to the allocation rates or counts specified. This new variable specifies how the data are allocated to the partitions.

Overfitting

The more variables we include in a model, the greater the risk of overfitting the particular data used for modeling. What is overfitting?

In Table 2.1 we show hypothetical data about advertising expenditures in one time period and sales in a subsequent time period (a scatterplot of the data is shown in Figure 2.5). We could connect up these points with a smooth but complicated function, one that interpolates all these data points perfectly and leaves no error (residuals). This can be seen in Figure 2.6. However, we can see that such a curve is unlikely to be accurate, or even useful, in predicting future sales on the basis of advertising expenditures (e.g., it is hard to believe that increasing expenditures from $400 to $500 will actually decrease revenue).

A basic purpose of building a model is to represent relationships among variables in such a way that this representation will do a good job of predicting future outcome values

TABLE 2.1

Advertising	Sales
239	514
364	789
602	550
644	1386
770	1394
789	1440
911	1354

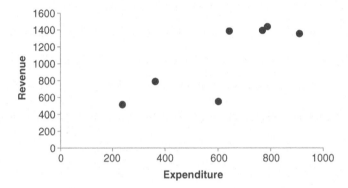

FIGURE 2.5   Scatterplot for advertising and sales data.

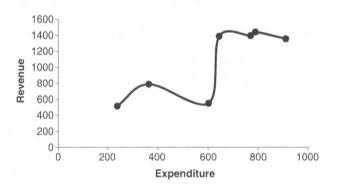

FIGURE 2.6   Overfitting: This function fits the data with no error.

on the basis of future predictor values. Of course, we want the model to do a good job of describing the data we have, but we are more interested in its performance with future data.

In the example above, a simple straight line might do a better job than the complex function in terms of predicting future sales on the basis of advertising. Instead, we devised a complex function that fit the data perfectly, and in doing so, we overreached. We ended up modeling some variation in the data that was nothing more than chance variation. We mislabeled the noise in the data as if it were a signal.

Similarly, we can add predictors to a model to sharpen its performance with the data at hand. Consider a database of 100 individuals, half of whom have contributed to a charitable cause. Information about income, family size, and zip code might do a fair job of predicting whether or not someone is a contributor. If we keep adding more predictors, we can improve the performance of the model with the data at hand and reduce the misclassification error to a negligible level. However, this low error rate is misleading because it probably includes spurious effects, which are specific to the 100 individuals but not beyond that sample.

For example, one of the variables might be height. We have no basis in theory to suppose that tall people might contribute more or less to charity, but if there are several tall people in our sample and they just happened to contribute heavily to charity, our model might include a term for height—the taller you are, the more you will contribute. Of course, when the model is applied to additional data, it is likely that this will not turn out to be a good predictor.

If the dataset is not much larger than the number of predictor variables, it is very likely that a spurious relationship like this will creep into the model. Continuing with our charity example, with a small sample just a few of whom are tall, whatever the contribution level of tall people may be, the algorithm is tempted to attribute it to their being tall. If the dataset is very large relative to the number of predictors, this is less likely to occur. In such a case each predictor must help predict the outcome for a large number of cases, so the job it does is much less dependent on just a few cases, which might be flukes.

Somewhat surprisingly, even if we know for a fact that a higher degree curve is the appropriate model, if the model-fitting dataset is not large enough, a lower degree function (that is not as likely to fit the noise) may perform better. Overfitting can also result from the application of many different models, from which the best performing model is selected.

2.6   BUILDING A PREDICTIVE MODEL WITH JMP PRO

Let us go through the steps typical to many data mining tasks using a familiar procedure: multiple linear regression. This will help you understand the overall process before we begin tackling new algorithms.

Predicting Home Values in a Boston Neighborhood

The Internet has revolutionized the real estate industry. Realtors now list houses and their prices on the web, and estimates of house and condominium prices have become widely available, even for units not on the market. Zillow (www.zillow.com) is the most popular online real estate information site,[1] and in 2014 they purchased their major rival, Trulia. By 2015 Zillow had become the dominant platform for checking house prices and, as such, the dominant online advertising venue for realtors. What used to be a comfortable 6% commission structure for realtors, affording them a handsome surplus (and an oversupply of realtors) was being rapidly eroded by an increasing need to pay for advertising on Zillow. (This, in fact, is the key to Zillow's business model—redirecting the 6% commission away from realtors and to itself.)

Zillow gets much of the data for its "Zestimates" of home values directly from publicly available city housing data, used to estimate property values for tax assessment. A

---

[1] "Zestimates may not be as right as you'd like" *Washington Post*, Feb. 7, 2015, p. T10, by K. Harney.

TABLE 2.2   Description of Variables in West Roxbury (Boston) Home Value Dataset

TOTAL VALUE	Total assessed value for property, in thousands of USD
TAX	Tax bill amount based on total assessed value multiplied by the tax rate, in USD
LOT SQ FT	Total lot size of parcel in square feet
YR BUILT	Year property was built
GROSS AREA	Gross floor area
LIVING AREA	Total living area for residential properties (ft^2)
FLOORS	Number of floors
ROOMS	Total number of rooms
BEDROOMS	Total number of bedrooms
FULL BATH	Total number of full baths
HALF BATH	Total number of half baths
KITCHEN	Total number of kitchens
FIREPLACE	Total number of fireplaces
REMODEL	When house was remodeled (Recent/Old/None)

competitor seeking to get into the market would likely take the same approach. So might realtors seeking to develop an alternative to Zillow.

A simple approach would be a naive, model-less method—just use the assessed values as determined by the city. Those values, however, do not necessarily include all properties, and they might not include changes warranted by remodeling, additions, and the like. Moreover, the assessment methods used by cities may not be transparent or always reflect true market values. However, the city property data can be used as a starting point to build a model, to which additional data (e.g., as collected by large realtors) can be added later.

Let us look at how Boston property assessment data, available from the City of Boston, might be used to predict home values. The data in `West Roxbury Housing.jmp` includes data for single family owner-occupied homes in West Roxbury, neighborhood southwest of Boston, MA, in 2014. The data include values for various predictor variables, and for a target—assessed home value ("total value"). This dataset has 14 variables, and a description of each variable is given in Table 2.2. (The full data dictionary provided by the City of Boston is at http://goo.gl/QBRIYF; we have modified a few variable names, and have omitted some of the variables.) The dataset includes 5802 homes. A sample of the data is shown in Figure 2.3.

Below the header row, each row in the data represents a home. For example, the first home was assessed at a total value of $344.2 thousand (TOTAL VALUE). Its tax bill was $4330. It has a lot size of 9965 square feet (ft^2), was built in year 1880, has two floors, 6 rooms, and so on.

Modeling Process

We now describe in detail the various model stages using the Boston home values example.

1. *Purpose*. We will assume that the purpose of our data mining project is to predict the value of homes in West Roxbury.
2. *Obtain the data*. We will use the 2014 West Roxbury Housing data. The dataset in question is small enough that we do not need to sample from it—we can use it in its entirety.

TABLE 2.3 First Ten Records in the West Roxbury home values dataset

TOTAL VALUE	TAX	LOT SQFT	YR BUILT	GROSS AREA	LIVING AREA	FLOORS	ROOMS	BED ROOMS	FULL BATH	HALF BATH	KIT CHEN	FIRE PLACE	REMODEL
344.2	4330	9965	1880	2436	1352	2	6	3	1	1	1	0	None
412.6	5190	6590	1945	3108	1976	2	10	4	2	1	1	0	Recent
330.1	4152	7500	1890	2294	1371	2	8	4	1	1	1	0	None
498.6	6272	13773	1957	5032	2608	1	9	5	1	1	1	1	None
331.5	4170	5000	1910	2370	1438	2	7	3	2	0	1	0	None
337.4	4244	5142	1950	2124	1060	1	6	3	1	0	1	1	Old
359.4	4521	5000	1954	3220	1916	2	7	3	1	1	1	0	None
320.4	4030	10000	1950	2208	1200	1	6	3	1	0	1	0	None
333.5	4195	6835	1958	2582	1092	1	5	3	1	0	1	1	Recent
409.4	5150	5093	1900	4818	2992	2	8	4	2	0	1	0	None

3. *Explore, clean, and preprocess the data.* Let us look first at the description of the variables, also known as the "data dictionary," to be sure that we understand them all. These descriptions are available as column notes in the data table (hold your mouse over the variable names in the *Columns Panel* to display) and in Table 2.3. The variable names and descriptions in this dataset all seem fairly straightforward, but this is not always the case. Often, variable names are cryptic and their descriptions may be unclear or missing.

It is useful to pause and think about what the variables mean and whether they should be included in the model. Consider the variable TAX. At first glance, we consider that the tax on a home is usually a function of its assessed value, so there is some circularity in the model—we want to predict a home's value using TAX as a predictor, yet TAX itself is determined by a home's value. TAX might be a very good predictor of home value in a numerical sense, but would it be useful if we wanted to apply our model to homes whose assessed value might not be known? For this reason we will exclude TAX from the analysis.

It is also useful to check for outliers that might be errors. For example, suppose that the column FLOORS (number of floors) looked like the one in Table 2.4, after sorting the data in descending order based on floors.

We can tell right away that the 15 is in error— it is unlikely that a home has 15 floors. All other values are between 1 and 2. Probably, the decimal was misplaced and the value should be 1.5.

Last, we may create dummy variables for categorical variables. Here we have one categorical variable: REMODEL, which has three categories. Recall that, in JMP, creating dummy variables for categorical variables is not required. So we can use REMODEL as is, without using dummy coding.

4. *Reduce the data dimension.* Our dataset has been prepared for presentation with fairly low dimension—it has only 12 potential predictor variables (not including TAX), and the single categorical variable considered has only three categories. If we had many more variables, at this stage we might want to apply a variable reduction technique such as condensing multiple categories into a smaller number, or applying principal components analysis to consolidate multiple similar numerical variables (e.g., LIVING AREA, ROOMS, BEDROOMS, BATH, HALF BATH) into a smaller number of variables.

5. *Determine the data mining task.* In this case, as noted, the specific task is to predict the value of TOTAL VALUE using the predictor variables.

6. *Partition the data (for supervised tasks).* Our task is to predict the house value and then assess how well that prediction performs. We will partition the data into a training set to build the model and a validation set to see how well the model does.

TABLE 2.4   Outlier in West Roxbury data

FLOORS	ROOMS
15	8
2	10
1.5	6
1	6

This technique is part of the "supervised learning" process in classification and prediction problems. These are problems in which we know the class or value of the outcome variable for some data, and we want to use those data in developing a model that can then be applied to other data where that value is unknown.

In JMP Pro, select *Modeling Utilities > Make Validation Column* from the *Cols* menus, and the dialog box shown in Figure 2.7 appears.[2] In JMP 12, the partitioning can be handled in one of two ways:

a. The partitioning can be done purely at random (*Purely Random*).

b. The partitioning can be based on a selected stratifying variable (or multiple stratifying variables) that governs the division into training and validation partitions (*Stratified Random*).

In this case we divide the data into two partitions, training and validation, using the *Purely Random* option and a 50-50 split. This produces a new column in the data table, named *Validation*, which specifies the partition each record is assigned to. The training partition is used to build the model, and the validation partition is used to see how well the model does when applied to new data. Although not used in our example, a test partition might also be used.

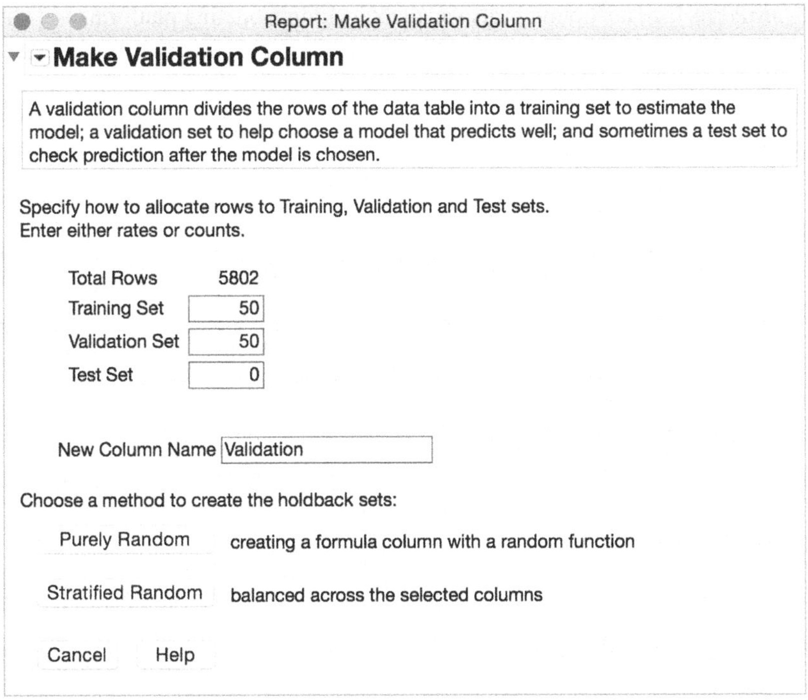

FIGURE 2.7 Partitioning the data using the JMP PRO *Make Validation Column* utility. In this example, a partition of 50% training, 50% validation and 0% test is used.

---

[2]The *Make Validation Column* utility has additional options in JMP Pro 13.

When partitioning data, a random seed can be set to duplicate the same random partition later should we need to. Here, we use the Validation column previously created and saved in the data table (this is the last column).

Note: The *Make Validation Column* utility and built-in model crossvalidation are only available in JMP Pro—these features are not available in the standard version of JMP. However, a validation column can be created manually in JMP using the *Column Info > Initialize Data > Random, Indicator* function. And, crossvalidation can be done manually for most modeling methods using the *Hide and Exclude* option from the *Rows* menu. *Throughout this book, we will be using JMP Pro for its automatic crossvalidation functionality and other model-building features. Many of the modeling techniques, however, are native in the standard version of JMP.*

SETTING THE RANDOM SEED IN JMP

The *Random Seed Reset* add-in is available from the JMP User Community (search for *Random Seed Reset* at community.jmp.com). To install the add-in, navigate to add-in page on the JMP community, read the instructions, and click on the attached file to open the add-in in JMP. Agree to install the file, and a new menu item will be added in JMP with the *Random Seed Reset* option.

The option to set the random seed when creating a validation column is available in JMP Pro 13. This option is also available within modeling platforms in JMP Pro 13.

---

7. *Choose the technique.* In this case it is multiple linear regression. Having divided the data into training and validation partitions, we can use JMP to build a multiple linear regression model with the training data. We want to predict the value of a house in West Roxbury on the basis of all the other values (except TAX).

8. *Use the algorithm to perform the task.* In JMP, we select *Fit Model* platform from the *Analyze* menu. The Fit Model dialog is shown in Figure 2.8. The variable TOTAL VALUE is selected as the *Y* (dependent) variable. All the other variables, except TAX and Validation, are selected as model effects (input or predictor variables). We use the partition column, *Validation*, to fit a model on the training data (we put this in the *Validation* field). JMP produces standard regression output, and provides both training and validation statistics. For now, we defer discussion of the output, along with other available options within the analysis window. (See Chapter 6, Multiple Linear Regression, or the book *Fitting Linear Models* under the *Help* menu in JMP for more information.)

Here, we focus on the model predictions and the prediction error (the residuals). These can be saved to the data table using the *red triangle* in the analysis window, and selecting the options from the *Save Columns* menu (as shown in Figure 2.9).

Figure 2.10 shows the actual values for the first ten records (observations that are in the validation set are selected), along with the predicted values and the residual values (the prediction error). Note that the predicted values are often called the *fitted values*, since they are for the records to which the model was fit.

The prediction error for the training and validation data can be compared using standard JMP platforms, such as *Analyze > Distribution* and *Analyze > Tabulate*.

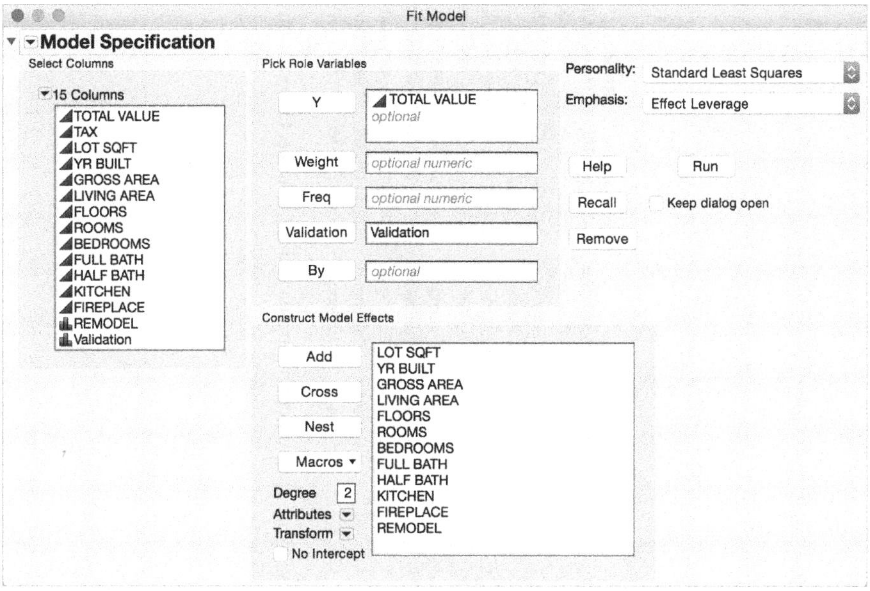

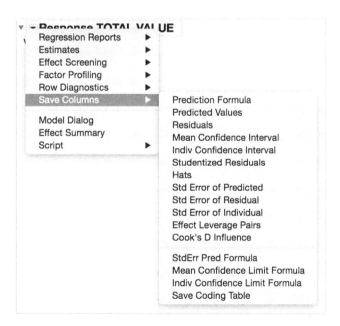

FIGURE 2.8   Using the JMP Pro Fit Model platform for multiple linear regression.

FIGURE 2.9   Use the Red Triangle in the Fit Least Squares analysis window to save the Prediction formula and residuals to the data table.

	TOTAL VALUE	TAX	LOT SQFT	YR BUILT	GROSS AREA	LIVING AREA	FLOORS	ROOMS	BEDROOMS	FULL BATH	HALF BATH	KITCHEN	FIREPLACE	REMODEL	Validation	Pred Formula TOTAL VALUE	Residual TOTAL VALUE
1	344.2	4330	9965	1880	2436	1352	2	6	3	1	1	1	1	0 None	Validation	383.93561633	-39.73561633
2	412.6	5190	6590	1945	3108	1976	2	10	4	2	1	1	1	0 Recent	Validation	461.42424888	-48.82424888
3	330.1	4152	7500	1890	2294	1371	2	8	4	1	1	1	1	0 None	Validation	360.46212838	-30.36212838
4	498.6	6272	13773	1957	5032	2608	1	9	5	1	1	1	1	1 None	Validation	548.5362171	-49.9362171
5	331.5	4170	5000	1910	2370	1438	2	7	3	2	0	1		0 None	Validation	347.43565162	-15.93565162
6	337.4	4244	5142	1950	2124	1060	1	6	3	1	0	1	1	1 Old	Training	286.521008	50.878991995
7	359.4	4521	5000	1954	3220	1916	2	7	3	1	1	1		0 None	Validation	402.3366772	-42.9366772
8	320.4	4030	10000	1950	2208	1200	1	6	3	1	0	1		0 None	Training	316.04156996	4.3584300434
9	333.5	4195	6835	1958	2582	1092	1	5	3	1	0	1	1	1 Recent	Training	339.45476038	-5.954760377
10	409.4	5150	5093	1900	4818	2992	2	8	4	2	0	1		0 None	Validation	503.29534518	-93.89534518

FIGURE 2.10 Predictions for the first 10 observations in the West Roxbury Housing data, made by saving the prediction formula to the data table. Observations in the validation set are highlighted.

▼ ⊡ **Tabulate**

	Residual TOTAL VALUE		
Validation	**Mean**	**Min**	**Max**
Training	-6.28683e-15	-276.1021151	232.06289994
Validation	0.4195725867	-209.0365292	290.74393395

FIGURE 2.11   Summary statistics for error rates for training (top) and validation (bottom) data produced using *Analyze > Tabulate* (error figures are in thousands of $).

▼ **Crossvalidation**

Source	RSquare	RASE	Freq
Training Set	0.8031	43.018	2821
Validation Set	0.8220	42.701	2981

FIGURE 2.12   Crossvalidation statistics from the JMP PRO *Fit Least Squares* analysis window for the regression model.

For example, in Figure 2.11 we see the result of using *Tabulate* to summarize the errors for the training and validations sets. Many different statistical summaries are available in *Tabulate*—here, we've selected the *Mean* (the average error), the *Min*, and the *Max*.

We see that the average error is quite small relative to the units of TOTAL VALUE, indicating that, in general, predictions average about right—our predictions are "unbiased." Of course, this simply means that the positive and negative errors balance out. It tells us nothing about how large these errors are. The *Min* and *Max* (see Figure 2.11) show the range of prediction errors, but there are several other possible measures of prediction error, many of which will be discussed in Chapter 5.

Two measures of prediction error are provided by default in the *Fit Least Squares* analysis window when a validation column is used (see Figure 2.12): *RSquare* and *RASE*. *RSquare*, which is based on the sum of the squared errors, is a measure of the percent of variation in the output variable explained by the inputs. RASE (or RMSE) is a far more useful measure of prediction error. RASE takes the square root of the average squared errors, so it gives an idea of the typical error (whether positive or negative) in the same scale as that used for the original data.

In this example, the RASE for the validation data ($42.7 thousand), which the model is seeing for the first time in making these predictions, is close to the RASE for the training data ($43.02 thousand), which were used in training the model.

9. *Interpret the results.* At this stage we would typically try other prediction algorithms (e.g., regression trees) and see how they do error-wise. We might also try different "settings" on the various models (e.g., we could use the *Stepwise* option in multiple linear regression to chose a reduced set of variables that might perform better with the validation data, or we might try a different algorithm, such as a *regression tree*). For each model we would save the prediction formula to the data table, and then use the *Model Comparison* platform in JMP Pro to compare validation statistics for the different models.

An additional measure of prediction error, *AAE*, or *average absolute error* is provided in the *Model Comparison* platform (under *Analyze > Modeling*). This is sometimes reported as *MAD (mean absolute deviation)* or *MAE (mean absolute error)*. (Note that additional measures can also be computed manually using the JMP *Formula Editor*.)

After choosing the best model (typically, the model with the lowest error on the validation data while also recognizing that "simpler is better"), we use that model to predict the output variable using fresh data. These steps are covered in more detail in the analysis of cases.

10. *Deploy the model.* After the best model is chosen, it is applied to new data to predict TOTAL VALUE for homes where this value is unknown. This was, of course, the overall purpose.

Predicting the output value for new records is called *scoring*. For predictive tasks, scoring produces predicted numerical values. For classification tasks, scoring produces classes and/or propensities.

In JMP, we can score new records using the models that we developed. To do this, we simply add new rows to the data table and enter values for the predictors into the appropriate cells. When the prediction formula for a model is saved to the data table, predicted values are automatically computed for these new rows added. Figure 2.13 shows an example of the JMP data table with two homes scored using our saved linear regression model. Note that all the required predictor columns are present, and the output column is absent.

Additional information for these predicted values, such as confidence intervals, can be added by using the *Save Columns* option from the top red triangle in the *Fit Least Squares* analysis window for the model. Data in a separate JMP data table, with the same predictor variables, can also be scored by copying the saved prediction formula into a column in the data table via the *Formula Editor*.

2.7   USING JMP PRO FOR DATA MINING

An important aspect of this process to note is that the heavy-duty analysis does not necessarily require a huge number of records. The dataset to be analyzed might have millions of records, and JMP can generally handle this. However, in applying multiple linear regression or applying a classification tree, the use of a sample of 20,000 is likely to yield as accurate an answer as that obtained when using the entire dataset. The principle involved is the same as the principle behind polling: if sampled judiciously, 2000 voters can give an estimate of the entire population's opinion within one or two percentage points. (See "How Many Variables and How Much Data" in Section 2.4 for further discussion.)

Therefore, in most cases, the number of records required in each partition (training, validation, and test) can be accommodated by JMP. Of course, we need to get those records into JMP, and for this purpose JMP provides the ability to import data in a number of different formats, to connect to external databases using the *Query Builder*, and to randomly sample records from an external database or SAS dataset.

	TOTAL VALUE	TAX	LOT SQFT	YR BUILT	GROSS AREA	LIVING AREA	FLOORS	ROOMS	BEDROOMS	FULL BATH	HALF BATH	KITCHEN	FIREPLACE	REMODEL	Validation	Pred Formula TOTAL VALUE
5795	371.2	4669	6116	1950	2522	1440	2	8	3	3	1	1	1	None	Training	385.55386053
5796	386.8	4865	6350	1950	2316	1326	2	7	3	3	2	1	1	None	Validation	393.38023876
5797	413....	5200	9150	1950	2324	1326	2	7	3	3	1	1	1	None	Validation	399.20831255
5798	404.8	5092	6762	1938	2594	1714	2	9	3	3	2	1	1	Recent	Validation	452.02284225
5799	407.9	5131	9408	1950	2414	1333	2	6	3	3	1	1	1	None	Training	403.72565544
5800	406.5	5113	7198	1987	2480	1674	2	7	3	3	1	1	1	None	Training	408.46216395
5801	308.7	3883	6890	1946	2000	1000	1	5	2	2	0	1	0	None	Validation	271.83275825
5802	447.6	5630	7406	1950	2510	1600	2	7	3	3	1	1	1	None	Validation	403.66586088
5803	•	•	6000	2000	2500	2000	1	6	3	3	0	1	0	None	Validation	354.89686925
5804	•	•	8000	2010	3000	2200	2	8	4	4	2	1	1	None	Validation	479.79538035

FIGURE 2.13   Two records scored using the saved linear regression model.

2.8   AUTOMATING DATA MINING SOLUTIONS

In most supervised data mining applications, the goal is not a static, one-time analysis of a particular dataset. Rather, we want to develop a model that can be used on an ongoing basis to predict or classify new records. Our initial analysis will be in prototype mode, while we explore and define the problem and test different models. We will follow all the steps outlined earlier in this chapter.

At the end of that process, we will typically want our chosen model to be deployed in automated fashion. For example, the US Internal Revenue Service (IRS) receives several hundred million tax returns per year—it does not want to have to pull each tax return out into an Excel sheet or other environment separate from its main database to determine the predicted probability that the return is fraudulent. Rather, it would like that determination to be made as part of the normal tax filing environment and process. Pandora or Spotify need to determine "recommendations" for next songs quickly for each of millions of users; there is no time to extract the data for manual analysis.

In practice, this is done by building the chosen algorithm into the computational setting in which the rest of the process or database lies. A tax return is entered directly into the IRS system by a tax preparer, a predictive algorithm is immediately applied to the new data in the IRS system, and a predicted classification is decided by the algorithm. Business rules would then determine what happens with that classification. In the IRS case, the rule might be "if no predicted fraud, continue routine processing; if fraud is predicted, alert an examiner for possible audit."

This flow of the tax return from data entry, into the IRS system, through a predictive algorithm, then back out to a human user is an example of a "data pipeline." The different components of the system communicate with one another via Application Programming Interfaces (API's) that establish locallyvalid rules for transmitting data and associated communciations. An API for a data mining algorithm would establish the required elements for a predictive algorithm to work—the exact predictor variables, their order, data formats, and so on. It would also establish the requirements for communciating the results of the algorithm. Algorithms to be used in an automated data pipeline will need to be compliant with the rules of the API's where they operate.

Finally, once the computational environment is set and functioning, the data miner's work is not done. The environment in which a model operates is typically dynamic, and predictive models often have a short shelf life—one leading consultant finds they rarely continue to function effectively for more than a year. So, even in a fully deployed state, models must be periodically checked and re-evaluated. Once performance lags, it is time to return to prototype mode and see if a new model can be developed.

In this book, our focus will be on the prototyping phase—all the steps that go into properly defining the model and developing and selecting a model. You should be aware, though, that most of the actual work of implementing a data mining model lies in the automated deployment phase. Much of this work is not in the analytic domain, rather it lies in the domains of databases and computer science, to ensure that detailed nuts and bolts of an automated dataflow all work properly.

DATA MINING SOFTWARE TOOLS: THE STATE OF THE MARKET BY HERB EDELSTEIN [3]

The data mining market has altered in some important ways in the last decade.

The most significant trends have been the increasing volume of information available and the growing use of the cloud for data storage and analytics. Data mining and analysis have evolved to meet these new demands.

The term "Big Data" reflects the surge in the amount and types of data collected. There is still an enormous amount of transactional data, data warehousing data, scientific data, and clickstream data. However, adding to the massive storage requirements of traditional data is the influx of information from unstructured sources (e.g., customer service calls and images), social media, and most recently the Internet of Things, which produces a flood of sensor data. The number of organizations collecting such data has greatly increased as virtually every business is expanding in these areas.

Rapid technological change has been an important factor. The price of data storage has dropped precipitously. At this writing, hard disk storage has fallen to about $50 per terabyte and solid state drives are about $200 per terabyte. Concomitant with the decrease in storage costs has been a dramatic increase in bandwidth at ever lower costs. This has enabled the spread of cloud-based computing.

Cloud-based computing refers to using remote platforms for storage, data management, and now analysis. Because scaling up the hardware and software infrastructure for Big Data is so complex, many organizations are entrusting their data to outside vendors. The largest cloud players (Amazon, Google, IBM, and Microsoft) are each reporting annual revenues in excess of US$5 billion, according to Forbes magazine.

Managing this much data is a challenging task. While traditional relational DBMSs—such as Oracle, Microsoft's SQL Server, IBM's DB2, and SAP's Adaptive Server Enterprise (formerly Sybase)—are still among the leading data management tools, open source DBMSs such as Oracle's MySQL are becoming increasingly popular. In addition, non-relational data storage is making headway in storing extremely large amounts of data. For example, Hadoop, an open source tool for managing large distributed database architectures, has become an important player in the cloud database space. However, Hadoop is a tool for the application developer community rather than end users. Consequently, many of the data mining analytics vendors such as SAS have built interfaces to Hadoop.

All the major database management system vendors offer data mining capabilities, usually integrated into their DBMS. Leading products include Microsoft SQL Server Analysis Services, Oracle Data Mining, and Teradata Warehouse Miner. The target user for embedded data mining is a database professional. Not surprisingly, these products take advantage of database functionality, including using the DBMS to transform variables, storing models in the database, and extending the data access language to include model-building and scoring the database. A few products also supply a separate graphical interface for building data mining models. Where the DBMS has parallel processing capabilities, embedded data mining tools will generally take advantage of it, resulting in greater performance. As with the data mining suites described below, these tools offer an assortment of algorithms. Not only does IBM have embedded analytics in DB2, but following its acquisition of SPSS, IBM has incorporated Clementine and SPSS into IBM Modeler.

There are still a large number of stand-alone data mining tools based on a single algorithm or on a collection of algorithms called a suite. Target users include both statisticians and business intelligence analysts. The leading suites include SAS Enterprise Miner, SAS JMP, IBM Modeler, Salford Systems SPM, Statistica, XLMiner, and RapidMiner. Suites are characterized by providing a wide range of functionality, frequently accessed via a graphical user interface designed to enhance model-building productivity. A popular approach for many of these GUIs is to provide a workflow interface in which the data mining steps and analysis are linked together.

Many suites have outstanding visualization tools and links to statistical packages that extend the range of tasks they can perform. They also provide interactive data transformation tools as well as a procedural scripting language for more complex data transformations. The suite vendors are working to link their tools more closely to underlying DBMSs; for example, data transformations might be handled by the DBMS. Data mining models can be exported to be incorporated into the DBMS through generating SQL, procedural language code (e.g., C++ or Java), or a standardized data mining model language called Predictive Model Markup Language (PMML).

In contrast to the general-purpose suites, application-specific tools are intended for particular analytic applications such as credit scoring, customer retention, or product marketing. Their focus may be further sharpened to address the needs of specialized markets such as mortgage lending or financial services. The target user is an analyst with expertise in the application domain. Therefore the interfaces, the algorithms, and even the terminology are customized for that particular industry, application, or customer. While less flexible than general-purpose tools, they offer the advantage of already incorporating domain knowledge into the product design, and can provide very good solutions with less effort. Data mining companies including SAS, IBM, and RapidMiner offer vertical market tools, as do industry specialists such as Fair Isaac. Other companies, such as Domo, are focusing on creating dashboards with analytics and visualizations for business intelligence.

Another technological shift has occurred with the spread of open source model building tools and open core tools. A somewhat simplified view of open source software is that the source code for the tool is available at no charge to the community of users and can be modified or enhanced by them. These enhancements are submitted to the originator or copyright holder, who can add them to the base package. Open core is a more recent approach in which a core set of functionality remains open and free, but there are proprietary extensions that are not free.

The most important open source statistical analysis software is R. R is descended from a Bell Labs program called S, which was commercialized as S+. Many data mining algorithms have been added to R, along with a plethora of statistics, data management tools, and visualization tools. Because it is essentially a programming language, R has enormous flexibility, but a steeper learning curve than many of the GUI-based tools. Although there are some GUIs for R, the overwhelming majority of use is through programming.

Some vendors, as well as the open source community, are adding statistical and data mining tools to Python, a popular programming language that is generally easier to use than C++ or Java, and faster than R.

As mentioned above, the cloud computing vendors have moved into the data mining/predictive analytics business by offering AaaS (Analytics as a Service) and pricing their products on a transaction basis. These products are oriented more toward application developers than business intelligence analysts. A big part of the attraction of mining data in the cloud is the ability to store and manage enormous amounts of data without requiring the expense and complexity of building an in-house capability. This can also enable a more rapid implementation of large distributed multi-user applications. Cloud based data can be used with non-cloud based analytics if the vendor's analytics do not meet the user's needs.

Amazon has added Amazon Machine Learning to its Amazon Web Services (AWS), taking advantage of predictive modeling tools developed for Amazon's internal use. AWS supports both relational databases and Hadoop data management. Models cannot be exported because they are intended to be applied to data stored on the Amazon cloud.

Google is very active in cloud analytics with its BigQuery and Prediction API. BigQuery allows the use of Google infrastructure to access large amounts of data using a SQL-like interface. The Prediction API can be accessed from a variety of languages including R and Python. It uses a variety of machine learning algorithms and automatically selects the best results. Unfortunately, this is not a transparent process. Furthermore, as with Amazon, models cannot be exported.

Google is very active in cloud analytics with its BigQuery and Prediction API. BigQuery allows the use of Google infrastructure to access large amounts of data using a SQL-like interface. The Prediction API can be accessed from a variety of languages including R and Python. It uses a variety of machine learning algorithms and automatically selects the best results. As with Amazon, models cannot be exported.

Microsoft is an active player in cloud analytics with its Azure Machine Learning Studio and Stream Analytics. Azure works with Hadoop clusters as well as with traditional relational databases. Azure ML offers a broad range of algorithms such as boosted trees and support vector machines as well as supporting R scripts and Python. Azure ML also supports a workflow interface making it more suitable for the nonprogrammer data scientist. The real-time analytics component is designed to allow streaming data from a variety of sources to be analyzed on the fly. Microsoft also acquired Revolution Analytics, a major player in the R analytics business, with a view to integrating Revolution's "R Enterprise" with SQL Server and Azure ML. R Enterprise includes extensions to R that eliminate memory limitations and take advantage of parallel processing.

One drawback of the cloud-based analytics tools is a relative lack of transparency and user control over the algorithms and their parameters. In some cases, the service will simply select a single model that is a black box to the user. Another drawback is that for the most part cloud based tools are aimed at more sophisticated data scientists who are systems savvy.

Data science is playing a central role in enabling many organizations to optimize everything from production to marketing. New storage options and analytical tools promise even greater capabilities. The key is to select technology that's appropriate for an organization's unique goals and constraints. As always, human judgment is the most important component of a data mining solution.

This book's focus is on a comprehensive understanding of the different techniques and algorithms used in data mining, and less on the data management requirements of real-time deployment of data mining models. JMP's short learning curve makes it ideal for this purpose, and for exploration, prototyping, and piloting of solutions.

---

[3] Herb Edelstein is president of Two Crows Consulting (www.twocrows.com), a leading data mining consulting firm near Washington, DC. He is an internationally recognized expert in data mining and data warehousing, a widely published author on these topics, and a popular speaker. @ 2015 Herb Edelstein.

PROBLEMS

2.1 Assuming that data mining techniques are to be used in the following cases, identify whether the task required is supervised or unsupervised learning.

    a. Deciding whether to issue a loan to an applicant based on demographic and financial data (with reference to a database of similar data on prior customers).

    b. In an online bookstore, making recommendations to customers concerning additional items to buy based on the buying patterns in prior transactions.

    c. Identifying a network data packet as dangerous (virus, hacker attack) based on comparison to other packets whose threat status is known.

    d. Identifying segments of similar customers.

    e. Predicting whether a company will go bankrupt based on comparing its financial data to those of similar bankrupt and nonbankrupt firms.

    f. Estimating the repair time required for an aircraft based on a trouble ticket.

    g. Automated sorting of mail by zip code scanning.

    h. Printing of custom discount coupons at the conclusion of a grocery store checkout based on what you just bought and what others have bought previously.

2.2 Describe the difference in roles assumed by the validation partition and the test partition.

2.3   Consider the sample from a database of credit applicants in Table 2.5. Comment on the likelihood that it was sampled randomly, and whether it is likely to be a useful sample.

2.4   Consider the sample from a bank database shown in Table 2.6; it was selected randomly from a larger database to be the training set. *Personal Loan* indicates whether a solicitation for a personal loan was accepted and is the response variable. A campaign is planned for a similar solicitation in the future and the bank is looking for a model that will identify likely responders. Examine the data carefully and indicate what your next step would be.

2.5   Using the concept of overfitting, explain why when a model is fit to training data, zero error with those data is not necessarily good.

2.6   In fitting a model to classify prospects as purchasers or nonpurchasers, a certain company drew the training data from internal data that include demographic and purchase information. Future data to be classified will be lists purchased from other sources, with demographic (but not purchase) data included. It was found that "refund issued" was a useful predictor in the training data. Why is this not an appropriate variable to include in the model?

2.7   A dataset has 1000 records and 50 variables with 5% of the values missing, spread randomly throughout the records and variables. An analyst decides to remove records that have missing values. About how many records would you expect would be removed?

2.8   Normalize the data in Table 2.7, showing calculations. Confirm your results in JMP (create a JMP data table, then use the *Formula Editor* or the dynamic transformation feature).

2.9   Statistical distance between records can be measured in several ways. Consider Euclidean distance, measured as the square root of the sum of the squared differences. For the first two records in Table 2.7, it is

$$\sqrt{(25 - 56)^2 + (49,000 - 156,000)^2}.$$

Can normalizing the data change which two records are farthest from each other in terms of Euclidean distance?

2.10 Two models are applied to a dataset that has been partitioned. Model A is considerably more accurate than model B on the training data, but slightly less accurate than model B on the validation data. Which model are you more likely to consider for final deployment?

2.11 The dataset `ToyotaCorolla.jmp` contains data on used cars on sale during the late summer of 2004 in the Netherlands. It has 1436 records containing details on 38 attributes, including Price, Age, Kilometers, HP, and other specifications.

   a. Explore the data using the data visualization (e.g., *Graph > Scatterplot Matrix* and *Graph > Graph Builder*) capabilities of JMP. Which of the pairs among the variables seem to be correlated? (Refer to the guides and videos at jmp.com/learn, under *Graphical Displays and Summaries*, for basic information on how to use these platforms.)

TABLE 2.5 Sample from a database of credit applications

OBS	CHEK ACCT	DURATION	HISTORY	NEW CAR	USED CAR	FURNITURE	RADIO TV	EDUC	RETRAIN	AMOUNT	SAVE ACCT	RESPONSE
1	0	6	4	0	0	0	1	0	0	1169	4	1
8	1	36	2	0	1	0	0	0	0	6948	0	1
16	0	24	2	0	0	0	1	0	0	1282	1	0
24	1	12	4	0	1	0	0	0	0	1804	1	1
32	0	24	2	0	0	1	0	0	0	4020	0	1
40	1	9	2	0	0	0	1	0	0	458	0	1
48	0	6	2	0	1	0	0	0	0	1352	2	1
56	3	6	1	1	0	0	0	0	0	783	4	1
64	1	48	0	0	0	0	0	0	1	14421	0	0
72	3	7	4	0	0	0	1	0	0	730	4	1
80	1	30	2	0	0	1	0	0	0	3832	0	1
88	1	36	2	0	0	0	0	1	1	12612	1	0
96	1	54	0	0	0	0	0	0	1	15945	0	0
104	1	9	4	0	0	1	0	0	0	1919	0	1
112	2	15	2	0	0	0	0	1	0	392	0	1

TABLE 2.6 Sample from a bank database

OBS	AGE	EXPERIENCE	INCOME	ZIP	FAMILY	CC AVG	EDUC	MORTGAGE	PERSONAL LOAN	SECURITIES ACCT
1	25	1	49	91107	4	1.6	1	0	0	1
4	35	9	100	94112	1	2.7	2	0	0	0
5	35	8	45	91330	4	1	2	0	0	0
9	35	10	81	90089	3	0.6	2	104	0	0
10	34	9	180	93023	1	8.9	3	0	1	0
12	29	5	45	90277	3	0.1	2	0	0	0
17	38	14	130	95010	4	4.7	3	134	1	0
18	42	18	81	94305	4	2.4	1	0	0	0
21	56	31	25	94015	4	0.9	2	111	0	0
26	43	19	29	94305	3	0.5	1	97	0	0
29	56	30	48	94539	1	2.2	3	0	0	0
30	38	13	119	94104	1	3.3	2	0	1	0
35	31	5	50	94035	4	1.8	3	0	0	0
36	48	24	81	92647	3	0.7	1	0	0	0
37	59	35	121	94720	1	2.9	1	0	0	0
38	51	25	71	95814	1	1.4	3	198	0	0
39	42	18	141	94114	3	5	3	0	1	1
41	57	32	84	92672	3	1.6	3	0	0	1

TABLE 2.7

Age	Income ($)
25	49,000
56	156,000
65	99,000
32	192,000
41	39,000
49	57,000

b. We plan to analyze the data using various data mining techniques described in future chapters. Prepare the dataset for data mining techniques of supervised learning by creating partitions using the JMP Pro *Make Validation Column* utility (from the *Cols* menu). Use the following partitioning percentages: training (50%), validation (30%), and test (20%). Describe the roles that these partitions will play in modeling.

PART II

DATA EXPLORATION AND DIMENSION REDUCTION

3

DATA VISUALIZATION

In this chapter we describe a set of plots that can be used to explore the multi-dimensional nature of a dataset. We present basic plots (bar charts, line graphs, and scatterplots), distribution plots (boxplots and histograms), and different enhancements that expand the capabilities of these plots to visualize more information. We focus on how the different visualizations and operations can support data mining tasks, from supervised (prediction, classification, and time series forecasting) to unsupervised tasks, and provide some guidelines on specific visualizations to use with each data mining task. We also describe the advantages of interactive visualization over static plots. The chapter concludes with a presentation of specialized plots that are suitable for data with special structure (hierarchical and geographical).

3.1 USES OF DATA VISUALIZATION

The popular saying "a picture is worth a thousand words" refers to the ability to condense diffused verbal information into a compact and quickly understood graphical image. In the case of numbers, data visualization and numerical summarization provide us with both a powerful tool to explore data and an effective way to present results and communicate findings.

Where do visualization techniques fit into the data mining process, as described so far? Their use is primarily in the preprocessing portion of the data mining process. Visualization supports data cleaning by finding incorrect values (e.g., patients whose age is 999 or −1), missing values, duplicate rows, columns with all the same value, and the like. Visualization techniques are also useful for variable derivation and selection: they can help determine which variables to include in the analysis and which might be redundant.

Data Mining for Business Analytics: Concepts, Techniques, and Applications with JMP Pro®, First Edition.
Galit Shmueli, Peter C. Bruce, Mia L. Stephens, and Nitin R. Patel.
© 2017 John Wiley & Sons, Inc. Published 2017 by John Wiley & Sons, Inc.

They can also help with determining appropriate bin sizes, should binning of numerical variables be needed (e.g., a numerical outcome variable might need to be converted to a binary variable if a yes/no decision is required). They can also play a role in combining categories as part of the data reduction process. Finally, if the data have yet to be collected and collection is expensive (as with the Pandora project at its outset, see Chapter 7), visualization methods can help determine, using a sample, which variables and metrics may be useful.

In this chapter we focus on the use of graphical presentations for the purpose of *data exploration*, in particular with relation to predictive analytics. Although our focus is not on visualization for the purpose of data reporting, this chapter offers ideas as to the effectiveness of various graphical displays for the purpose of *data presentation*. The graphical displays we introduce offer a wealth of information beyond tabular summaries and basic bar charts, and are currently the most popular form of data presentation in the business environment. For an excellent comprehensive discussion on using graphs to report business data, see Few (2004). In terms of reporting data mining results graphically, we describe common graphical displays elsewhere in the book, some of which are technique specific (e.g., dendrograms for hierarchical clustering in Chapter 14, and tree charts for classification and regression trees in Chapter 9) while others are more general (e.g., receiver operating characteristic (ROC) curves and lift curves for classification in Chapter 5).

Data exploration is a mandatory initial step whether or not more formal analysis follows. Graphical exploration can support free-form exploration for the purpose of understanding the data structure, cleaning the data (e.g., identifying unexpected gaps or "illegal" values), identifying outliers, discovering initial patterns (e.g., correlations among variables and surprising clusters), and generating interesting questions. Graphical exploration can also be more focused, geared toward specific questions of interest. In the data mining context a combination is needed: free-form exploration performed with the purpose of supporting a specific goal.

Graphical exploration can range from generating very basic plots to using operations such as filtering and zooming interactively to explore a set of interconnected plots that include advanced features such as color and multiple panels. This chapter is not meant to be an exhaustive guidebook on visualization techniques, but rather to introduce main principles and features that support data exploration in a data mining context. We start by describing varying levels of sophistication in terms of visualization, and proceed to show the advantages of different features and operations. Our discussion is from the perspective of how visualization supports the subsequent data mining goal. In particular, we distinguish between supervised and unsupervised learning; within supervised learning, we further distinguish between classification (categorical Y) and prediction (numerical Y).

3.2 DATA EXAMPLES

To illustrate data visualization, we use two datasets. The same two datasets are used in later chapters of the book to allow the reader to compare some of the basic graphs that are improved in the later chapters and to easily see the merit of advanced visualization.

TABLE 3.1   Description of Variables in the Boston Housing Dataset

CRIM	Crime rate
ZN	Percentage of residential land zoned for lots over 25,000 ft^2
INDUS	Percentage of land occupied by nonretail business
CHAS	Charles River dummy variable (= 1 if tract bounds river; = 0 otherwise)
NOX	Nitric oxide concentration (parts per 10 million)
RM	Average number of rooms per dwelling
AGE	Percentage of owner-occupied units built prior to 1940
DIS	Weighted distances to five Boston employment centers
RAD	Index of accessibility to radial highways
TAX	Full-value property-tax rate per $10,000
PTRATIO	Pupil/teacher ratio by town
LSTAT	% Lower status of the population
MEDV	Median value of owner-occupied homes in $1000s
CAT.MEDV	Binary coding of MEDV (= 1 if MEDV > $30,000; = 0 otherwise)

Example 1: Boston Housing Data

The Boston housing data (`Boston Housing.jmp`) contain information on census tracts in Boston[1] for which several measurements are taken (e.g., crime rate, pupil/teacher ratio). The dataset has 14 variables in all. A description of each variable is given in Table 3.1, and these descriptions have been added as notes in the JMP data table (hold your mouse over the variables in the *Columns Panel* in the data table to view the descriptions). The first ten records are shown in Figure 3.1. In addition to the original 13 variables, the dataset contains the variable CAT.MEDV, which was created by categorizing median value (MEDV) into two categories, high (1) and low (0).

We consider three possible tasks:

1. A supervised predictive task, where the outcome variable of interest is the median value of a home in the tract (MEDV),
2. A supervised classification task, where the outcome variable of interest is the binary variable CAT.MEDV that indicates whether the median home value is above or below $30,000.
3. An unsupervised task, where the goal is to cluster census tracts.

(MEDV and CAT.MEDV are not used together in any of the three cases).

Example 2: Ridership on Amtrak Trains

Amtrak, a US railway company, routinely collects data on ridership. Here we focus on forecasting future ridership using the series of monthly ridership data between January

---

[1]The Boston Housing dataset was originally published by Harrison and Rubinfeld in "Hedonic prices and the demand for clean air," *Journal of Environmental Economics & Management*, vol. 5, pp. 81–102, 1978.

	CRIM	ZN	INDUS	CHAS	NOX	RM	AGE	DIS	RAD	TAX	PTRATIO	LSTAT	MEDV	CAT.MEDV
1	0.00632	18	2.31	0	0.538	6.575	65.2	4.09	1	296	15.3	4.98	24	0
2	0.02731	0	7.07	0	0.469	6.421	78.9	4.9671	2	242	17.8	9.14	21.6	0
3	0.02729	0	7.07	0	0.469	7.185	61.1	4.9671	2	242	17.8	4.03	34.7	1
4	0.03237	0	2.18	0	0.458	6.998	45.8	6.0622	3	222	18.7	2.94	33.4	1
5	0.06905	0	2.18	0	0.458	7.147	54.2	6.0622	3	222	18.7	5.33	36.2	1
6	0.02985	0	2.18	0	0.458	6.43	58.7	6.0622	3	222	18.7	5.21	28.7	0
7	0.08829	12.5	7.87	0	0.524	6.012	66.6	5.5605	5	311	15.2	12.43	22.9	0
8	0.14455	12.5	7.87	0	0.524	6.172	96.1	5.9505	5	311	15.2	19.15	27.1	0
9	0.21124	12.5	7.87	0	0.524	5.631	100	6.0821	5	311	15.2	29.93	16.5	0
10	0.17004	12.5	7.87	0	0.524	6.004	85.9	6.5921	5	311	15.2	17.1	18.9	0

FIGURE 3.1 First ten records in the Boston Housing dataset. Each row represents one census tract.

1991 and March 2004. The data and their source are described in Chapter 15. Hence our task here is (numerical) time series forecasting. Data are in Amtrak.jmp.

3.3 BASIC CHARTS: BAR CHARTS, LINE GRAPHS, AND SCATTERPLOTS

The three most effective basic plots—bar charts, line graphs, and scatterplots—are easily created using the JMP *Graph Builder* (*Graph > Graph Builder*). These plots are the commonly used in the current business world, in both data exploration and presentation (unfortunately, pie charts are also popular, although they are usually ineffective visualizations). Basic charts support data exploration by displaying one or more columns of data (variables) at a time. This is useful in the early stages of getting familiar with the data structure, the amount and types of variables, the volume and type of missing values, and so on. The initial *Graph Builder* screen is shown in Figure 3.3.

The nature of the data mining task and domain knowledge about the data will affect the use of basic charts in terms of the amount of time and effort allocated to different variables. In supervised learning, there will be more focus on the outcome variable. In scatterplots, the outcome variable is typically associated with the *y*-axis. In unsupervised learning (for the purpose of data reduction or clustering), basic plots that convey relationships (e.g., scatterplots) are useful.

USING THE JMP GRAPH BUILDER

The JMP *Graph Builder* is a versatile graphing platform for creating basic (and more advanced) graphs for one, two, or many variables in two-dimensional space. Drag variables to zones (e.g., *Y* or *X*), and click the available graph icons at the top to change the graph type. Drag a graph icon onto a graph pane to add an additional graph element, and change display options at the bottom left for selected graph types. To close the control panel and produce the finished graph, click *Done*. (See Figure 3.2.)

Examples of basic plots created using the *Graph Builder* are shown in Figure 3.3. The top left panel in Figure 3.3 displays a line chart for the time series of monthly railway

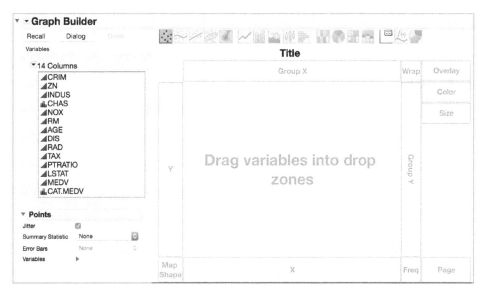

FIGURE 3.2   The JMP Graph Builder initial display.

passengers on Amtrak. Line graphs are used primarily for showing time series. The choice of the time frame to plot, as well as the temporal scale, should depend on the horizon of the forecasting task and on the nature of the data.

Bar charts are useful for comparing a single statistic (e.g., average, count, percentage) across groups. The height of the bar (or length, in a horizontal display) represents the value of the statistic, and different bars correspond to different groups. The bottom left panel in Figure 3.3 shows a bar chart for a numerical variable (MEDV), which uses the average MEDV on the y-axis. Separate bars are used to denote homes in Boston that are near the Charles River (CHAS = 1) as opposed to those that are not (CHAS = 0), thereby comparing the two categories of CHAS. This supports the second predictive task mentioned earlier: the numerical outcome is on the y-axis and the x-axis is used for a potential categorical predictor.[2] (Note that the x-axis on a bar chart must be used only for categorical variables, since the order of bars in a bar chart should be interchangeable.)

For the classification task, CAT.MEDV is on the y-axis in the bottom right panel in Figure 3.3, but its aggregation is a percentage (the alternative would be a count). This graph is a form of a stacked bar graph called a *mosaic plot*. The width of the bars relative to the x-axis shows us that the vast majority (nearly 85%) of the tracts do not border the Charles River (the bar for CHAS = 0 is much wider than for CHAS = 1). Note that the labeling of the y-axis can be confusing in this case: the y-axis is simply a % of records in each category of CAT.MEDV for CHAS = 0 and CHAS = 1.

The top right panel in Figure 3.3 displays a scatterplot of MEDV vs. LSTAT. This is an important plot in the prediction task. Note that the output MEDV is again on the

---

[2]We refer here to a bar chart with vertical bars. The same principles apply if using a bar chart with horizontal lines, except that the x-axis is now associated with the numerical variable and the y-axis with the categorical variable.

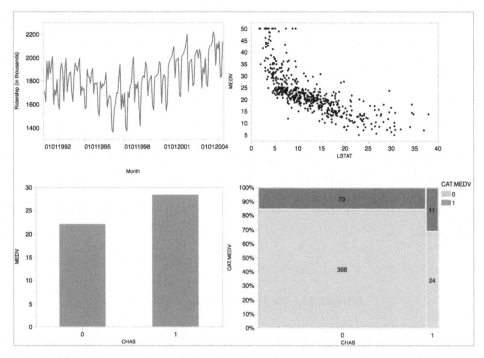

FIGURE 3.3 Basic plots: line graph (top left), scatterplot (top right), bar chart for numerical variable (bottom left), and mosaic plot for categorical variable (bottom right).

y-axis (and LSTAT on the x-axis is a potential predictor). Because both variables in a basic scatterplot must be numerical, the scatterplot cannot be used to display the relation between CAT.MEDV and potential predictors for the classification task (but we can enhance it to do so—see Section 3.4). For unsupervised learning, this particular scatterplot helps us study the association between two numerical variables in terms of information overlap as well as identifying clusters of observations.

All four of these plots highlight global information, such as the overall level of ridership or MEDV, as well as changes over time (line chart), differences between subgroups (bar chart and mosaic plot), and relationships between numerical variables (scatterplot).

Distribution Plots: Boxplots and Histograms

Before moving on to more sophisticated visualizations that enable multidimensional investigation, we note two important plots that are usually not considered "basic charts" but are very useful in statistical and data mining contexts. The *histogram* and the *boxplot* are two plots that display the entire distribution of a numerical variable. Although averages are very popular and useful summary statistics, there is usually much to be gained by looking at additional statistics such as the median and standard deviation of a variable, and even more so by examining the entire distribution. Whereas bar charts can only use a single aggregation, boxplots and histograms display the entire distribution of a numerical variable.

Boxplots are also effective for comparing subgroups. Side-by-side (comparative) box-plots display a different boxplot for each of the values of a second variable. A series of boxplots can also be used for looking at distributions over time. In JMP, histograms and boxplots for a single numeric variable are generated in the *Graph Builder* (or in *Analyze > Distribution*, which also provides summary statistics). Comparative boxplots can be produced using the *Graph Builder* or *Analyze > Fit Y by X* (which also produces summary statistics for the groups).

Distribution plots (histograms and boxplots) are useful in supervised learning for deter-mining potential data mining methods and variable transformations. For example, skewed numerical variables might warrant transformation (e.g., moving to a logarithmic scale) if used in methods that assume normality (e.g., linear regression, discriminant analysis).

A histogram represents the frequencies of all *x* values with a series of connected bars. For example, in the top left panel of Figure 3.4, there are about 25 tracts where the median value (MEDV) is between 5 to 10 thousand dollars.

A boxplot represents the variable being plotted on the *y*-axis, although the plot can potentially be turned in a 90 degrees angle so that the boxes are parallel to the *x*-axis (as we see above the histogram in Figure 3.4). In the top right panel of Figure 3.4 there are two side-by-side boxplots. The box encloses 50% of the data—for example, in the boxplot for CHAS = 1 half of the tracts have median values (MEDV) between 20 and 33 thousand dollars. The horizontal line inside the box represents the median (50th percentile). The top and bottom of the box represent the 75th and 25th percentiles, respectively. Lines extending above and below the box cover the rest of the data range; potential outliers are depicted as points beyond the lines.

In the JMP *Distribution* platform a diamond is displayed in the boxplot, where the center of the diamond is the mean (see the boxplot above the histogram in Figure 3.4). In *Fit Y by X* (not shown), the mean may display as a line or as a diamond (depending on the red triangle option you select). Comparing the mean and the median helps in assessing how skewed the data are. If the data are highly right-skewed, the mean will generally be much larger than the median.

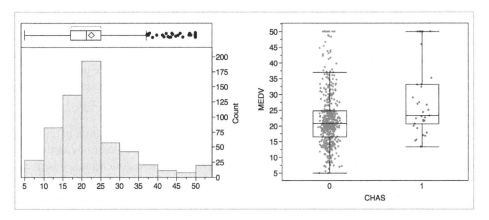

FIGURE 3.4 Distribution charts for numerical variable MEDV. Left: Histogram and boxplot, Right: Comparative boxplots.

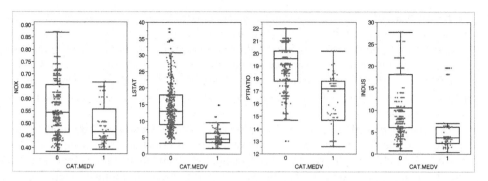

FIGURE 3.5   Side-by-side boxplots for exploring the CAT.MEDV output variable by different numerical predictors created in the *Graph Builder* using the *Column Switcher*.

Because histograms and boxplots are geared toward numerical variables, in their basic form they are useful for prediction tasks. Boxplots can also support unsupervised learning by displaying relationships between a numerical variable (*y*-axis) and a categorical variable (*x*-axis). To illustrate these points, look again at Figure 3.4. The left panel shows a histogram and boxplot of MEDV, revealing a skewed distribution. Transforming the output variable to log(MEDV) might improve results of a linear regression prediction model.

The right panel in Figure 3.4 shows side-by-side boxplots comparing the distribution of MEDV for homes that border the Charles River (1) or not (0). We see that not only is the median MEDV for river bounding homes higher than the non–river bounding homes, the entire distribution is higher (median, quartiles, min and max). We also see that all river-bounding homes have MEDV above 10 thousand dollars, unlike non–riverbounding homes. This information is useful for identifying the potential importance of this predictor (CHAS), and for choosing data mining methods that can capture the nonoverlapping area between the two distributions (e.g., trees). Boxplots and histograms applied to numerical variables can also provide directions for deriving new variables. For example, they can indicate how to bin a numerical variable, or in the Boston Housing example, they can help choose the cutoff to convert MEDV to CAT.MEDV.

Finally, side-by-side boxplots are useful in classification tasks for evaluating the potential of numerical predictors. This is done by using the *x*-axis for the categorical outcome and the *y*-axis for a numerical predictor. An example is shown in Figure 3.5, where we can see the effects of four numerical predictors on CAT.MEDV. These were generated in the *Graph Builder* using the *Column Switcher* (under the red triangle > *Script*) to swap out the predictor variables one at a time. The pairs that are most separated (e.g., LSTAT and INDUS) indicate potentially useful predictors.

The main weakness of basic charts in their basic form (i.e., using position in relation to the axes to encode values) is that they only display two variables and therefore cannot reveal high-dimensional information. Each of the basic charts has two dimensions, where each dimension is dedicated to a single variable. In data mining, the data are usually multivariate by nature, and the analytics are designed to capture and measure multivariate information. Visual exploration should therefore incorporate this important aspect. In the next section we describe how to extend basic and distribution charts to multidimensional data visualization by adding features, employing manipulations, and incorporating interactivity. We then present several specialized charts that are geared toward displaying special data structures (Section 3.5).

TOOLS FOR DATA VISUALIZATION IN JMP

The three most popular tools in JMP for creating visualizations are *Distribution* and *Fit Y by X* (from the *Analyze* menu) and the *Graph Builder* (from the *Graph* menu). *Distribution* is used for creating univariate graphs and statistics (bar charts, histograms, frequency distributions, and summary statistics). The graphs and statistics available are based on the types of the variables, and additional options are available under the red triangle for the variable. *Fit Y by X* is used for bivariate graphs and statistics (scatterplots, boxplots and summary statistics, mosaic plots and contingency tables, etc.). Again, the graphs and statistical methods provided are based on the variables selected, and the red triangle offers additional options. The *Graph Builder* is primarily an interactive graphing platform, providing a wide array of graphs for one, two, or many variables.

Heatmaps (Color Maps and Cell Plots): Visualizing Correlations and Missing Values

Heatmaps are graphical displays of numerical data where color is used to represent value magnitude. In a data mining context, these charts are especially useful for two purposes: for visualizing correlation tables and for visualizing missing values in the data. In both cases the information is conveyed in a two-dimensional table. We can either color the text of the values according to their magnitude, or even the cell background. Two types of heatmaps in JMP for visualizing correlations and missing values are *color maps* and *cell plots*.

Heatmaps are based on the fact that it is much easier and faster to scan the color-coding rather than the values. A heatmap is useful when examining a large number of values, but it is not a replacement for more precise graphical displays, such as bar charts or scatterplots, since color differences cannot be perceived accurately.

The text-colored correlation table and its corresponding color map for the Boston Housing data are shown in Figure 3.6. All of the pairwise correlations between 13 variables (MEDV and 12 predictors) are given in the correlation table (top). In the color map (bottom), a white to black color theme has been applied to highlight negative (white) and positive correlations (black). Note that the y-axis labels are not displayed—the rows correspond to the variables listed across the top of the color map. It is easy to quickly spot the high and low correlations. This output was produced in the *Multivariate* platform (under *Analyze > Multivariate Methods*).

Next we can examine the data table (with n rows and p variables) using a color map to search for missing values. In JMP, a missing value analysis can be conducted using *Tables > Missing Data Pattern*. This option produces a *missing data pattern* data table, which uses a binary coding of the original dataset where 1 denotes a missing value for a variable and 0 otherwise. The pattern indicates which variables, or columns in the data table, are missing values. For example, a pattern of 00000 for five variables indicates that there are no missing values, while a pattern of 10100 indicates that there are missing values for the first and third variables. The count indicates how many observations are missing values for those particular variables.

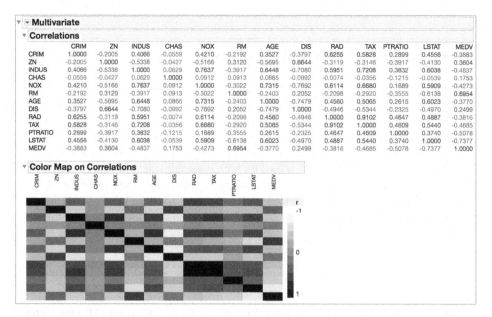

FIGURE 3.6   Color-coded correlation table and heatmap (also called a color map) of correlations with a white to black color theme.

Missing Data Pattern

▾ 🗄 Missing Data Pattern		▾ Source		▾ Treemap		▾ Cell Plot

	Count	Number of columns missing	Patterns	Banding?	timestamp	cylin
1	277	0	00000000000000000000000000000000000000000	0	0	
2	1	1	00000000000000000000000000000000100000	0	0	
3	2	1	00000000000000000000000000000010000000	0	0	
4	3	1	00000000000000000000000000010000000000	0	0	
5	36	1	0000000000000000000000001000000000000	0	0	
6	1	3	0000000000000000000000001000000110000	0	0	
7	1	2	0000000000000000000000000101000000000	0	0	
8	8	1	0000000000000000000000001000000000000	0	0	
9	1	4	0000000000000000000000010010000110000	0	0	
10	8	2	0000000000000000000000011000000000000	0	0	
11	1	3	0000000000000000000000011010000000000	0	0	
12	2	1	0000000000000000000001000000000000	0	0	
13	9	1	0000000000000000000100000000000000	0	0	
14	2	1	0000000000000000010000000000000	0	0	
15	5	1	0000000000000000100000000000000	0	0	
16	1	2	0000000000000001000001000000000	0	0	
17	1	2	0000000000000010000010000000000	0	0	
18	1	4	0000000000000010010000000000110	0	0	
19	1	5	0000000000000100100100000000110	0	0	
20	1	6	0000000000000110100000000000111	0	0	
21	72	1	0000000000001000000000000000	0	0	
22	1	2	0000000000001000000010000000000	0	0	
23	9	2	0000000000001000000010000000000	0	0	
24	4	2	0000000000001000001000000000000	0	0	

▾ Columns (41/0)
◢ Count
◢ Number of columns missing
▦ Patterns ✔
▦ Banding?
▦ timestamp
▦ cylinder number
▦ customer
▦ job number
▦ grain screened
▦ proof on ctd ink
▦ blade mfg
▦ paper type
▦ ink type
▦ direct steam
▦ solvent type
▦ type on cylinder
▦ press type

▾ Rows
All rows   68
Selected   0
Excluded   0
Hidden   0
Labelled   0

FIGURE 3.7   Missing data pattern table. The pattern indicates which columns are missing values. Run the scripts in the top right corner to produce a treemap or cell plot of missing values.

For illustration, Figure 3.7 shows a partial missing data table for a dataset with over 40 columns and 539 rows (the file Bands Data.jmp from the JMP *Sample Data Library*). The variables were merged from multiple sources, and for each source information was not always available. In Figure 3.7 we see that only 277 rows (see the first row in the missing data pattern table) are not missing values for any the columns, and that several rows are missing values for just one or two columns.

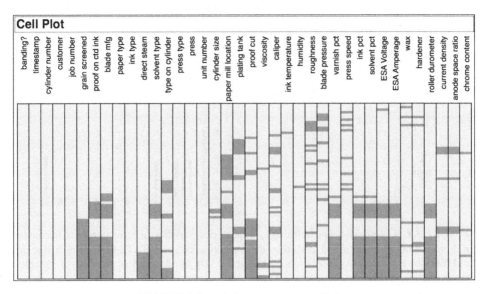

FIGURE 3.8   Missing data pattern cell plot.

The next step is to create a *cell plot*, which highlights the missing values in the data matrix. Figure 3.7 shows the result for our example. The cell plot is created via a shortcut provided as a script in the top left corner of the missing data pattern table. To create the cell plot, right-click on the script and select *Run Script*.

The cell plot in Figure 3.8 graphically displays missing rows for particular variables as shaded regions, along with rows missing values for multiple variables (shaded regions for multiple variables). The cell plot helps us visualize the amount of "missingness" in the merged data file. Some patterns of missingness easily emerge: variables that are missing values for nearly all observations, as well as rows that are missing many values. Variables with little missingness are also visible. This information can then be used for determining how to handle the missingness (e.g., dropping some variables, dropping some records, imputing values, or via other techniques).

Note: Additional options for exploring missing values, including options to impute missing values, are available in JMP 12 Pro from *Cols > Modeling Utilities > Explore Missing Values*.

3.4   MULTIDIMENSIONAL VISUALIZATION

Basic plots in JMP can convey richer information with features such as color, size, and multiple panels, and by enabling operations such as rescaling, aggregation, interactivity, dynamic linking between graphs and the data table, and dynamic filtering of data based on values of other variables. These additions allow us to look at more than one or two variables at a time. The beauty of these additions is their effectiveness in displaying complex information in an easily understandable way. Effective features are based on understanding how visual perception works (see Few, 2009). The goal is to make the information more understandable, not just represent the data in higher dimensions (e.g., three-dimensional

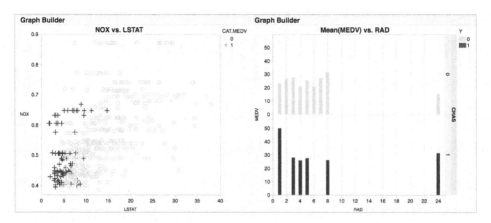

FIGURE 3.9 Color-coding and trellising by a categorical variable. Left: Scatterplot of two numerical predictors, color-coded and marked by the categorical outcome CAT.MEDV. Right: Bar chart of MEDV by two categorical predictors (CHAS and RAD), using CHAS as the *Group Y* variable to create multiple panels.

plots, which are usually ineffective visualizations), and to allow the exploration of multiple variables *in flow* and on the fly.

Adding Variables: Color, Size, Shape, Multiple Panels, and Animation

In order to include more variables in a plot, we must consider the types of variables to include. To represent additional categorical information, the best way is to use hue, shape, or multiple panels. For additional numerical information, we can use color intensity or size. Temporal information can be added via animation.

Incorporating additional categorical and/or numerical variables into the standard plots means that we can now use all of them for both prediction and classification tasks! For example, we mentioned earlier that a basic scatterplot cannot be used for studying the relationship between a categorical outcome and predictors (in the context of classification). However, a very effective plot for classification is a scatterplot of two numerical predictors color-coded by the categorical outcome variable. An example is shown in the left panel of Figure 3.9, with color and marker denoting CAT.MEDV (using *Graph Builder*, with CAT.MEDV as both the *overlay* and the *color* variable).

In the context of prediction, color-coding supports the exploration of the conditional relationship between the numerical outcome (on the *y*-axis) and a numerical predictor. Color-coded scatterplots then help us assess the need for including interaction terms in the model. For example, is the relationship between MEDV and LSTAT different for homes near vs. away from the river? Simply use the *Graph Builder* to graph MEDV versus LSTAT, and drag CHAS to the *Overlay* zone. For a more formal exploration of this relationship, click on the *Line of Fit* graph icon. If the resulting lines are not parallel, this is a good indication that an interaction term for LSTAT and CHAS is needed (i.e., the relationship between LSTAT and MEDV depends on the value of CHAS).

Color can also be used to add categorical variables to a bar chart, as long as the number of categories is small. When the number of categories is large, a better alternative is to

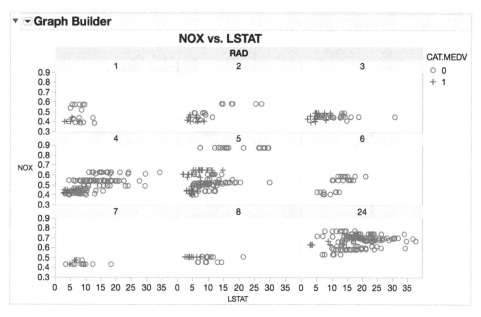

FIGURE 3.10   Scatterplot of NOX vs. LSTAT, trellised by RAD.

use *multiple panels*. Creating multiple panels (also called *trellising*) is done by splitting the observations according to a categorical variable, and creating a separate plot (of the same type) for each category. An example of trellising is shown in the right-hand panel of Figure 3.9, where a bar chart of average MEDV by highway accessibility (RAD) is broken down into two panels by CHAS. In this example, CHAS is used as the *Group Y* variable in the *Graph Builder*. We see that the average MEDV for different highway accessibility levels (RAD) behaves differently for homes near the river (lower panel) compared to homes away from the river (upper panel). This is especially salient for RAD = 1. We also see that there are no near-river homes at RAD levels 2, 6, and 7. Such information might lead us to create an interaction term between RAD and CHAS, and to consider combining some of the bins in RAD. All these explorations are useful for prediction and classification.

The *Group* zones creates multiple panels that are aligned either vertically or horizontally. For categorical variables with multiple levels, it is sometimes better to align the panels in multiple rows by columns. To achieve that, use the *Wrap* zone (in the top right corner of the *Graph Builder*; see Figure 3.2). This will create a *trellis* plot by the variable in the wrap zone, which automatically wraps to efficiently display multiple panels. Figure 3.10 shows an example of a trellis plot that uses wrap. Specifically, we created a scatterplot of NOX versus LSTAT, color-coded and marked the points by CAT.MEDV, and then trellised the scatterplot by adding RAD into the wrap zone.

Note: When numeric variables are placed in the wrap zone, or in either of the group zones, the variables are automatically binned to create trellis plots.

A special plot that uses scatterplots with multiple panels is the scatterplot matrix. In this plot, all pairwise scatterplots are shown in a single display. The panels in a scatterplot

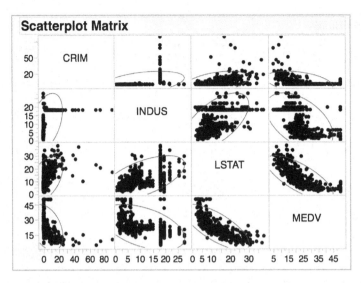

FIGURE 3.11   Scatterplot matrix for MEDV and three numerical predictors. The *density ellipses* graphically show the nature and strength of correlation among variables.

matrix are organized in a special way, such that each column corresponds to a variable and each row corresponds to a variable; thereby the intersections create all the possible pairwise scatterplots. The scatterplot matrix is useful in unsupervised learning for studying the associations between pairs of numerical variables, detecting outliers, and identifying clusters. For supervised learning, it can be used for examining pairwise relationships (and their nature) between predictors to support variable transformations and variable selection (see the correlation analysis in Chapter 4). For prediction, it can also be used to depict the relationship between the outcome and the numerical predictors.

An example of a scatterplot matrix is shown in Figure 3.11, with MEDV and three predictors. Variable names in each row indicate the *y*-axis variable, and variable names in each column indicate the *x*-axis variable. For example, the plots in the bottom row all have MEDV on the *y*-axis (which allows studying the individual outcome–predictor relations). We can see different types of relationships from the different shapes (e.g., an exponential relationship between MEDV and LSTAT and a highly skewed relationship between CRIM and INDUS) that can indicate potentially useful transformations. Note that the plots above and to the right of the diagonal are mirror images of those below and to the left.

Once hue is used, further categorical variables can be added via shape and multiple panels. However, one must proceed cautiously in adding multiple variables, as the display can become overcluttered and then visual perception is lost.

Adding a numerical variable via size is especially useful in scatterplots (thereby creating "bubble plots"), since, in a scatterplot, points typically represent individual observations. Size variables can be added in the *Graph Builder* and the *Bubble Plot* (under the *Graph* menu). In plots that aggregate across observations (e.g., boxplots, histograms, bar charts) size and hue are not normally incorporated.

Finally, adding a temporal dimension to a plot to show how the information changes over time can be achieved via animation. A famous example is Rosling's animated scatterplots showing how world demographics changed over the years (www.gapminder.org). In JMP, this form of animation is available with the *Bubble Plot*. Additional animation is available with the *Data Filter* and the *Column Switcher*. However, while animations of this type work for "statistical storytelling," they are usually not very effective for data exploration.

Manipulations: Rescaling, Aggregation and Hierarchies, Zooming, Filtering

Most of the time spent in data mining projects is spent in preprocessing data. Typically, considerable effort is expended getting all the data in a format that can actually be used in data mining. Additional time is spent processing the data in ways that improve the performance of the data mining procedures. The preprocessing step includes variable transformation and derivation of new variables to help models perform more effectively. Transformations include changing the numeric scale of a variable, binning numerical variables, combining categories in categorical variables, and so on. The following manipulations support the preprocessing step as well the choice of adequate data mining methods. They do so by revealing hidden patterns in the data.

Rescaling   Changing the scale in a display can enhance the plot and illuminate relationships. For example, in Figure 3.12 we see the effect of changing both axes of the scatterplot (top) and the y-axis of a boxplot (bottom) to logarithmic (log) scale. Whereas the original plots (left) are hard to understand, the patterns become visible in log scale (right). In the scatterplots the nature of the relationship between MEDV and CRIM is hard to determine in the original scale because too many of the points are "crowded" near the y-axis. The rescaling removes this crowding and allows a better view of the linear relationship between the two log-scaled variables (indicating a log-log relationship). In the boxplots the crowding toward the x-axis in the original units does not allow us to compare the two box sizes, their locations, lower outliers, and most of the distribution information. Rescaling removes the "crowding to the x-axis" effect, thereby allowing a comparison of the two boxplots. (Double-click on an axis in JMP to change the scale.)

Aggregation and Hierarchies   Another useful manipulation of scaling is changing the level of aggregation. For a temporal scale, we can aggregate by using different granularity (e.g., monthly, daily, hourly) or even by a "seasonal" factor of interest such as month-of-year or day-of-week. A popular aggregation for time series is a moving average, where the average of neighboring values within a given window of time is plotted. Moving average plots enhance global trend visualization (see Chapter 15).

Nontemporal variables can be aggregated if some meaningful hierarchy exists: geographical (tracts within a zip code in the Boston Housing example), organizational (people within departments within units), and so forth. Figure 3.13 illustrates two types of aggregation for the railway ridership time series. The original monthly series is shown in the top left panel. Seasonal aggregation (by month-of-year) is shown in the top right panel, where it is easy to see the peak in ridership in August and the dip in February and March. The bottom right panel shows temporal aggregation, where the series is now displayed in yearly aggregates. This plot reveals the global long-term trend in ridership and the generally increasing trend from 1996 on.

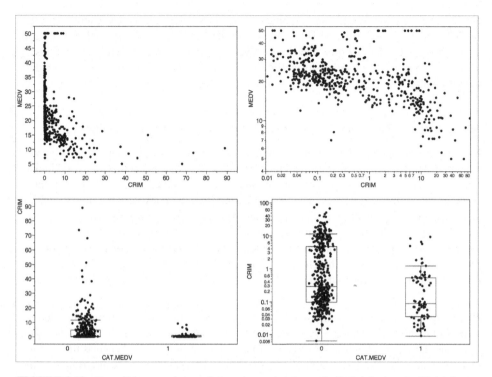

FIGURE 3.12   Rescaling can enhance plots and reveal patterns. Left: Original scale. Right: Logarithmic scale.

Examining different scales, aggregations, or hierarchies supports both supervised and unsupervised tasks in that it can reveal patterns and relationships at various levels, and can suggest new sets of variables with which to work.

In JMP, different aggregations and transformations can be applied from the data table, the *Graph Builder*, or in any analysis dialog window. From the data table, right-click on the variable of interest, select *New Formula Column*, and select from the available functions. To create transformed variables from a graphing or analysis platform, right-click on the variable in the dialog, and select the function of interest. This produces a temporary variable that can be used in the current graph or analysis. This transformed variable (and its formula) can then be saved to the data table for future use.

Zooming and Panning   The ability to zoom in and out of certain areas of the data on a plot is important for revealing patterns and outliers. Zooming is particularly useful with geographic maps (e.g., produced via the *Graph Builder*). We are often interested in more detail on areas of dense information or of special interest. Panning refers to moving the zoom window to other areas or time frames. An example of zooming is shown in the bottom left panel of Figure 3.13, where the ridership series is zoomed in to the first two

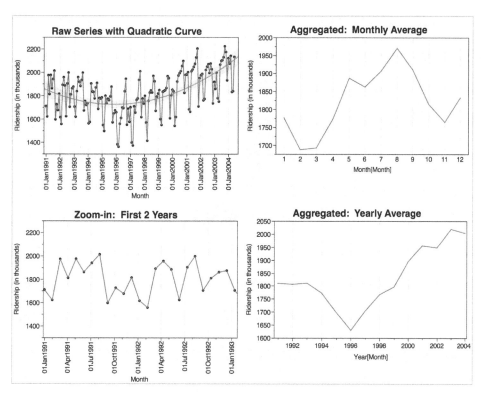

FIGURE 3.13   Time series line graphs using different aggregations (right panels), adding curves (top left panel), and zooming in (bottom left panel).

years of the series. Zooming and panning support supervised and unsupervised methods by detecting areas of different behavior, which may lead to creating new interaction terms, new variables, or even separate models for data subsets. In addition, zooming and panning can help us choose between methods that assume global behavior (e.g., regression models) and data-driven methods (e.g., exponential smoothing forecasters and k-nearest neighbors classifiers), and indicate the level of global/local behavior, as manifested by parameters such as k in k-nearest neighbors, the size of a tree, or the smoothing parameters in exponential smoothing.

Filtering   Filtering means highlighting or removing some of the observations from the plot (or the analysis). The purpose of filtering is to focus the attention on certain data while eliminating "noise" created by other data. Filtering supports supervised and unsupervised learning in a similar way to zooming and panning: it assists in identifying different or unusual local behavior. In JMP, the *Data Filter* (from the *Rows* menu) can be used to select, hide, and/or exclude values. Hiding removes observations from graphical displays, while excluding removes observations from analyses and calculations. (*Note:* In analysis

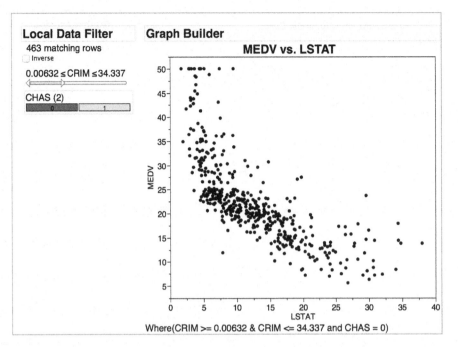

FIGURE 3.14 Using the *Local Data Filter* in the *Graph Builder* to view only data in specific ranges of CRIM and CHAS values.

platforms you need to re-run the analysis after excluding observations.) The Data Filter is a *global* filter, and it applies to the data table and all output windows. The *Local Data Filter* (available under the *red triangle > Scripts* menu in any output window, or as a shortcut on the toolbar) selects and includes values of the specified columns for only the current (or active) graphs and analysis. In Figure 3.14 we explore the relationship between MEDV and LSTAT at specified values of two other variables, CRIM and CHAS. The graph dynamically updates as values of the filtering variables are changed.

Reference: Trend Lines and Labels

Using trend lines and in-plot labels also helps us detect patterns and outliers. Trend lines serve as a reference, and allow us to more easily assess the shape of a pattern. Although linearity is easy to visually perceive, more elaborate relationships such as exponential and polynomial trends are harder to assess by eye. Trend lines are useful in line graphs as well as in scatterplots. An example is shown in the top left panel of Figure 3.13, where a polynomial curve is overlaid on the original line graph (see also Chapter 15).

ADDING TRENDLINES IN THE GRAPH BUILDER

To create a graph with a quadratic curve (as shown in Figure 3.15):

- Drag a continuous variable to the *Y* zone.
- Drag a continuous variable to the *X* zone.
- Click on the *Line of Fit* icon (the third icon above the graph). This fits a trendline to the data (the shaded bands are confidence bands.) A series of options for this line of fit will appear on the bottom left, below the list of variables.
- Under the *Line of Fit* outline, change the *Degree* from Linear to Quadratic (the option for a Cubic fit is also available).
- To connect the points, drag the *Line* icon onto the graph.

In Figure 3.15, three graph elements (points, line of fit, and line) have been used—these icons all appear shaded. Options for selected graph elements are provided on the bottom left of the graph. To close the control panel on the left and hide these options, click *Done*.

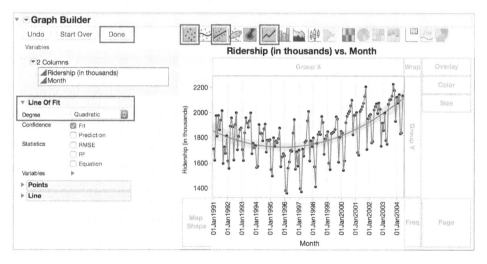

FIGURE 3.15 Adding a trendline in Graph Builder.

In displays that are not cluttered or overcrowded, the use of in-plot labels can be useful for better exploration of outliers and clusters. An example is shown in Figure 3.16. The figure shows different utilities on a scatterplot that compares fuel cost with total sales. We might be interested in clustering the data, and using clustering algorithms to identify clusters that differ markedly with respect to fuel cost and sales. Figure 3.16 helps us visualize these clusters and their members (e.g., Nevada and Puget are part of a clear cluster with low fuel costs and high sales). For more on clustering, and this example, see Chapter 14. Labels are added as column and row properties in the data table. Within a graph, labels can be repositioned (dragged manually) for optimal spacing and visibility.

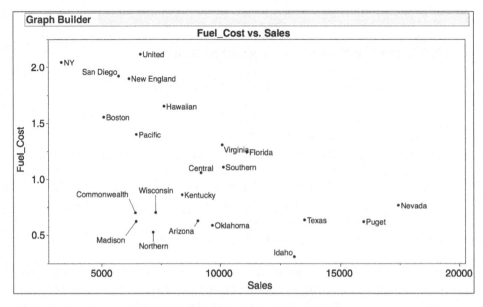

FIGURE 3.16   Scatterplot with labeled points.

Points can be labeled in JMP using tags and column/row labels. To display a tag with the plotted values for a point in a graph, hold your mouse over the point—this will display a temporary tag. To pin (retain) this tag on the graph, right-click on the tag and select *Pin*. To add the values for a variable to the tag, right-click on the variable name in the data table and select *Label/Unlabel*. To add permanent variable labels for particular points, select the points and use *Rows* > Label/*Unlabel*.

Scaling Up: Large Datasets

When the number of observations (rows) is large, plots that display each individual observation (e.g., scatterplots) can become ineffective. Aside from using aggregated charts such as boxplots, some alternatives are as follows:

1. Sampling (by drawing a random sample and using it for plotting). JMP has a sampling utility, and sampling can be done directly in the *Graph Builder*.
2. Reducing marker size.
3. Using more transparent marker colors and removing fill.
4. Breaking down the data into subsets (e.g., by creating multiple panels).
5. Using aggregation (e.g., bubble plots where size corresponds to number of observations in a certain range).
6. Using jittering (slightly moving the marker by adding a small amount of noise so that individual markers can be seen).

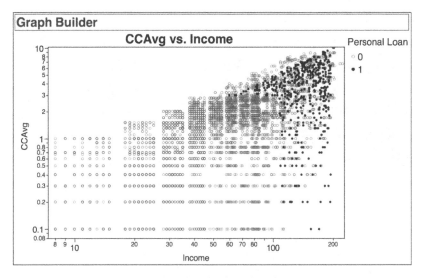

FIGURE 3.17   Scatterplot of 5000 records with reduced marker size, different markers, and more transparent coloring.

An example of the advantage of plotting a sample over the large dataset is shown in Chapter 12 (Figure 12.2), where a scatterplot of 5000 records is plotted along with a scatterplot of a sample. The dataset, `UniversalBank.jmp`, is described in a future chapter. In Figure 3.17 we illustrate techniques for plotting the large datasets described above. In this scatterplot, we use smaller markers, different markers for each class, and more transparent colors (a gray scale was used for printing). We can see that larger areas of the plot are dominated by the gray class (hollow markers). The black class (solid markers) is mainly on the right, while there is a lot of overlap in the top right area.

Multivariate Plot: Parallel Coordinates Plot

Another way to present multidimensional information in a two-dimensional plot is via specialized plots such as the *parallel coordinates plot*. In this plot a vertical axis is drawn for each variable. Then each observation is represented by drawing a line that connects its values on the different axes, thereby creating a "multivariate profile." An example is shown in Figure 3.18 for the Boston Housing data. In this display separate panels are used for the two values of CAT.MEDV, in order to compare the profiles of tracts in the two classes (for a classification task). We see that tracts with more expensive homes (right panel) consistently have low CRIM, low LSAT, and high RM compared to tracts with less expensive homes (left panel), which are more mixed on CRIM and LSAT, and have a medium level of RM. This observation gives indication of useful predictors and suggests possible binning for some numerical predictors.

Parallel coordinate plots are also useful in unsupervised tasks. They can reveal clusters, outliers, and information overlap across variables. A useful manipulation is to re-order the columns to better reveal clustering of observations. Parallel coordinates plots are available in JMP under *Graph > Parallel Plots*, and in the *Graph Builder* in JMP 13.

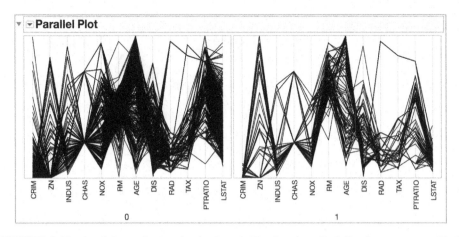

FIGURE 3.18 Parallel coordinates plot for Boston Housing data. Each line is census tract. Each of the variables (shown on the horizontal axis) is scaled to 0–100%. Panels are used to distinguish CAT.MEDV (left panel = tracts with median home value below $30,000).

Interactive Visualization

Similar to the interactive nature of the data mining process, interactivity is key to enhancing our ability to gain information from graphical visualization and is fundamental in JMP. In the words of Stephen Few (Few, 2009), an expert in data visualization:

> We can only learn so much when staring at a static visualization such as a printed graph.
> . . . If we can't interact with the data . . . , we hit the wall.

By interactive visualization we mean an interface that supports the following principles:

1. Making changes to a plot is *easy, rapid, and reversible*.
2. Multiple concurrent plots can be easily combined and displayed on a single screen.
3. A set of visualizations can be linked so that operations in one display are reflected in the other displays.

JMP is designed, from the ground up, to support all these operations. Let us consider a few examples where we contrast a static plot generator (e.g., Excel) with an interactive visualization interface (i.e., JMP).

Histogram Re-binning   Consider the need to bin a numerical variable and using a histogram for that purpose. A static histogram would require re-plotting for each new binning choice (in Excel it would even require creating the new bins manually). If the user generates multiple plots, then the screen becomes cluttered. If the same plot is recreated, then it is hard to compare to other binning choices. In contrast, an interactive visualization provides an easy way to change bin width interactively, and then the histogram automatically and rapidly replots as the user changes the bin width. The hand tool in JMP (aka the *Grabber* tool) on the toolbar makes this effortless (select the *Grabber*, click on the histogram and drag up or down to rebin, or drag left or right to rescale.)

Aggregation and Zooming   Consider a time series forecasting task, given a long series of data. Temporal aggregation at multiple levels is needed for determining short- and long-term patterns. Zooming and panning are used to identify unusual periods. A static plotting software requires the user to create new data columns for each temporal aggregation (e.g., aggregate daily data to obtain weekly aggregates). Zooming and panning in Excel requires manually changing the min and max values on the axis scale of interest (thereby losing the ability to quickly move between different areas without creating multiple charts). In contrast, JMP provides immediate temporal hierarchies that the user can easily switch between (via dynamic column transformations, discussed previously). The cursor, or arrow tool, is also used to quickly zoom, stretch, or reposition axes, thereby allowing direct manipulation and rapid reaction.

Combining Multiple Linked Plots That Fit in a Single Screen   In earlier sections we used charts to illustrate the advantages of visualization, because "a picture is worth a thousand words." The advantages of an interactive visualization are even harder to convey in words. As Ben Shneiderman, a well-known researcher in information visualization and interfaces, puts it:

> A picture is worth a thousand words. An interface is worth a thousand pictures.

In the exploration process we typically create multiple visualizations. These can include side-by-side boxplots, color-coded scatterplots with trend lines, multi-panel bar charts, and so on. We might want to detect possible multidimensional relationships (and identify possible outliers) by selecting a certain subset of the data (e.g., a single category of some variable) and locating the observations on the other plots. In a static interface, we would have to manually organize the plots of interest and resize them in order to fit them within a single screen. Moreover a static interface would usually not support inter-plot linkage, and even if so, the entire set of plots would have to be regenerated each time a selection is made.

In contrast, JMP provides an easy way to automatically organize and re-size the set of plots to fit within a screen. All plots are dynamically linked, and plots are dynamically linked to the data table. You can also create dashboards (using the *Application Builder* in JMP 12), with related graphs, analyses, and data filters. Multiple observations are selected by drawing a box around the points with the cursor (or using a data filter), and selections in one plot are automatically highlighted in the other plots. A simple example is shown in Figure 3.19. For more information on building applications in JMP, search for *Applications* in the *JMP Help*.

Note: The *Dashboard Builder* in JMP 13 provides a more interactive and flexible interface for creating dashboards.

3.5   SPECIALIZED VISUALIZATIONS

In this section we mention a few specialized visualizations that are able to capture data structures beyond the standard time series and cross-sectional structures—special types of relationships that are usually hard to capture with ordinary plots. In particular, we address hierarchical data, network data, and geographical data—three types of data that are becoming increasingly more available.

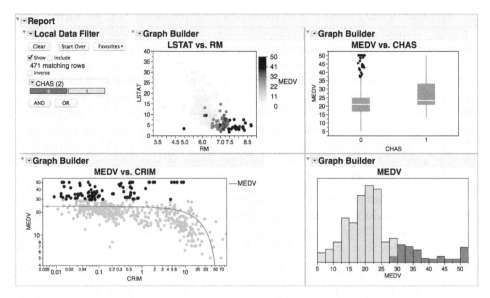

FIGURE 3.19 Multiple inter-linked plots in a single dashboard (a JMP *application*). All graphs were created using the *Graph Builder*, and these graphs were combined in one window to create a JMP Application. The *Local Data Filter* was used to select observations corresponding to Chas = 0. These observations are automatically selected in all of the graphs.

Visualizing Networked Data

Network analysis techniques were spawned by the explosion of social and product network data. Examples of social networks are networks of sellers and buyers on eBay and networks of people on Facebook. An example of a product network is the network of products on Amazon (linked through the recommendation system). Network data visualization is available in various network-specialized software, and in some general-purpose software (although it is not directly available in JMP).

A network diagram consists of actors and the relations between them. "Nodes" are the actors (e.g., people in a social network or products in a product network), and are represented by circles. "Edges" are the relations between nodes, and are represented by lines connecting nodes. For example, in a social network such as Facebook, we can construct a list of users (nodes) and all the pairwise relations (edges) between users who are "Friends." Alternatively, we can define edges as a posting that one user posts on another user's Facebook page. In this setup we might have more than a single edge between two nodes. Networks can also have nodes of multiple types. A common structure is networks with two types of nodes.

An example of a two-type node network is shown in Figure 3.20, where we see a set of transactions between a network of sellers and buyers on the online auction site www.eBay.com (the data are for auctions selling Swarovski beads, and these auctions took place over a period of several months; see Jank and Yahav, 2010). The circles on the left side represent sellers and on the right side buyers. Circle size represents the number of transactions that the node (seller or buyer) was involved in within this network. Line width represents the number of auctions that the bidder–seller pair interacted in (in this case we use arrows to denote the directional relationship from buyer to seller). We can see that this

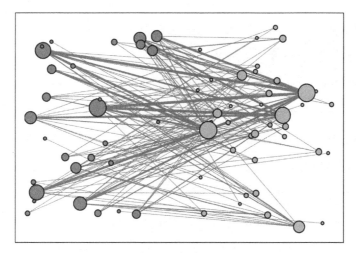

FIGURE 3.20   Network graph of eBay sellers (left side) and buyers (right side) of Swarovski beads. Circle size represents the node's number of transactions. Line width represents the number of transactions between that pair of seller–buyer. Produced using a JMP add-in.

marketplace is dominated by three or four high-volume sellers. We can also see that many buyers interact with a single seller. The market structures for many individual products could be reviewed quickly in this way. Network providers could use the information, for example, to identify possible partnerships to explore with sellers.

Figure 3.20 was produced using a JMP add-in, the *Transaction Visualizer*. This add-in was developed using the JMP Scripting Language (JSL), and it is freely availble at the JMP User Community (*community.jmp.com*). For more information on JMP add-ins, search for *Add-ins* in the *JMP Help*.

Visualizing Hierarchical Data: More on Treemaps

Treemaps (available from the *Graph* menu and within the *Graph Builder*) are useful visualizations for exploring large datasets that are hierarchically structured (tree-structured). They allow exploration of various dimensions of the data while maintaining the hierarchical nature of the data. An example is shown in Figure 3.21, which displays a large set of eBay auctions,[3] hierarchically ordered by item category, subcategory, and brand. The levels in the hierarchy of the treemap are visualized as rectangles containing sub-rectangles. Categorical variables can be included in the display by using hue. Numerical variables can be included via rectangle size and color intensity (ordering of the rectangles is sometimes used to reinforce size).

In the example in Figure 3.21 size is used to represent the average closing price (which reflects item value), and color intensity represents the percentage of sellers with negative feedback (a negative seller feedback indicates buyer dissatisfaction in past transactions and often indicative of fraudulent seller behavior). Consider the task of classifying ongoing auctions in terms of a fraudulent outcome. From the treemap we see that the highest

---

[3]We thank Sharad Borle for sharing this dataset.

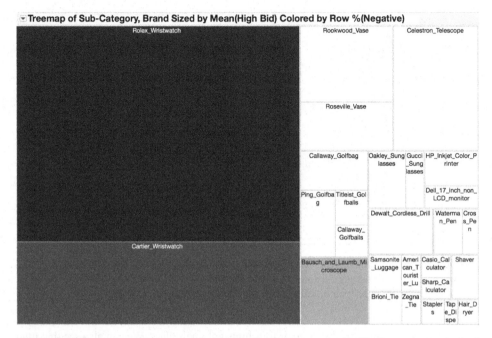

FIGURE 3.21   Treemap showing nearly 11,000 eBay auctions, organized by item subcategory and brand. Rectangle size represents average closing price (reflecting item value). Shade represents % of sellers with negative feedback (darker = higher %).

proportion of sellers with negative ratings (black) is concentrated in expensive item auctions (Rolex and Cartier wristwatches).

One example of an interactive online application of treemaps is currently available at www.drasticdata.nl/drastictreemap.htm. One of these examples displays player-level data from the 2014 World Cup, aggregated to team level. The user can choose to explore players and team data. Similar treemaps can be produced in JMP 13.

Visualizing Geographical Data: Maps

Many data sets used for data mining now include geographical information. Zip codes are one example of a categorical variable with many categories, where it is not straightforward to create meaningful variables for analysis. Plotting the data on a geographic map can often reveal patterns that are harder to identify otherwise. A map chart uses a geographical map as its background, and then color, hue, and other features to include categorical or numerical variables. The ability to plot data on background maps and overlay data on images is readily available in JMP (see www.jmp.com/learn) or search for "background maps" in the JMP documentation).

Figure 3.22 shows two world map charts (created in the JMP *Graph Builder*), comparing education levels in the top map, and gross domestic product per capita in US dollars (GDP) in the bottom map (from the `World Demographic.jmp` dataset in the JMP *Sample Data Library*). Darker shades mean higher values. Tools like the JMP *Column Switcher* and *Data Filter*, when used with background maps, bring added dimensions to the visualizations.

Google Maps provides API's (application programming interfaces) that allow organizations to overlay their data on a Google map. While Google Maps is readily available, the resulting map charts (Figure 3.23) are somewhat inferior in terms of effectiveness compared to map charts in dedicated interactive visualization software.

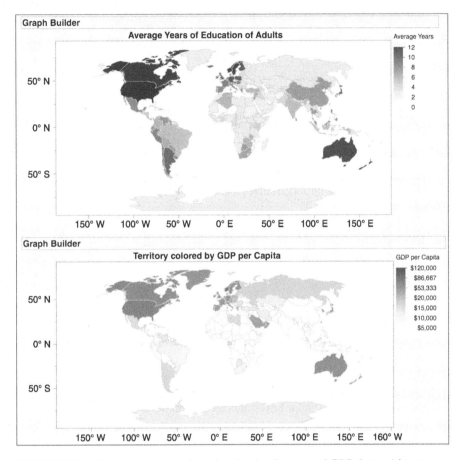

FIGURE 3.22   World maps comparing education levels (top) and GDP (bottom) by country.

3.6   SUMMARY OF MAJOR VISUALIZATIONS AND OPERATIONS, ACCORDING TO DATA MINING GOAL

Prediction

- Plot outcome on the numerical axis of boxplots, bar charts, scatterplots.
- Study relation of outcome to categorical predictors via side-by-side boxplots, bar charts, and multiple panels.
- Study relation of outcome to numerical predictors via scatterplots.
- Use distribution plots (boxplot, histogram) for determining potentially useful transformations of the outcome variable (and/or numerical predictors).

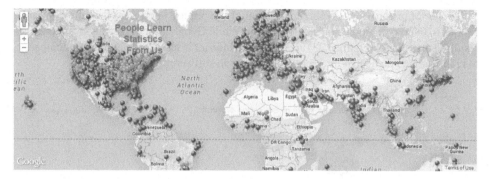

FIGURE 3.23   Map chart of students (blue and orange) and instructors' (red) locations on a Google Map (from Statistics.com).

- Examine scatterplots with added color/panels/size to determine the need for interaction terms.
- Use various aggregation levels, zooming, and filtering to determine areas of the data with different behavior, and to evaluate the level of global vs. local patterns.

Classification

- Study relation of outcome to categorical predictors using bar charts with the outcome frequencies/percentages on the numerical axis.
- Study relation of outcome to pairs of numerical predictors via color-coded scatterplots (color denotes the outcome).
- Study relation of outcome to numerical predictors via side-by-side boxplots: plot boxplots of a numerical variable by outcome. Create similar displays for each numerical predictor. The most dissimilar boxes indicate potentially useful predictors.
- Use color to represent the outcome variable on a parallel coordinate plot.
- Use distribution plots (boxplot, histogram) for determining potentially useful transformations of numerical predictor variables.
- Examine scatterplots of numerical predictor variables with added color/panels for outcome to determine the need for interaction terms.
- Use various aggregation levels, zooming, and filtering to determine areas of the data with different behavior, and to evaluate the level of global vs. local patterns.

Time Series Forecasting

- Create line graphs at different temporal aggregations to determine types of patterns.
- Use zooming and panning to examine shorter periods of the series to determine areas of the data with different behavior.
- Use different aggregation levels to identify global and local patterns.
- Identify missing values in the series.
- Overlay trend lines of different types to determine adequate modeling choices.

Unsupervised Learning

- Create scatterplot matrices to identify pairwise relationships and clustering of observations.
- Use color maps to examine the correlation table.
- Use various aggregation levels and zooming to determine areas of the data with different behavior.
- Generate a parallel coordinate plot to identify clusters of observations.

PROBLEMS

3.1 Shipments of Household Appliances: Line Graphs. The file `ApplianceShipments.jmp` contains the series of quarterly shipments (in millions of dollars) of US household appliances between 1985 and 1989.

    a. Create a well-formatted time plot of the data using the JMP *Graph Builder*.

    b. Does there appear to be a quarterly pattern? For a closer view of the patterns, try zooming in on different ranges of data on the two axes.

    c. Create one chart with four separate lines, one for each of Q1, Q2, Q3, and Q4. This can be achieved by creating a transformed variable (right-click on Quarter in the *Graph Builder*, and select *Date Time*, *Quarter*). Then drag this new column to the *Overlay*, *Wrap* or *Group Y* zone. Does there appear to be a difference between quarters?

    d. Create a line graph of the series at a yearly aggregated level (i.e., the total shipments in each year). Again, you'll need to create a transformed variable in *Graph Builder*.

3.2 Sales of Riding Mowers: Scatterplots. A company that manufactures riding mowers wants to identify the best sales prospects for an intensive sales campaign. In particular, the manufacturer is interested in classifying households as prospective owners or nonowners on the basis of income (in $1000s) and lot size (in 1000 ft^2). The marketing expert looked at a random sample of 24 households, given in the file `RidingMowers.jmp`.

    a. Create a scatterplot of lot size versus income, color-coded by the outcome variable Ownership. Make sure to obtain a well-formatted plot. The result should be similar to Figure 9.2. Describe the potential relationship(s) of ownership to lot size and income.

    b. Explore different methods for saving your work in JMP. Search for "saving work" in the JMP documentation or see the Using JMP section of the *JMP Learning Library* at jmp.com/learn. Mention two methods.

3.3 Laptop Sales at a London Computer Chain: Bar Charts and Boxplots. The file `LaptopSalesJanuary2008.jmp` contains data for all sales of laptops at a computer chain in London in January 2008. This is a subset of the full dataset that includes data for the entire year.

    a. Create a bar chart, showing the average retail price by store (Store Postcode). Adjust the *y*-axis scaling to magnify differences. Which store has the highest average? Which has the lowest? Use the *Local Data Filter* (under the *red triangle > Scripts*) to isolate these two stores.

b. To better compare retail prices across stores, create side-by-side boxplots of retail price by store. Now compare the prices in the two stores above. Does there seem to be a difference between their price distributions?

3.4 **Laptop Sales at a London Computer Chain: Interactive Visualization.** The file `LaptopSales.txt` is a comma-separated file with nearly 300,000 rows. ENBIS (the European Network for Business and Industrial Statistics) provided these data as part of a contest organized in the fall of 2009.

Scenario: Imagine that you are a new analyst for a company called Acell (a company selling laptops). You have been provided with data about products and sales. You need to help the company with their business goal of planning a product strategy and pricing policies that will maximize Acell's projected revenues in 2009. Import the data into JMP (for details on importing data into JMP, search for "import text files" in the JMP documentation or at jmp.com/learn). Check to ensure that the data and modeling types in the data table are correct for each of the variables, and answer the following questions.

a. Price Questions:

    i. At what price are the laptops actually selling?

    ii. Does price change with time? (Hint: Make sure that the date column is recognized as such. JMP will then allow dynamic transformations and allow you to plot the data by weekly or monthly aggregates, or even by day of week.)

    iii. Are prices consistent over retail outlets?

    iv. How does price change with configuration?

b. Location Questions:

    i. Where are the stores and customers located?

    ii. Which stores are selling the most?

    iii. How far would customers travel to buy a laptop?

        o *Hint 1:* You should be able to aggregate the data, such as by a plot of the sum or average of the prices.

        o *Hint 2:* Use the coordinated highlighting between multiple visualizations in the same page, for example, select a store in one view to see the matching customers in another visualization.

        o *Hint 3:* Explore the use of filters to see differences. Be sure to filter in the zoomed-out view. For example, try to use "store location" as an alternative way to dynamically compare store locations. This might be more useful to spot outlier patterns if there were 50 store locations to compare.

    iv. Try an alternative way of looking at how far customers traveled. Do this by creating a new data column that computes the distance between customer and store. (For information on creating formulas in JMP, search for "Creating Formulas" in the JMP documentation or see jmp.com/learn.)

c. Revenue Questions:

    i. How do the sales volume in each store relate to Acell's revenues?

    ii. How does this depend on the configuration?

d. Configuration Questions:

    i. What are the details of each configuration? How does this relate to price?

    ii. Do all stores sell all configurations?

4

DIMENSION REDUCTION

In this chapter we describe the important step of dimension reduction. The dimension of a dataset, which is the number of variables, often must be reduced for the data mining algorithms to operate efficiently. This process is part of the pilot/prototype phase of data mining and is done before deploying a model. We present and discuss several dimension reduction approaches: (1) incorporating domain knowledge to remove or combine categories, (2) using data summaries to detect information overlap between variables (and remove or combine redundant variables or categories), (3) using data conversion techniques such as converting categorical variables into numerical variables, and (4) employing automated reduction techniques such as principal components analysis (PCA), where a new set of variables (which are weighted averages of the original variables) is created. These new variables are uncorrelated and a small subset of them usually contains most of their combined information (hence we can reduce dimension by using only a subset of the new variables). Finally, we mention data mining methods such as regression models and classification and regression trees, which can be used for removing redundant variables and for combining "similar" categories of categorical variables.

4.1 INTRODUCTION

In data mining one often encounters situations where there is a large number of variables in the database. Even when the initial number of variables is small, this set quickly expands in the data preparation step, where new derived variables are created (e.g., recoded or transformed variables). In such situations it is likely that subsets of variables are highly correlated with each other. Including highly correlated variables in a classification or prediction model, or including variables that are unrelated to the outcome of interest, can lead to overfitting, and accuracy and reliability can suffer. A large number of variables also poses computational problems for some supervised as well as unsupervised algorithms (aside from questions of correlation). In model deployment, superfluous variables can

Data Mining for Business Analytics: Concepts, Techniques, and Applications with JMP Pro®, First Edition.
Galit Shmueli, Peter C. Bruce, Mia L. Stephens, and Nitin R. Patel.
© 2017 John Wiley & Sons, Inc. Published 2017 by John Wiley & Sons, Inc.

increase costs due to the collection and processing of these variables. The *dimensionality* of a model is the number of independent or input variables used by the model.

4.2 CURSE OF DIMENSIONALITY

The *curse of dimensionality* is the affliction caused by adding variables to multivariate data models. As variables are added, the data space becomes increasingly sparse, and classification and prediction models fail because the available data are insufficient to provide a useful model across so many variables. An important consideration is the fact that the difficulties posed by adding a variable increase exponentially with the addition of each variable. One way to think of this intuitively is to consider the location of an object on a chessboard, which has two dimensions and 64 squares or choices. If you expand the chessboard to a cube, you increase the dimensions by 50%—from 2 dimensions to 3 dimensions. However, the location options increase by 800%, to 512 ($8 \times 8 \times 8$). In statistical distance terms, the proliferation of variables means that nothing is close to anything else anymore—too much noise has been added and patterns and structure are no longer discernable. The problem is particularly acute in big data applications, including genomics, where, for example, an analysis might have to deal with values for thousands of different genes. One of the key steps in data mining therefore is finding ways to reduce dimensionality with minimal sacrifice of information content or accuracy. In the artificial intelligence literature, dimension reduction is often referred to as *factor selection* or *feature extraction*.

4.3 PRACTICAL CONSIDERATIONS

Although data mining relies heavily on automated methods versus domain knowledge, it is important at the first step of data exploration to make sure that the variables measured are reasonable for the task at hand. The integration of expert knowledge through a discussion with the data provider (or user) will probably lead to better results. Practical considerations include: Which variables are most important for the task at hand, and which are most likely to be useless? Which variables are likely to contain much error in the measurements? Which variables will be available for measurement (and what will it cost to measure them) in the future if the analysis is repeated? Which variables can actually be measured before the outcome occurs? (For example, if we want to predict the closing price of an ongoing online auction, we cannot use the number of bids as a predictor because this will not be known until the auction closes.)

Example 1: House Prices in Boston

We return to the Boston housing example introduced in Chapter 3 (the dataset is Boston Housing.jmp.) For each neighborhood, a number of variables are given, such as the crime rate, the student/teacher ratio, and the median value of a housing unit in the neighborhood. A description of all 14 variables is given in Table 4.1. The first ten records of the data are shown in Figure 4.1. The first row represents the first neighborhood, which had an average per capita crime rate of 0.006, 18% of the residential land zoned for lots over 25,000 ft^2, 2.31% of the land devoted to nonretail business, no border on the Charles River, and so on.

TABLE 4.1   Description of Variables in the Boston HousingDataset

CRIM	Crime rate
ZN	Percentage of residential land zoned for lots over 25,000 ft^2
INDUS	Percentage of land occupied by nonretail business
CHAS	Does tract bound Charles River (= 1 if tract bounds river, = 0 otherwise)
NOX	Nitric oxide concentration (parts per 10 million)
RM	Average number of rooms per dwelling
AGE	Percentage of owner-occupied units built prior to 1940
DIS	Weighted distances to five Boston employment centers
RAD	Index of accessibility to radial highways
TAX	Full-value property tax rate per $10,000
PTRATIO	Pupil-to-teacher ratio by town
LSTAT	Percentage of lower status of the population
MEDV	Median value of owner-occupied homes in $1000s
CAT.MEDV	Is median value of owner-occupied homes in tract above $30,000 (CAT.MEDV = 1) or not (CAT.MEDV = 0)

	CRIM	ZN	INDUS	CHAS	NOX	RM	AGE	DIS	RAD	TAX	PTRATIO	LSTAT	MEDV	CAT.MEDV
1	0.00632	18	2.31	0	0.538	6.575	65.2	4.09	1	296	15.3	4.98	24	0
2	0.02731	0	7.07	0	0.469	6.421	78.9	4.9671	2	242	17.8	9.14	21.6	0
3	0.02729	0	7.07	0	0.469	7.185	61.1	4.9671	2	242	17.8	4.03	34.7	1
4	0.03237	0	2.18	0	0.458	6.998	45.8	6.0622	3	222	18.7	2.94	33.4	1
5	0.06905	0	2.18	0	0.458	7.147	54.2	6.0622	3	222	18.7	5.33	36.2	1
6	0.02985	0	2.18	0	0.458	6.43	58.7	6.0622	3	222	18.7	5.21	28.7	0
7	0.08829	12.5	7.87	0	0.524	6.012	66.6	5.5605	5	311	15.2	12.43	22.9	0
8	0.14455	12.5	7.87	0	0.524	6.172	96.1	5.9505	5	311	15.2	19.15	27.1	0
9	0.21124	12.5	7.87	0	0.524	5.631	100	6.0821	5	311	15.2	29.93	16.5	0
10	0.17004	12.5	7.87	0	0.524	6.004	85.9	6.5921	5	311	15.2	17.1	18.9	0

FIGURE 4.1   First Ten Records in the Boston Housing Data.

4.4   DATA SUMMARIES

As we saw in the preceding chapter on data visualization, an important initial step of data exploration is getting familiar with the data and their characteristics through summaries and graphs. The importance of this step cannot be overstated. The better you understand the data, the better the results from the modeling or data mining process.

Numerical summaries and graphs of the data are very helpful for data reduction. The information that they convey can assist in combining categories of a categorical variable, in choosing variables to remove, in assessing the level of information overlap between variables, and more. Before discussing such strategies for reducing the dimension of a data set, let us consider useful summaries and tools.

Summary Statistics

JMP has several platforms for summarizing data (most are under the *Analyze* menu), and provides a variety of summary statistics for learning about the characteristics of each variable. The most common of these statistics are the sample *average* (or *mean*), the *median*, the *standard deviation*, the *minimum* and *maximum values*, and for categorical data the *number of levels* and the *counts* for each level. These statistics give us information about the scale and type of values that a variable takes. The minimum and maximum values

Boston Housing.jmp (506 rows, 14 columns)

▶ **Columns View Selector**

▼ ▼**Summary Statistics**

13 Columns   Clear Select   Distribution

Columns	N Categories	Min	Max	Mean	Std Dev	Median	Lower Quartile	Upper Quartile	Interquartile Range
CRIM	.	0.01	88.98	3.61	8.60	0.26	0.08	3.68	3.60
ZN	.	0.00	100.00	11.36	23.32	0.00	0.00	12.50	12.50
INDUS	.	0.46	27.74	11.14	6.86	9.69	5.18	18.10	12.93
CHAS	2.00	.	.	.	.	.	.	.	.
NOX	.	0.39	0.87	0.55	0.12	0.54	0.45	0.62	0.18
RM	.	3.56	8.78	6.28	0.70	6.21	5.88	6.63	0.74
AGE	.	2.90	100.00	68.57	28.15	77.50	44.85	94.10	49.25
DIS	.	1.13	12.13	3.80	2.11	3.21	2.10	5.21	3.12
RAD	.	1.00	24.00	9.55	8.71	5.00	4.00	24.00	20.00
TAX	.	187.00	711.00	408.24	168.54	330.00	279.00	666.00	387.00
PTRATIO	.	12.60	22.00	18.46	2.16	19.05	17.38	20.20	2.83
LSTAT	.	1.73	37.97	12.65	7.14	11.36	6.93	16.99	10.07
MEDV	.	5.00	50.00	22.53	9.20	21.20	16.95	25.00	8.05

FIGURE 4.2   Summary statistics the Boston Housing data from the *Columns Viewer*.

can be used to detect extreme values that might be errors. The mean and median give a sense of the central values of a variable, and a large difference between the two also indicates skew. The standard deviation gives a sense of how dispersed the data are (relative to the mean). Other statistics, such as *N Missing*, can tell us about missing values.

To quickly summarize all of the variables in a dataset, use the *Columns Viewer* (under the *Cols* menu). Select the variables, then click *Show Summary*. Categorical variables show a value under *N Categories*, while summary statistics are provided for numeric (continuous) variables. If data are missing, the table will contain a column summarizing missing data across the variables (*N Missing*). Checking the *Show Quartiles* box before clicking *Show Summary* provides the statistics in the last four columns shown in Figure 4.2.

We immediately see that the different variables have very different ranges of values. We will soon see how variation in scale across variables can distort analyses if not treated properly. Another observation that can be made is that the average of the first variable, CRIM (as well as several others), is much larger than the median, indicating right skew. None of the variables have missing values. There also do not appear to be indications of extreme values that might result from data entry errors (the Min and Max values for all of the variables look reasonable).

For a more detailed look at the distributions and summary statistics for specific variables, select the variables and click the *Distribution* button in the *Columns Viewer*. The *Distribution* platform (also accessible from the *Analyze* menu) provides summary statistics and basic graphical displays. Figure 4.3 shows default output for CRIM and CHAS, from the Boston Housing data. Additional summary statistics and graphical options are available under the red triangles (e.g., we used the Stack option from the top red triangle to produce the horizontal layout).

Next we summarize relationships between two or more variables. For numerical variables, we can compute pairwise correlations using the *Analyze > Multivariate Methods > Multivariate* platform (as shown in Chapter 3). This produces a complete matrix of correlations between each pair of variables. Figure 4.4 shows the correlation matrix for a subset of the Boston Housing variables. We see that most correlations are low and that many

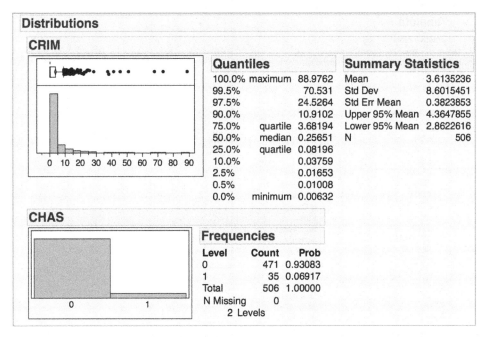

FIGURE 4.3   Basic graphs and summary statistics CRIM and CHAS from the *Distribution* platform.

are negative. Recall also the visual display of a correlation matrix via a color map (see Figure 3.6 in Chapter 3 for the color map corresponding to this correlation table). We will return to the importance of the correlation matrix soon, in the context of correlation analysis.

Tabulating Data (Pivot Tables)

Another very useful tool for summarizing data in JMP is *Tabulate* (under the *Analyze* menu). *Tabulate* provides interactive tables that can combine information from multiple variables and compute a range of summary statistics (count, mean, percentage, etc.). The initial *Tabulate* window is shown in Figure 4.5. Like *Graph Builder*, *Tabulate* has drop zones for dragging and dropping variables in an interactive table, but a more traditional dialog window is also available.

▽ **Multivariate**
▽ **Correlations**

	CRIM	ZN	INDUS	CHAS	NOX	RM	AGE	DIS	RAD	TAX	PTRATIO	LSTAT	MEDV
CRIM	1.0000	-0.2005	0.4066	-0.0559	0.4210	-0.2192	0.3527	-0.3797	0.6255	0.5828	0.2899	0.4556	-0.3883
ZN	-0.2005	1.0000	-0.5338	-0.0427	-0.5166	0.3120	-0.5695	0.6644	-0.3119	-0.3146	-0.3917	-0.4130	0.3604
INDUS	0.4066	-0.5338	1.0000	0.0629	0.7637	-0.3917	0.6448	-0.7080	0.5951	0.7208	0.3832	0.6038	-0.4837
CHAS	-0.0559	-0.0427	0.0629	1.0000	0.0912	0.0913	0.0865	-0.0992	-0.0074	-0.0356	-0.1215	-0.0539	0.1753
NOX	0.4210	-0.5166	0.7637	0.0912	1.0000	-0.3022	0.7315	-0.7692	0.6114	0.6680	0.1889	0.5909	-0.4273
RM	-0.2192	0.3120	-0.3917	0.0913	-0.3022	1.0000	-0.2403	0.2052	-0.2098	-0.2920	-0.3555	-0.6138	0.6954
AGE	0.3527	-0.5695	0.6448	0.0865	0.7315	-0.2403	1.0000	-0.7479	0.4560	0.5065	0.2615	0.6023	-0.3770
DIS	-0.3797	0.6644	-0.7080	-0.0992	-0.7692	0.2052	-0.7479	1.0000	-0.4946	-0.5344	-0.2325	-0.4970	0.2499
RAD	0.6255	-0.3119	0.5951	-0.0074	0.6114	-0.2098	0.4560	-0.4946	1.0000	0.9102	0.4647	0.4887	-0.3816
TAX	0.5828	-0.3146	0.7208	-0.0356	0.6680	-0.2920	0.5065	-0.5344	0.9102	1.0000	0.4609	0.5440	-0.4685
PTRATIO	0.2899	-0.3917	0.3832	-0.1215	0.1889	-0.3555	0.2615	-0.2325	0.4647	0.4609	1.0000	0.3740	-0.5078
LSTAT	0.4556	-0.4130	0.6038	-0.0539	0.5909	-0.6138	0.6023	-0.4970	0.4887	0.5440	0.3740	1.0000	-0.7377
MEDV	-0.3883	0.3604	-0.4837	0.1753	-0.4273	0.6954	-0.3770	0.2499	-0.3816	-0.4685	-0.5078	-0.7377	1.0000

FIGURE 4.4   Correlation table for Boston Housing data, generated using the *Multivariate* platform.

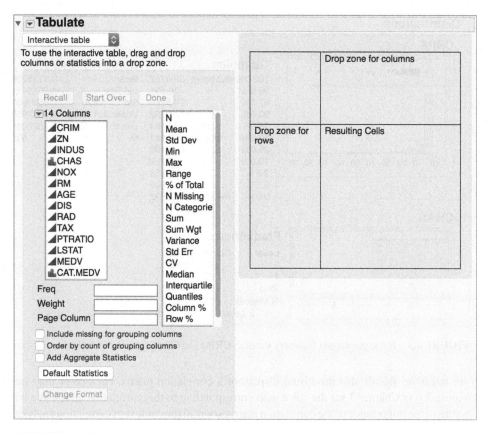

FIGURE 4.5 The initial *Tabulate* window. Drag and drop variables in the drop zones for columns or rows, and select the summary statistics of interest from the list provided.

A simple example is the number of neighborhoods that bound the Charles River versus those that do not, and the mean MEDV for each group. This is shown in the left-hand panel of Figure 4.6. This tabulation was obtained by dragging MEDV to the drop zone for columns, CHAS to the drop zone for rows, and dragging the statistics Mean and N on top of the existing statistic (the Sum). Clicking the *Done* button (on the left in Figure 4.5) closes the control panel and produces the table. For categorical variables we obtain a breakdown of the records by the combination of categories. For instance, the right-hand panel of Figure 4.6 shows the average of each of the numeric variables for the two levels of CAT.MEDV, along with a count of neighborhoods bordering the Charles River for each category of CAT.MEDV. There are many more possibilities and options for using *Tabulate*. We leave it to the reader to explore these options. For more information, search for *Tabulate* in the *JMP Help*.

In classification tasks, where the goal is to find predictor variables that do a good job of distinguishing between two classes, a good exploratory step is to produce summaries for each class. This can assist in detecting useful predictors that display some separation between the two classes. Data summaries are useful for almost any data mining task and

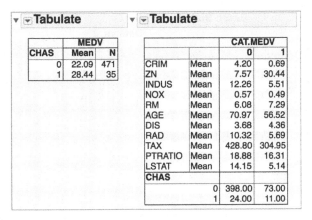

FIGURE 4.6   Tabulations for the Boston housing data, created using *Tabulate*.

are therefore an important preliminary step for cleaning and understanding the data before carrying out further analyses.

4.5   CORRELATION ANALYSIS

In datasets with a large number of variables (which are likely to serve as predictors), there is usually much overlap in the information covered by the set of variables. One simple way to find redundancies is to look at a correlation matrix (see the example in Figure 4.4). This shows all the pairwise correlations between variables. Pairs that have a very strong (positive or negative) correlation contain a lot of overlap in information and are good candidates for data reduction by removing one of the variables. Removing variables that are strongly correlated to others is useful for avoiding multicollinearity problems that can arise in various models. (*Multicollinearity* is the presence of two or more predictors sharing the same linear relationship with the outcome variable.)

Correlation analysis is also a good method for detecting duplications of variables in the data. Sometimes the same variable appears accidentally more than once in the dataset (under a different name) because the dataset was merged from multiple sources, the same phenomenon is measured in different units, and so on. Use of correlation table color map, as shown in Chapter 3, can make the task of identifying strong correlations easier.

4.6   REDUCING THE NUMBER OF CATEGORIES IN CATEGORICAL VARIABLES

When a categorical variable has many categories, and this variable is destined to be a predictor, many data mining methods will require converting it into many *dummy* or *indicator variables*. In particular, a variable with m categories will be transformed into either m or $m - 1$ dummy variables (depending on the method). Note that JMP does not

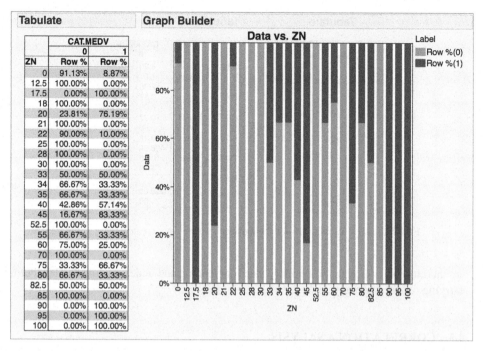

FIGURE 4.7   Distribution of CAT.MEDV (dark gray indicates CAT.MEDV = 1) by ZN. Similar bars indicate low separation of classes, suggesting those categories can be combined.

require converting data into dummy variables per se—it will do this behind the scenes in most modeling and data mining platforms. Nonetheless, even if we have very few original categorical variables, they can greatly inflate the dimensions of the dataset. One way to handle this is to reduce the number of categories by combining close or similar categories. To combine categories requires incorporating expert knowledge and common sense. Tabulations are useful for this task: We can examine the sizes of the various categories and how the response behaves at each category. Generally, categories that contain very few observations are good candidates for combining with other categories. A best practice is to use only the categories that are most relevant to the analysis, recode missing values as "missing," and recode the rest as "other."

In classification tasks (with a categorical output), a tabulation broken down by the output classes can help identify categories that do not separate the classes. Those categories are also candidates for inclusion in the "other" category. An example is shown on the left in Figure 4.7. We can see that the distribution of CAT.MEDV is identical for ZN = 17.5, 90, 95, and 100 (where all neighborhoods have CAT.MEDV = 1). These four categories can then be combined into a single category. Similarly categories ZN = 12.5, 25, 28, 30, and 70 (and others where CAT.MEDV = 0) can be combined. Further combination is also possible based on similar bars.

To create the tabulation and stacked bar graph in Figure 4.7:

- Change the modeling type for ZN to Nominal.
- In *Tabulate*, drag CAT.MEDV to the drop zone for columns, drag ZN to the drop zone for rows, and drag Row % to the results area.
- Select *Make into Data Table* from the top red triangle.
- In *Graph Builder*, drag ZN to the X zone, drag the Row % columns for both classes to the Y zone (at once), and click on the bar icon (above the graph). Under Bar Style (on the bottom left) select Stacked.

In a time series context where we might have a categorical variable denoting season (e.g., month or hour of day) that will serve as a predictor, reducing categories can be done by examining the time series plot and identifying similar periods. For example, the time plot in Figure 4.8 shows the quarterly revenues of Toys "R" Us between 1992 and 1995. Only quarter 4 periods appear different, and therefore we can combine quarters 1–3 into a single category.

One JMP tool for combining categories is *Recode* (under *Cols > Utilities* in JMP 12—first select the column in the data table). We can use recode, for example, to create a new numeric category combining the zones 12.5, 25, 28, 30, and 70, as shown in Figure 4.9. Selecting *Done > Formula Column* will result in the data being recoded in a new column, with a stored formula for recoding (so, if additional data are added to the data table, these data will be recoded as well). Note that additional options for data cleanup for categorical data, such as grouping similar values, trimming white space and addressing capitalization issues, are available under the red triangle in the *Recode* window (these are not available for numeric data).

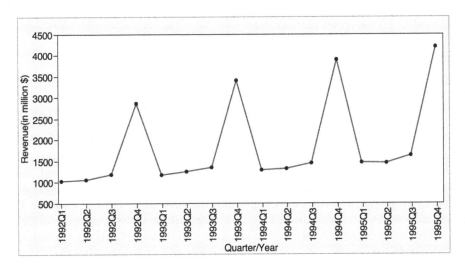

FIGURE 4.8   Quarterly Revenues of Toys "R" US, 1992–1995.

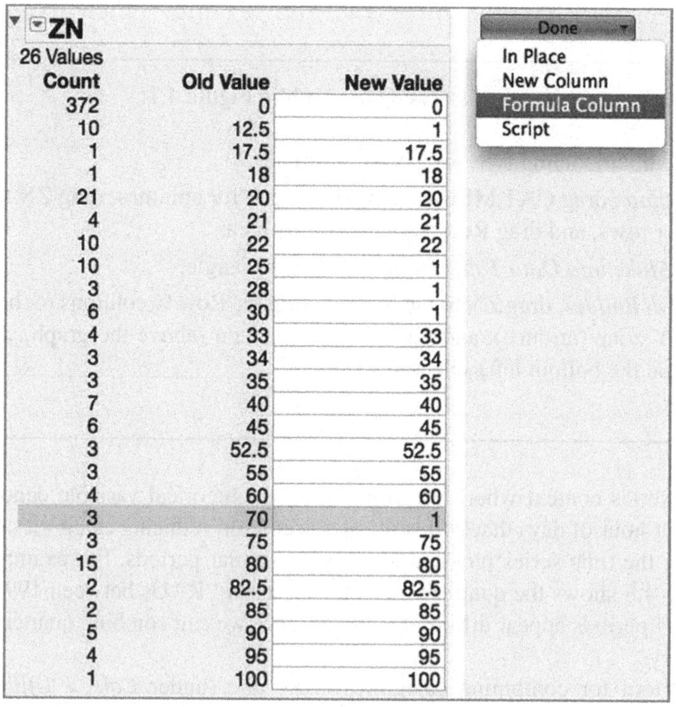

ZN			
26 Values			
Count	Old Value	New Value	
372	0	0	
10	12.5	1	
1	17.5	17.5	
1	18	18	
21	20	20	
4	21	21	
10	22	22	
10	25	1	
3	28	1	
6	30	1	
4	33	33	
3	34	34	
3	35	35	
7	40	40	
6	45	45	
3	52.5	52.5	
3	55	55	
4	60	60	
3	70	1	
3	75	75	
15	80	80	
2	82.5	82.5	
2	85	85	
5	90	90	
4	95	95	
1	100	100	

Done

In Place
New Column
Formula Column
Script

FIGURE 4.9 Combining categories using *Recode*.

4.7 CONVERTING A CATEGORICAL VARIABLE TO A CONTINUOUS VARIABLE

Sometimes the categories in a categorical variable represent intervals. Common examples are age group or income bracket. If the interval values are known (e.g., category 2 is the age interval 20–30), we can replace the categorical value ("2" in the example) with the mid-interval value (here "25"). The resulting numeric values can be treated as continuous data, thus simplifying analyses.

4.8 PRINCIPAL COMPONENTS ANALYSIS

Principal components analysis (PCA) is a useful method for dimension reduction, especially when the number of variables is large. PCA is especially valuable when we have subsets of measurements that are measured on the same scale and are highly correlated. In that case it provides a few variables (often as few as three) that are weighted linear combinations of the original variables, and that retain the majority of the information of the full original set of variables. PCA is intended for use with quantitative variables. For categorical variables, other methods, such as correspondence analysis, are more suitable. (*Multiple Correspondence Analysis* is available in JMP from the *Analyze > Consumer Research* platform.)

Cereal Name	mfr	type	calories	protein	fat	sodium	fiber	carbo	sugars	potass	vitamins
100% Bran	N	C	70	4	1	130	10	5	6	280	25
100% Natural Bran	Q	C	120	3	5	15	2	8	8	135	0
All-Bran	K	C	70	4	1	260	9	7	5	320	25
All-Bran with Extra Fiber	K	C	50	4	0	140	14	8	0	330	25
Almond Delight	R	C	110	2	2	200	1	14	8		25
Apple Cinnamon Cheerios	G	C	110	2	2	180	1.5	10.5	10	70	25
Apple Jacks	K	C	110	2	0	125	1	11	14	30	25
Basic 4	G	C	130	3	2	210	2	18	8	100	25
Bran Chex	R	C	90	2	1	200	4	15	6	125	25
Bran Flakes	P	C	90	3	0	210	5	13	5	190	25
Cap'n'Crunch	Q	C	120	1	2	220	0	12	12	35	25
Cheerios	G	C	110	6	2	290	2	17	1	105	25
Cinnamon Toast Crunch	G	C	120	1	3	210	0	13	9	45	25
Clusters	G	C	110	3	2	140	2	13	7	105	25
Cocoa Puffs	G	C	110	1	1	180	0	12	13	55	25
Corn Chex	R	C	110	2	0	280	0	22	3	25	25
Corn Flakes	K	C	100	2	0	290	1	21	2	35	25
Corn Pops	K	C	110	1	0	90	1	13	12	20	25
Count Chocula	G	C	110	1	1	180	0	12	13	65	25
Cracklin' Oat Bran	K	C	110	3	3	140	4	10	7	160	25

FIGURE 4.10   Sample from the 77 breakfast cereals dataset (Cereals.jmm).

Example 2: Breakfast Cereals

Data were collected on the nutritional information and consumer rating of 77 breakfast cereals (data are in `Cereals.jmp`).[1] The consumer rating is a rating of cereal "healthiness" for consumer information (not a rating by consumers). For each cereal the data include 13 numerical variables, and we are interested in reducing this dimension. For each cereal the information is based on a bowl of cereal rather than a serving size, since most people simply fill a cereal bowl (resulting in constant volume, but not weight). A snapshot of these data is given in Figure 4.10, and the description of the different variables is given in Table 4.2.

We focus first on two variables: *calories* and consumer *rating*. These are given in Table 4.3 for all cereals. The average calories across the 77 cereals is 106.88 and the average consumer rating is 42.67. The estimated covariance matrix between the two variables is

$$S = \begin{bmatrix} 379.63 & -188.68 \\ -188.68 & 197.32 \end{bmatrix}$$

Here, the variance of calories is 379.63, the variance of rating is 197.32, and the covariance is 188.68. It can be seen that the two variables are strongly correlated with a negative correlation of

$$-0.69 = \frac{-188.68}{\sqrt{(379.63)(197.32)}}$$

[1] The data are available at http://lib.stat.cmu.edu/DASL/Stories/HealthyBreakfast.html.

TABLE 4.2   Description of the Variables in the Breakfast Cereal Dataset

Variable	Description
mfr	Manufacturer of cereal (American Home Food Products, General Mills, Kellogg, etc.)
type	Cold or hot
calories	Calories per serving
protein	Grams of protein
fat	Grams of fat
sodium	Milligrams of sodium
fiber	Grams of dietary fiber
carbo	Grams of complex carbohydrates
sugars	Grams of sugars
potass	Milligrams of potassium
vitamins	Vitamins and minerals: 0, 25, or 100, indicating the typical percentage of FDA recommended
shelf	Display shelf (1, 2, or 3, counting from the floor)
weight	Weight in ounces of one serving
cups	Number of cups in one serving
rating	Rating of the cereal calculated by Consumer Reports

These numbers can be confirmed in the *Multivariate* platform. Roughly speaking, 69% of the total variation in both variables is actually "co-variation," or variation in one variable that is duplicated by similar variation in the other variable. Can we use this fact to reduce the number of variables, while making maximum use of their unique contributions to the overall variation? Since there is redundancy in the information that the two variables contain, it might be possible to reduce the two variables to a single variable without losing too much information. The idea in PCA is to find a linear combination of the two variables that contains most, even if not all, of the information, so that this new variable can replace the two original variables. Information here is in the sense of variability: What can explain the most variability *among* the 77 cereals? The total variability here is the sum of the variances of the two variables, which in this case is $379.63 + 197.32 = 576.95$. This means that *calories* accounts for $66\% = 379.63/577$ of the total variability, and *rating* the remaining 34%. If we drop one of the variables for the sake of dimension reduction, we lose at least 34% of the total variability. Can we redistribute the total variability between two new variables in a more polarized way? If so, it might be possible to keep only the one new variable that (hopefully) accounts for a large portion of the total variation.

Figure 4.11 shows a scatterplot of *rating* vs. *calories*. The line z_1 is the direction in which the variability of the points is largest. It is the line that captures the most variation in the data if we decide to reduce the dimensionality of the data from two to one. Among all possible lines, it is the line for which, if we project the points in the dataset orthogonally to get a set of 77 (one-dimensional) values, the variance of the z_1 values will be maximum. This is called the *first principal component*. It is also the line that minimizes the sum-of-squared perpendicular distances from the line. The z_2-axis is chosen to be perpendicular to the z_1-axis. In the case of two variables, there is only one line that is perpendicular to z_1, and it has the second largest variability, but its information is uncorrelated with z_1. This is called the *second principal component*. In general, when we have more than two variables, once we find the direction z_1 with the largest variability, we search among all the

TABLE 4.3 Cereal Calories and Ratings

Cereal	Calories	Rating	Cereal	Calories	Rating
100% Bran	70	68.40297	Just Right Fruit & Nut	140	36.471512
100% Natural Bran	120	33.98368	Kix	110	39.241114
All-Bran	70	59.42551	Life	100	45.328074
All-Bran with Extra Fiber	50	93.70491	Lucky Charms	110	26.734515
Almond Delight	110	34.38484	Maypo	100	54.850917
Apple Cinnamon Cheerios	110	29.50954	Muesli Raisins, Dates & Almonds	150	37.136863
Apple Jacks	110	33.17409	Muesli Raisins, Peaches & Pecans	150	34.139765
Basic 4	130	37.03856	Mueslix Crispy Blend	160	30.313351
Bran Chex	90	49.12025	Multi-Grain Cheerios	100	40.105965
Bran Flakes	90	53.31381	Nut&Honey Crunch	120	29.924285
Cap'n'Crunch	120	18.04285	Nutri-Grain Almond-Raisin	140	40.69232
Cheerios	110	50.765	Nutri-grain Wheat	90	59.642837
Cinnamon Toast Crunch	120	19.82357	Oatmeal Raisin Crisp	130	30.450843
Clusters	110	40.40021	Post Nat. Raisin Bran	120	37.840594
Cocoa Puffs	110	22.73645	Product 19	100	41.50354
Corn Chex	110	41.44502	Puffed Rice	50	60.756112
Corn Flakes	100	45.86332	Puffed Wheat	50	63.005645
Corn Pops	110	35.78279	Quaker Oat Squares	100	49.511874
Count Chocula	110	22.39651	Quaker Oatmeal	100	50.828392
Cracklin' Oat Bran	110	40.44877	Raisin Bran	120	39.259197
Cream of Wheat (Quick)	100	64.53382	Raisin Nut Bran	100	39.7034
Crispix	110	46.89564	Raisin Squares	90	55.333142
Crispy Wheat & Raisins	100	36.1762	Rice Chex	110	41.998933
Double Chex	100	44.33086	Rice Krispies	110	40.560159
Froot Loops	110	32.20758	Shredded Wheat	80	68.235885
Frosted Flakes	110	31.43597	Shredded Wheat 'n'Bran	90	74.472949
Frosted Mini-Wheats	100	58.34514	Shredded Wheat spoon size	90	72.801787
Fruit & Fibre Dates, Walnuts & Oats	120	40.91705	Smacks	110	31.230054
Fruitful Bran	120	41.01549	Special K	110	53.131324
Fruity Pebbles	110	28.02577	Strawberry Fruit Wheats	90	59.363993
Golden Crisp	100	35.25244	Total Corn Flakes	110	38.839746
Golden Grahams	110	23.80404	Total Raisin Bran	140	28.592785
Grape Nuts Flakes	100	52.0769	Total Whole Grain	100	46.658844
Grape-Nuts	110	53.37101	Triples	110	39.106174
Great Grains Pecan	120	45.81172	Trix	110	27.753301
Honey Graham Ohs	120	21.87129	Wheat Chex	100	49.787445
Honey Nut Cheerios	110	31.07222	Wheaties	100	51.592193
Honey-comb	110	28.74241	Wheaties Honey Gold	110	36.187559
Just Right Crunchy Nuggets	110	36.52368			

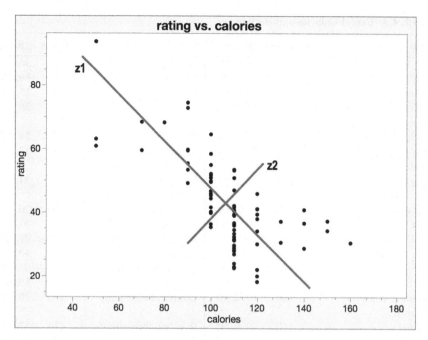

FIGURE 4.11   Scatterplot of *rating* vs. *calories* for 77 breakfast cereals, with the two principal component directions, created using the *Graph Builder* for illustration.

orthogonal directions to z_1 for the one with the next-highest variability. That is z_2. The idea is then to find the coordinates of these lines and to see how they redistribute the variability.

Figure 4.12 shows the JMP output from running PCA on these two variables (using *Analyze > Multivariate Methods > Principal Components*). Note that JMP can conduct PCA using either the correlation matrix or the covariance matrix (more on this distinction later). For this discussion, we have conducted the analysis on covariances (the default in JMP is to use the correlation matrix, but this can be changed using the first option under the red triangle in the platform).

Looking at the reallocated variances for the new derived variables (listed under *Eigenvalues* in Figure 4.12) we see that z_1 accounts for 86.3% of the total variability, and that z_2 accounts for the remaining 13.7%. Therefore, if we drop the second component, we still maintain 86.3% of the total variability in the data.

Eigenvectors are the weights that are used to project the original data onto the two new directions. The eigenvectors for the cereal data are shown in Figure 4.12. The weights for z_1 (which is labeled *Prin1*) are given by 0.847 for *calories* and −0.532 for *rating*. For z_2 they are given by 0.532 and 0.847, respectively. These weights are used to compute principal component scores for each observation, which are the projected values of *calories* and *rating* onto the new axes (after subtracting the means for each variables). Figure 4.13 shows the scores for the two dimensions. The first column is the projection onto z_1 using the weights (0.847, −0.532). The second column is the projection onto z_2 using the weights (0.532, 0.847). For example, the first score for the 100% Bran cereal (with 70 calories and a rating of 68.4) is $(0.847)(70 - 106.88) + (-0.532)(68.4 - 42.67) = -44.92$.

Note that the means of the new variables z_1 and z_2 are zero because we've subtracted the mean of each variable (106.88 for calories and 42.67 for rating—this can be confirmed using

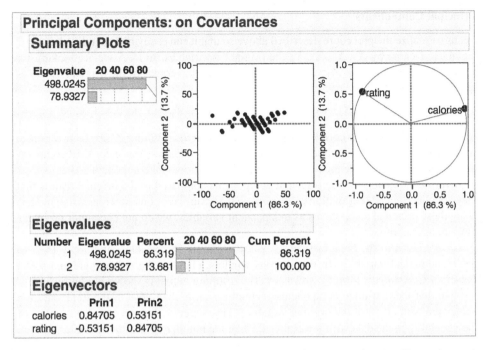

FIGURE 4.12 Output from principal components analysis of *calories* and *rating*. *Eigenvalues* and *Eigenvectors* are displayed using red triangle options.

the *Distribution* platform). The variances of z_1 and z_2 are 498.02 and 78.93, respectively, and the sum of these variances $\text{var}(z_1) + \text{var}(z_2)$ is equal to the sum of the variances of the original variables, $\text{var}(calories) + \text{var}(rating)$. So, the first principal component, z_1, accounts for 86% of the total variance. Since it captures most of the variability in the data, it seems reasonable to use one variable, the first principal score, to represent the two variables in the original data. Next, we generalize these ideas to more than two variables.

Graphical and numeric summaries of the principal components analysis are provided under *Summary Plots* (this is the default output). The bar chart displays the percentage of variation accounted for by each principal component (also shown in the *Eigenvalues* table). The *Score Plot* (top, center in Figure 4.12) helps interpret the data relative to the principal components–points that are close together are similar. The *Loadings Plot* is for interpreting the relationships among the variables and their "loadings," or projections, onto the principal components.

Cereals	calories	rating	Prin1	Prin2
1 100%_Bran	70	68.402973	-44.92152782	2.1971832552
2 100%_Natural_Bran	120	33.983679	15.725264881	-0.38241646
3 All-Bran	70	59.425505	-40.14993478	-5.407212287
4 All-Bran_with_Extra_Fiber	50	93.704912	-75.31077196	12.999125557
5 Almond_Delight	110	34.384843	7.0415083066	-5.35768573

FIGURE 4.13 Principal component scores from PCA of *calories* and *rating* for the first 5 cereals.

Principal Components

Let us formalize the procedure described above so that it can easily be generalized to $p > 2$ variables. Denote by $X_1, X_2, \ldots, X_p$ the original p variables. In PCA we are looking for a set of new variables $Z_1, Z_2, \ldots, Z_p$ that are weighted averages of the original variables (after subtracting their mean):

$$Z_i = a_{i,1}(X_1 - \bar{X}_1) + a_{i,2}(X_2 - \bar{X}_2) + \cdots + a_{i,p}(X_p - \bar{X}_p), \quad i = 1, \ldots, p$$

where each pair of Z's has correlation = 0. We then order the resulting Z's by their variance, with Z_1 having the largest variance and Z_p having the smallest variance. JMP computes the weights $a_{i,j}$, which are then used in computing the principal component scores.

A further advantage of the principal components compared to the original data is that they are uncorrelated (correlation coefficient = 0). Because we construct regression models using these principal components as independent variables, we do not encounter problems of multicollinearity.

Let us return to the breakfast cereal dataset with all 15 variables, and apply PCA to the 13 numerical variables. In the analysis window, click on the red triangle and select *Principal Components > on Covariances*. The resulting output is shown in Figure 4.14.

Note that the first three components account for more than 96% of the total variation associated with all 13 of the original variables. This suggests that we can capture most of the variability in the data with fewer than 25% of the number of original variables. In fact,

Eigenvalues

Number	Eigenvalue	Percent	20 40 60 80	Cum Percent
1	7207.777	55.001		55.001
2	4961.273	37.859		92.860
3	498.3640	3.803		96.663
4	357.0866	2.725		99.388
5	72.3054	0.552		99.939
6	3.8532	0.029		99.969
7	2.7603	0.021		99.990
8	0.6209	0.005		99.995
9	0.4361	0.003		99.998
10	0.1857	0.001		99.999
11	0.0595	0.000		100.000
12	0.0329	0.000		100.000
13	0.0038	0.000		100.000

Eigenvectors

	Prin1	Prin2	Prin3	Prin4	Prin5	Prin6	Prin7
calories	0.07629	-0.00959	0.61056	-0.61793	0.45563	0.11148	-0.11231
protein	-0.00153	0.00863	0.00054	0.00188	0.05494	0.07855	-0.27998
fat	-0.00016	0.00271	0.01598	-0.02603	-0.01894	-0.19240	-0.34032
sodium	0.98041	0.13452	-0.14069	-0.00292	0.01546	0.02185	-0.00728
fiber	-0.00504	0.03042	-0.01669	0.02122	0.00630	0.26427	-0.08320
carbo	0.01551	-0.01732	0.01227	0.02268	0.36294	-0.44395	0.73170
sugars	0.00407	-0.00005	0.09946	-0.11648	-0.30415	0.70344	0.48650
potass	-0.12850	0.98741	0.03598	-0.04251	-0.04625	-0.04820	0.02484
vitamins	0.10134	0.01697	0.70796	0.69787	-0.02556	0.01171	-0.01251
shelf	-0.00097	0.00440	0.01270	0.00568	-0.00868	-0.08087	-0.08151
weight	0.00049	0.00098	0.00370	-0.00267	0.00310	0.01093	0.02284
cups	0.00049	-0.00158	0.00059	0.00097	0.00296	-0.00726	0.02105
rating	-0.07681	0.07217	-0.30688	0.33775	0.74942	0.41590	-0.07587

FIGURE 4.14   PCA output using all 13 numerical variables in the breakfast cereals dataset. The eigenvectors table gives results for the first seven principal components.

the first two principal components alone capture 92.9% of the total variation. However, these results are influenced by the scales of the variables, as we describe next.

Normalizing the Data

A further use of PCA is to understand the structure of the data. This is done by examining the weights (the eigenvectors) to see how the original variables contribute to the different principal components. In our example it is clear that the first principal component is dominated by the sodium content of the cereal: it has the highest (in this case, positive) weight. This means that the first principal component is in fact measuring how much sodium is in the cereal. Similarly, the second principal component seems to be measuring the amount of potassium. Since both these variables are measured in milligrams, whereas the other nutrients are measured in grams, the scale is obviously leading to this result. The variances of potassium and sodium are much larger than the variances of the other variables, and thus the total variance is dominated by these two variances. A solution is to normalize the data before performing the PCA. Normalization (or standardization) means replacing each original variable by a standardized version of the variable that has unit variance. This is easily accomplished by dividing each variable by its standard deviation. The effect of this normalization (standardization) is to give all variables equal importance in terms of the variability.

When should we normalize the data like this? It depends on the nature of the data. When the units of measurement are common for the variables (e.g., dollars), and when their scale reflects their importance (sales of jet fuel, sales of heating oil), it is probably best not to normalize (i.e., not to rescale the data so that they have unit variance). If the variables are measured in quite differing units so that it is unclear how to compare the variability of different variables (e.g., dollars for some, parts per million for others), or if for variables measured in the same units, scale does not reflect importance (earnings per share, gross revenues), then it is generally advisable to normalize. In this way the changes in units of measurement do not change the principal components' weights. In the rare situations where we can give relative weights to variables, we multiply the normalized variables by these weights before doing the principal components analysis.

Thus far we have calculated principal components using the covariance matrix. An alternative to normalizing and then performing PCA is to perform PCA on the *correlation matrix* instead of the covariance matrix. Most software programs allow the user to choose between the two. In fact, the default in JMP is to use the correlation matrix. Using the correlation matrix means that you are operating on the normalized data.

Returning to the breakfast cereal data, we conduct PCA on the correlation matrix to normalize the 13 variables due to the different scales of the variables. The output is shown in Figure 4.15. Now we find that we need 7 principal components to account for more than 90% of the total variability. The first 2 principal components account for only 52% of the total variability, and thus reducing the number of variables to 2 would mean losing a lot of information. Examining the weights, we see that the first principal component measures the balance between 2 quantities: (1) fiber, potassium, protein, and rating (large positive weights) versus (2) calories and cups (large negative weights). High scores on principal component 1 mean that the cereal is high in fiber, protein, and potassium and low in calories and the amount per bowl. Unsurprisingly, this type of cereal is associated with a high consumer rating. The second principal component is most affected by the weight of a serving, and the third principal component by the carbohydrate content. We can continue

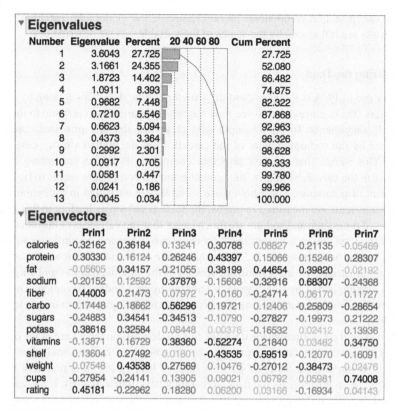

FIGURE 4.15 PCA output using all 13 numerical variables in the breakfast cereals dataset, *normalized* by using correlations. The Eigenvectors table gives results for the first 7 principal components.

labeling the next principal components in a similar fashion to learn about the structure of the data.

The *Score Plot*, which displays by default when conducting a PCA in JMP, shows the first two principal component scores for the different breakfast cereals (see Figure 4.16). *Score Plots* for additional principal components are available from the *red triangle*. The observations can be labeled if the dataset is not too large. We can see that as we move from right (bran cereals) to left, the cereals are less "healthy" in the sense of high calories, low protein and fiber, and so on. Also, moving from bottom to top, we get heavier cereals (moving from puffed rice to raisin bran). These plots are especially useful if interesting clusterings of observations can be found. For instance, we see here that children's cereals are close together on the middle-left part of the plot.

Reminder: To label observations in JMP, right-click on the variable(s) in the data table and select *Label/Unlabel*. When you move your mouse over these points in a graphical display, the labels will appear. To permanently label these points, select the rows in the data table and select *Label/Unlabel* (or select the points directly in the graph, right-click, and select *Row Label*).

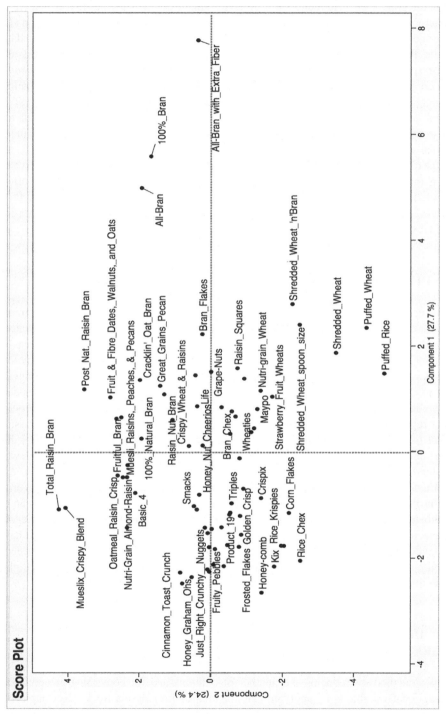

FIGURE 4.16 Score plot of the second vs. first principal components scores for PCA based on correlations.

Using Principal Components for Classification and Prediction

When the goal of the data reduction is to have a smaller set of variables that will serve as predictors, we can proceed as follows: Partition the data into training and validation sets (*hide* and *exclude* the validation data). Apply PCA to the predictors using the training data. Use the PCA output to determine the number of principal components to be retained, and save the principal components to the data table (*red triangle > Save Principal Components*). This creates new formula columns in the data table, with the weights for each principal component computed using only the training data. Principal scores for each observation are automatically calculated for the validation set using these formulas. Use the validation data to fit a model, using the principal component columns as the predictors rather than the original variables.

One disadvantage of using a subset of principal components as predictors in a supervised task is that we might lose predictive information that is nonlinear (e.g., a quadratic effect of a predictor on the outcome or an interaction between predictors). This is because PCA produces linear transformations, thereby capturing only linear relationships between the original variables.

4.9 DIMENSION REDUCTION USING REGRESSION MODELS

In this chapter we discussed methods for reducing the number of columns using summary statistics, plots, and principal components analysis. All these are considered exploratory methods. Some of them completely ignore the output variable (e.g., PCA), whereas in other methods we informally try to incorporate the relationship between the predictors and the output variable (e.g., combining similar categories, in terms of their behavior with y). Another approach to reducing the number of predictors, which directly considers the predictive or classification task, is by fitting a regression model. For prediction, a linear regression model is used (see Chapter 6) and for classification, a logistic regression model (see Chapter 10). In both cases we can employ subset selection procedures that algorithmically choose a subset of variables among the larger set (see details in the relevant chapters).

Fitted regression models can be used to further combine similar categories: categories that have coefficients that are not statistically significant (i.e., have a high p-value) can be combined with the reference category because their distinction from the reference category appears to have no significant effect on the output variable. Moreover, categories that have similar coefficient values (and the same sign) can often be combined because their effect on the output variable is similar. See the example in Chapter 10 on predicting delayed flights for an illustration of how regression models can be used for dimension reduction.

4.10 DIMENSION REDUCTION USING CLASSIFICATION AND REGRESSION TREES

Another method for reducing the number of columns and for combining categories of a categorical variable is by applying classification and regression trees (see Chapter 9). Classification trees are used for classification tasks and regression trees for prediction tasks. In both cases the algorithm creates binary splits on the predictors that best classify/predict the outcome (e.g., above/below age 30). Although we defer the detailed discussion to Chapter 9, we note here that the resulting tree diagram can be used for determining the important predictors (using *Column Contributions*). Predictors (numerical or categorical) that do not appear in the tree can be removed. Similarly, categories that do not appear in the tree can be combined.

PROBLEMS

4.1   Breakfast Cereals. Use the data for the breakfast cereals example in Section 4.8 (Cereals.jmp) to explore and summarize the data as follows:

    a. Which variables are continuous/numerical? Which are ordinal? Which are nominal?

    b. Calculate the following summary statistics: mean, median, min, max, and standard deviation for each of the continuous variables, and the count for each categorical variable. This can be done using *Cols > Columns Viewer*.

        i. Is there any evidence of extreme values?

        ii. Which, if any, of the variables is missing values?

    c. Use *Analyze > Distribution* to plot a histogram for each of the continuous variables and create summary statistics. Based on the histograms and summary statistics, answer the following questions:

        i. Which variables have the largest variability?

        ii. Which variables seem skewed?

        iii. Are there any values that seem extreme?

    d. Use the *Graph Builder* to plot a side-by-side boxplot comparing the calories in hot vs. cold cereals. What does this plot show us?

    e. Use the *Graph Builder* to plot a side-by-side boxplot of consumer rating as a function of the shelf height (the variable *shelf*). If we were to predict consumer rating from shelf height, does it appear that we need to keep all three categories of shelf height?

    f. Compute the correlation table and generate a scatterplot matrix for the continuous variables (use *Analyze > Multivariate Methods > Multivariate*).

        i. Which pair of variables is most strongly correlated?

        ii. How can we reduce the number of variables based on these correlations?

        iii. How would the correlations change if we normalized the data first?

    g. Consider the first column on the left under *Eigenvectors* in Figure 4.14. Describe briefly what this column represents.

4.2   Chemical Features of Wine. Figure 4.17 shows the PCA output for analyses conducted on normalized (correlations) and nonnormalized data (covariances). Each variable represent a chemical characteristics of wine, and each case is a different wine.

    a. The data are in the file Wine.jmp. Consider the variances in the columns labeled *Eigenvalue* for the principal components analysis conducted using covariances.

Principal Components: on Correlations				
Eigenvalues				
Number	Eigenvalue	Percent	20 40 60 80	Cum Percent
1	4.7059	36.199		36.199
2	2.4970	19.207		55.406
3	1.4461	11.124		66.530
4	0.9190	7.069		73.599
5	0.8532	6.563		80.162
6	0.6417	4.936		85.098
7	0.5510	4.239		89.337
8	0.3485	2.681		92.018
9	0.2889	2.222		94.240
10	0.2509	1.930		96.170
11	0.2258	1.737		97.907
12	0.1688	1.298		99.205
13	0.1034	0.795		100.000

Principal Components: on Covariances				
Eigenvalues				
Number	Eigenvalue	Percent	20 40 60 80	Cum Percent
1	99201.79	99.809		99.809
2	172.5353	0.174		99.983
3	9.4381	0.009		99.992
4	4.9912	0.005		99.997
5	1.2288	0.001		99.998
6	0.8411	0.001		99.999
7	0.2790	0.000		100.000
8	0.1514	0.000		100.000
9	0.1121	0.000		100.000
10	0.0717	0.000		100.000
11	0.0376	0.000		100.000
12	0.0211	0.000		100.000
13	0.0082	0.000		100.000

FIGURE 4.17   Principal Components of Normalized and Nonnormalized Wine Data.

Why is the variance for the first principal component so much larger than the others?

b. Comment on the use of correlations versus covariances. Would the results change dramatically if PCA (in this example) were conducted on the correlations instead?

4.3   **University Rankings.** The dataset on American college and university rankings (`Colleges.jmp`) contains information on 1302 American colleges and universities offering an undergraduate program. For each university there are 17 measurements that include continuous measurements (e.g., tuition and graduation rate) and categorical measurements (e.g., location by state and whether it is a private or a public school).

a. Make sure the variables are coded correctly in JMP (*Nominal*, *Ordinal*, or *Continuous*), then use the *Columns Viewer* to summarize the data. Are there any missing values? How many Nominal columns are there?

b. Conduct a principal components analysis on the data and comment on the results. Recall that, by default, JMP will conduct the analysis on correlations rather than covariances. Is this necessary? Do the data need to be normalized in this case? Discuss key considerations in this decision.

4.4   **Sales of Toyota Corolla Cars.** The file `ToyotaCorolla.jmp` contains data on used cars (Toyota Corollas) on sale during late summer of 2004 in the Netherlands. The data table has 1436 records containing details on 38 attributes, including *Price, Age, Kilometers, HP*, and some categorical and dummy-coded variables. The ultimate goal will be to predict the price of a used Toyota Corolla based on its specifications. Although special coding of categorical variables in JMP is generally not required, in this exercise we explore how to create dummy variables (and why).

a. Identify the categorical variables.

b. Which variables have been dummy coded?

c. Consider the variable *Fuel_Type*. How many binary dummy variables are required to capture the information for this variable?

d. Use the *Make Indicator Variables* option under *Cols > Utilities* to convert the *Fuel > Type* into dummy variables, then change the *Modeling Types* for these dummy variables to *Continuous*. Explain in words the values in the derived binary dummies.

PART III

PERFORMANCE EVALUATION

5

EVALUATING PREDICTIVE PERFORMANCE

In this chapter we discuss how the predictive performance of data mining methods can be assessed. This is a critical step in any analytics project. We point out the danger of overfitting to the training data, and the need for testing model performance on data that were not used in the training step. We discuss popular performance metrics. For prediction, metrics include average absolute error (*AAE*) and root mean squared error (RMSE) (based on the validation data). For classification tasks, metrics based on the classification matrix include overall accuracy, specificity, and sensitivity and metrics that account for misclassification costs. We also show the relation between the choice of cutoff value and method performance, and present the ROC curve, which is a popular chart for assessing method performance at different cutoff values. When the goal is to accurately classify the most interesting or important cases, called *ranking*, rather than accurately classify the entire sample (e.g., the 10% of customers most likely to respond to an offer, or the 5% of claims most likely to be fraudulent), lift curves are used to assess performance. We also discuss the need for oversampling rare classes and how to adjust performance metrics for the oversampling. Finally, we mention the usefulness of comparing metrics based on the validation data to metrics based on the training data for the purpose of detecting overfitting. While some differences are expected, extreme differences can be indicative of overfitting.

5.1 INTRODUCTION

In supervised learning we are interested in predicting the outcome variable for new records. There are main types of outcomes of interest:

Predicted numerical value—when the outcome variable is numerical (e.g., house price)

Predicted class membership—when the outcome variable is categorical (e.g., buyer/nonbuyer)

Data Mining for Business Analytics: Concepts, Techniques, and Applications with JMP Pro®, First Edition.
Galit Shmueli, Peter C. Bruce, Mia L. Stephens, and Nitin R. Patel.
© 2017 John Wiley & Sons, Inc. Published 2017 by John Wiley & Sons, Inc.

Propensity—the probability of class membership, when the outcome variable is categorical (e.g., the propensity to default)

Prediction methods are used for generating numerical predictions, while classification methods ("classifiers") are used for generating propensities, and using a cutoff value on the propensities, we generate predicted class memberships.

A subtle distinction to keep in mind is the two distinct predictive uses of classifiers: one use, *classification*, is aimed at predicting class membership for new records. The other, *ranking*, is detecting, among a set of new records, the ones most likely to belong to a class of interest.

In this chapter we examine the approach for judging the performance and overall usefulness of predictive models. We'll see measures of predictive performance for a model generating numerical predictions (Section 5.2), a classifier used for predicting class membership (Section 5.3), and a classifier used for ranking (Section 5.4). In Section 5.5 we'll look at evaluating performance under the scenario of oversampling.

The need for performance measures arises from the wide choice of classifiers and predictive methods. Not only do we have several different methods, but even within a single method there are usually many options that can lead to completely different results. A simple example is the choice of predictors used within a particular predictive algorithm. Before we study these various algorithms in detail and face decisions on how to set these options, we need to know how we will measure success.

5.2   EVALUATING PREDICTIVE PERFORMANCE

First let us emphasize that predictive accuracy is not the same as goodness-of-fit. Classical measures of performance are aimed at finding a model that fits well to the data on which the model was trained. In data mining we are interested in models that have high predictive accuracy when applied to *new* records. Measures such as R^2 and standard error of estimate are common metrics in classical regression modeling, and residual analysis is used to gauge goodness-of-fit in that situation. However, these measures, do not tell us much about the ability of the model to predict new cases.

For assessing prediction performance, several measures are used. In all cases the measures are based on the validation set, which serves as a more objective ground than the training set to assess predictive accuracy. This is because records in the validation set are more similar to the future records to be predicted, in the sense that they are not used to select predictors or to estimate the model coefficients. Models are trained on the training data, applied to the validation data, and measures of accuracy then use the prediction errors on that validation set.

Benchmark: The Average

The benchmark criterion in prediction is using the average outcome (thereby ignoring all predictor information). In other words, the prediction for a new record is simply the average outcome of the records in the training set ($\bar{y}$). A good predictive model should outperform the benchmark criterion in terms of predictive accuracy.

Prediction Accuracy Measures

The *prediction error*, or *residual* for record i is defined as the difference between its actual y value and its predicted y value: $e_i = y_i - \hat{y}_i$. There are a few popular numerical measures of predictive accuracy, calculated on the validation data:

- *MAE, MAD*, or *AAE* (mean absolute error/deviation, or average absolute error) = $1/n \sum_{i=1}^{n} |e_i|$. This gives the magnitude of the average error.
- *Average error* = $1/n \sum_{i=1}^{n} e_i$. This measure is similar to *MAE* except that it retains the sign of the errors, so that negative errors cancel out positive errors of the same magnitude. It therefore gives an indication of whether the predictions are on average over- or underpredicting the response.
- *MAPE* (mean absolute percentage error) = $100 \times 1/n \sum_{i=1}^{n} |e_i/y_i|$. This measure gives a percentage score of how predictions deviate (on average) from the actual values.
- *RMSE* (root mean squared error) or *RASE* (root average squared error) = $\sqrt{1/n \sum_{i=1}^{n} e_i^2}$. This measure is similar to the estimate of prediction error in linear regression, except that it is computed on the validation data rather than on the training data. It has the same units as the variable predicted.
- Total *SSE* (sum of squared errors) = $\sum_{i=1}^{n} e_i^2$.

Such measures can be used to compare models and to assess their degree of prediction accuracy. Of these, JMP Pro reports the following measures of predictive accuracy for the validation data (additional measures are available for the training data, and R^2 is also reported):

- *RASE* (or RMSE in some platforms)
- *AAE* (in the *Model Comparison* platform)

Note that all of these measures are influenced by outliers. To check outlier influence, we can compute median-based measures (and compare to the mean-based measures) or simply plot a histogram or boxplot of the errors. Plotting the distribution of prediction errors is in fact very useful and can highlight more information than the metrics alone. To illustrate the use of predictive accuracy measures and charts of the distribution of prediction error, consider the error metrics and charts shown in Figures 5.1 and 5.2. These are the result of fitting a certain predictive model to prices of used Toyota Corolla cars. The training set includes 600 cars and the validation set includes 400 cars. The prediction errors (residuals) for the validation set are summarized in the bottom row in Figure 5.1. Residuals for this model were saved to the data table. The histogram, boxplot, and summary statistics corresponding to the residuals for the validation set (at the bottom of Figure 5.2) show a

Crossvalidation

Source	RSquare	RASE	Freq
Training Set	0.7621	11.109	18
Validation Set	0.3381	16.771	22

FIGURE 5.1 Prediction error metrics from a model for Toyota car prices. Training (top row) and validation (bottom row).

small number of outliers, with over half of the errors in the [−900, 800] range. On average, the model overpredicts (the average residual value for the validation set is −80.7).

Comparing Training and Validation Performance

Prediction errors (residuals) based on the training set tell us about model fit, whereas errors based on the validation set measure the model's ability to predict new data (predictive performance). We expect training errors to be smaller than the validation errors (because the model was fit using the training set), and the more complex the model, the greater is the likelihood that it will *overfit* the training data (indicated by a greater difference between the training and validation errors). In an extreme case of overfitting, the training errors would

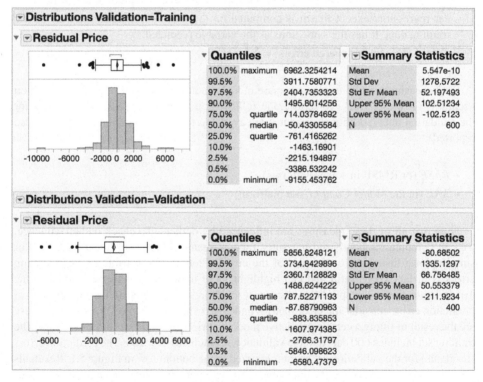

FIGURE 5.2 Histograms, boxplots, and summary statistics for Toyota price prediction errors, training, and validation sets.

be zero (perfect fit of the model to the training data), and the validation errors would be nonzero and nonnegligible. For this reason it is important to compare the error plots and metrics (RMSE, average error, etc.) of the training and validation sets. Figure 5.1 illustrates this comparison: the RASE for the training set is lower than for the validation set, while the RSquare is much higher.

5.3   JUDGING CLASSIFIER PERFORMANCE

A natural criterion for judging the performance of a classifier is the probability of making a *misclassification error*. Misclassification means that the observation belongs to one class but the model classifies it as a member of a different class. A classifier that makes no errors would be perfect, but we do not expect to be able to construct such classifiers in the real world due to "noise" and to not having all the information needed to classify cases precisely. Is there a minimal probability of misclassification that we should require of a classifier?

Benchmark: The Naive Rule

A very simple rule for classifying a record into one of m classes, ignoring all predictor information $(x_1, x_2, \ldots, x_p)$ that we might have, is to classify the record as a member of the majority class. In other words, "classify as belonging to the most prevalent class." The *naive rule* is used mainly as a baseline or benchmark for evaluating the performance of more complicated classifiers. Clearly, a classifier that uses external predictor information (on top of the class membership allocation) should outperform the naive rule.

Similar to using the sample mean ($\bar{y}$) as the naive benchmark in the numerical outcome case, the naive rule for classification relies solely on the y information and excludes any additional predictor information.

Class Separation

If the classes are well separated by the predictor information, even a small dataset will suffice in finding a good classifier, whereas if the classes are not separated at all by the predictors, even a very large dataset will not help. Figure 5.3 illustrates this for a two-class case. The top panel includes a small dataset ($n = 24$ observations) where two predictors (income and lot size) are used for separating owners from nonowners (we thank Dean Wichern for this example, described in Johnson and Wichern, 2002). In this case the predictor information seems useful in that it separates the two classes (owners/nonowners). The bottom panel shows a much larger dataset ($n = 5000$ observations) where the two predictors (income and average credit card spending) do not separate the two classes well in most of the higher ranges (loan acceptors/nonacceptors).

The Classification Matrix

In practice, most accuracy measures are derived from the *classification matrix*, also called the *confusion matrix*. This matrix summarizes the correct and incorrect classifications that

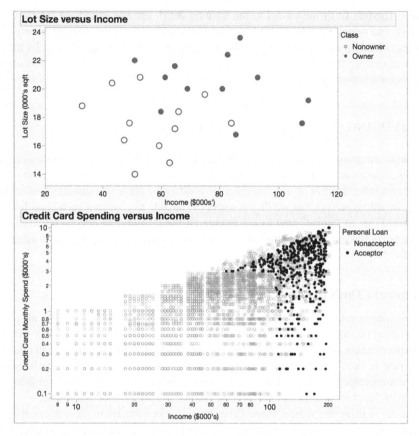

FIGURE 5.3 High (top) and low (bottom) levels of separation between two classes, using two predictors.

a classifier produced for a certain dataset. Rows and columns of the classification matrix correspond to the true and predicted classes, respectively. Figure 5.4 shows an example of a classification (confusion) matrix for a two-class (0/1) problem resulting from applying a certain classifier to 3000 observations. The two diagonal cells (upper left, lower right) give the number of correct classifications, where the predicted class coincides with the actual class of the observation. The off-diagonal cells give counts of misclassification. The top right cell gives the number of class 0 members that were misclassified as 1's (in this example, there were 25 such misclassifications). Similarly, the lower left cell gives the number of class 1 members that were misclassified as 0's (85 such observations).

	Predicted Class		
Count	0	1	Total
0	2689	25	2714
1	85	201	286
Total	2774	226	3000

FIGURE 5.4 Classification matrix based on 3000 observations and two classes.

TABLE 5.1   Classification Matrix: Meaning of Each Cell

Actual Class	Predicted Class	
	C_0	C_1
C_0	$n_{0,0}$ = number of C_0 cases classified correctly	$n_{0,1}$ = number of C_0 cases classified incorrectly as C_1
C_1	$n_{1,0}$ = number of C_1 cases classified incorrectly as C_0	$n_{1,1}$ = number of C_1 cases classified correctly

The classification matrix gives estimates of the true classification and misclassification rates. Of course, these are estimates and they can be incorrect, but if we have a large enough dataset and neither class is very rare, our estimates will be reliable. Sometimes, we may be able to use public data such as US Census data to estimate these proportions. However, in most business settings, we will not know them.

Using the Validation Data

To obtain an honest estimate of future classification error, we use the classification matrix that is computed from the *validation data*. In other words, we first partition the data into training and validation sets by random selection of cases. We then construct a classifier using the training data, and then apply it to the validation data. This will yield the predicted classifications for observations in the validation set (see Figure 2.4 in Chapter 2). We then summarize these classifications in a classification matrix. Although we can summarize our results in a classification matrix for training data as well, the resulting classification matrix is not useful for getting an honest estimate of the misclassification rate for new data due to the danger of overfitting.

In addition to examining the validation data classification matrix to assess the classification performance on new data, we compare the training data classification matrix to the validation data classification matrix in order to detect overfitting: although we expect somewhat inferior results on the validation data, a large discrepancy in training and validation performance might be indicative of overfitting.

Accuracy Measures

Different accuracy measures can be derived from the classification matrix. Consider a two-class case with classes C_0 and C_1 (e.g., nonbuyer/buyer). The schematic classification matrix in Table 5.1 uses the notation $n_{i,j}$ to denote the number of cases that are class C_i members, and were classified as C_j members. Of course, if $i \neq j$, these are counts of misclassifications. The total number of observations is $n = n_{0,0} + n_{0,1} + n_{1,0} + n_{1,1}$.

A main accuracy measure is the *estimated misclassification rate*, also called the *overall error rate*. It is given by

$$\mathrm{err} = \frac{n_{0,1} + n_{1,0}}{n}$$

where n is the total number of cases in the validation dataset. In the example in Figure 5.4, we get err $= (25 + 85)/3000 = 3.67\%$.

We can measure accuracy by looking at the correct classifications—the full half of the cup—instead of the misclassifications. The *overall accuracy* of a classifier is estimated by

$$\text{accuracy} = 1 - \text{err} = \frac{n_{0,0} + n_{1,1}}{n}$$

In the example we have $(2689 + 201)/3000 = 96.33\%$.

Propensities and Cutoff for Classification

The first step in most classification algorithms is to estimate the probability that a case belongs to each of the classes. These probabilities are also called *propensities*. Propensities are typically used either as an interim step for generating predicted class membership (classification), or for rank-ordering the records by their probability of belonging to a class of interest. Let us consider their first use in this section. The second use is discussed in Section 5.4.

If overall classification accuracy (involving all the classes) is of interest, the case can be assigned to the class with the highest probability. In many cases a single class is of special interest, so we will focus on that particular class and compare the propensity of belonging to that class to a *cutoff value* set by the analyst. This approach can be used with two classes or more than two classes, though it may make sense in such cases to consolidate classes so that you end up with two: the class of interest and all other classes. If the probability of belonging to the class of interest is above the cutoff, the case is assigned to that class.

CUTOFF VALUES FOR TRIAGE

In some cases it is useful to have two cutoffs, and allow a "cannot say" option for the classifier. In a two-class situation, this means that for a case we can make one of three predictions: The case belongs to C_0, or the case belongs to C_1, or we cannot make a prediction because there is not enough information to pick C_0 or C_1 confidently. Cases that the classifier cannot classify are subjected to closer scrutiny either by using expert judgment or by enriching the set of predictor variables by gathering additional information that is perhaps more difficult or expensive to obtain. An example is classification of documents found during legal discovery (reciprocal forced document disclosure in a legal proceeding). Under traditional human-review systems, qualified legal personnel are needed to review what might be tens of thousands of documents to determine their relevance to a case. Using a classifier and a triage outcome, documents could be sorted into clearly relevant, clearly not relevant, and the gray area documents requiring human review. This substantially reduces the costs of discovery.

	Actual Class	Prob[owner]	Most Likely Ownership	Predicted Class at 0.25	Predicted Class at 0.75
1	owner	0.102697297	nonowner	nonowner	nonowner
2	owner	0.561213154	owner	owner	nonowner
3	owner	0.941747935	owner	owner	owner
4	owner	0.7323180727	owner	owner	nonowner
5	owner	0.9998892877	owner	owner	owner
6	owner	0.9803565717	owner	owner	owner
7	owner	0.9448657753	owner	owner	owner
8	owner	0.9978033093	owner	owner	owner
9	owner	0.6683372877	owner	owner	nonowner
10	owner	0.9859447076	owner	owner	owner
11	owner	0.8794488943	owner	owner	owner
12	owner	0.8830365999	owner	owner	owner
13	nonowner	0.7088413043	owner	owner	nonowner
14	nonowner	0.4863742324	nonowner	owner	nonowner
15	nonowner	0.0917943278	nonowner	nonowner	nonowner
16	nonowner	0.1213584078	nonowner	nonowner	nonowner
17	nonowner	0.5788381	owner	owner	nonowner
18	nonowner	0.0133127658	nonowner	nonowner	nonowner
19	nonowner	0.0331493122	nonowner	nonowner	nonowner
20	nonowner	0.2010408317	nonowner	nonowner	nonowner
21	nonowner	0.0057710408	nonowner	nonowner	nonowner
22	nonowner	0.0033442195	nonowner	nonowner	nonowner
23	nonowner	0.0138203993	nonowner	nonowner	nonowner
24	nonowner	0.0646960932	nonowner	nonowner	nonowner

FIGURE 5.5   Riding Mower Example—24 Records with the Actual Class and the Probability (Propensity) of Them Being Owners, as Estimated by a Classifier in JMP.

▼ ⊡ Tabulate

	Most Likely Ownership				Predicted Class at 0.25				Predicted Class at 0.75			
	nonowner		owner		nonowner		owner		nonowner		owner	
Actual Class	Row %	N	Row %	N	Row %	N	Row %	N	Row %	N	Row %	N
nonowner	83.33%	10	16.67%	2	75.00%	9	25.00%	3	100.00%	12	0.00%	0
owner	8.33%	1	91.67%	11	8.33%	1	91.67%	11	33.33%	4	66.67%	8

FIGURE 5.6   Classification matrices based on cutoffs of 0.5, 0.25, and 0.75 for the Riding Mower data.

The default cutoff value in two-class classifiers is 0.5. Thus, if the probability of a record being a class 1 member is greater than 0.5, that record is classified as a 1. Any record with an estimated probability of less than 0.5 would be classified as a 0. It is possible, however, to use a cutoff that is either higher or lower than 0.5. A cutoff greater than 0.5 will end up classifying fewer records as 1's, whereas a cutoff less than 0.5 will end up classifying more records as 1. Typically, the misclassification rate will rise in either case.

CHANGING THE CUTOFF VALUES FOR A CONFUSION MATRIX IN JMP

To see how the accuracy or misclassification rates change as a function of the cutoff, we can change the default cutoff value (from 0.50) as follows:

1. Fit a model in JMP.
2. Save the probability formula to the data table. This will also save the predicted probability for each class, and a *Most Likely Class* using a cutoff of 0.50.
3. Add a new column, and use the formula editor to create a conditional *IF* formula using the desired cutoff (see the formula shown in Figure 5.7).
4. Repeat the above step for different cutoff values.
5. Use *Analyze > Tabulate* to create confusion matrices at each cutoff value (see Figure 5.6.)

Another approach is to use the JMP *Alternate Cutoff Confusion Matrix Add-in* to create confusion matrices for a range of cutoff values. This add-in is freely available from the JMP user community at community.jmp.com.

Consider the data in Figure 5.5, showing the actual class for 24 records (based on the riding mowers example), along with propensities (*Prob[owner]*) and classifications (*Most Likely Ownership*) based on some model. To get the propensities and classifications, a model was fit, and the probability formula was saved to the data table (note that some columns are hidden, and that for illustration the *Actual Class* column was changed from its original name, *Ownership*).

The misclassification rate for the *Most Likely Ownership*, which uses the standard 0.5 as the cutoff, is 3/24. This is summarized in a confusion matrix on the left in Figure 5.6. If we instead adopt a cutoff of 0.25, we classify more records as nonowners as owners and the misclassification rate goes up to 4/24 (see column the column *Predicted Class at 0.25* in Figure 5.5). If we adopt a cutoff of 0.75, more owners are classified as nonowners and the misclassification rate again goes up to 4/24, as shown in the last column in Figure 5.5. All of this can be seen in the confusion matrices in Figure 5.6 (produced using *Tabulate*).

Why would we want to use cutoff values different from 0.5 if they increase the misclassification rate? The answer is that it might be more important to classify 1's properly than 0's, and we would tolerate a greater misclassification of the latter. Or the reverse might be true: the costs of misclassification might be asymmetric. We can adjust the cutoff value in such a case to classify more records as the high-value class, that is, accept more misclassifications where the misclassification cost is low. Keep in mind that we are doing so after the data mining model has already been selected—we are not changing that model. It is also possible to incorporate costs into the picture before deriving the model. These subjects are discussed in greater detail below.

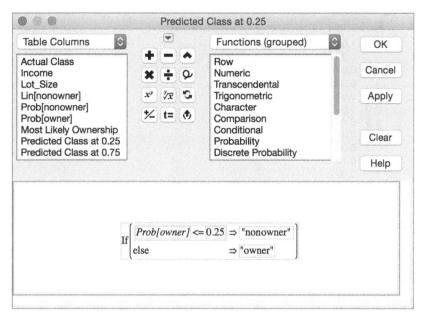

FIGURE 5.7   Using the *Formula Editor* with *Conditional* and *Comparison* functions to change the cutoff for classification.

Performance in Unequal Importance of Classes

Suppose that it is more important to predict membership correctly in class 1 than in class 0. An example is predicting the financial status (bankrupt/solvent) of firms. It may be more important to correctly predict a firm that is going bankrupt than to correctly predict a firm that is going to remain solvent. The classifier is essentially used as a system for detecting or signaling bankruptcy. In such a case the overall accuracy is not a good measure for evaluating the classifier. Suppose that the important class is C_1. The following pair of accuracy measures are the most popular:

> The **sensitivity** of a classifier is its ability to detect the important class members correctly. This is measured by $n_{1,1}/(n_{1,0} + n_{1,1})$, the percentage of C_1 members classified correctly.

> The **specificity** of a classifier is its ability to rule out C_0 members correctly. This is measured by $n_{0,0}/(n_{0,0} + n_{0,1})$, the percentage of C_0 members classified correctly.

It can be useful to plot these measures versus the cutoff value in order to find a cutoff value that balances these measures.

ROC Curve   A popular method for plotting the sensitivity and specificity of a classifier is through *ROC* (receiver operating characteristic) *curves*. The ROC curve plots the pairs {sensitivity, 1−specificity} as the cutoff value increases from 0 and 1. Better performance is reflected by curves that are closer to the top-left corner. The line drawn at an angle tangent to the ROC Curve marks a good cutoff value (under the assumption that false negatives and false positives have similar costs). The ROC curve for the riding lawn mower example is

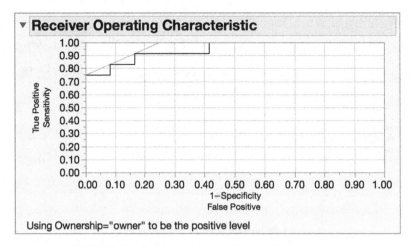

FIGURE 5.8   ROC curve for the Riding Mower example.

shown in Figure 5.8. ROC curves are available as a red triangle option in JMP. A common metric is "area under the curve (AUC)," which ranges from 1 (perfect discrimination between classes) to 0.5 (no better than the naive rule).

FALSE-POSITIVE AND FALSE-NEGATIVE RATES

Sensitivity and specificity measure the performance of a classifier from the point of view of the "classifying agency" (e.g., a company classifying customers or a hospital classifying patients). They answer the question "how well does the classifier segregate the important class members?". It is also possible to measure accuracy from the perspective of the entity that is being predicted (e.g., the customer or the patient), who asks "what is my chance of belonging to the important class?" Then again, this question is usually less relevant in a data mining application. The terms "false-positive rate" and "false-negative rate," which are sometimes used erroneously to describe $1-$sensitivity and $1-$specificity, are measures of performance from the perspective of the individual entity. If C_1 is the important class, they are defined as:

The *false-positive rate* is the proportion of C_1 predictions that are wrong: $n_{0,1}/(n_{0,1} + n_{1,1})$. Notice that this is a ratio within the column of C_1 predictions (i.e., it uses only records that were classified as C_1).

The *false-negative rate* is the proportion of C_0 predictions that are wrong: $n_{1,0}/(n_{0,0} + n_{1,0})$. Notice that this is a ratio within the column of C_0 predictions (i.e., it uses only records that were classified as C_0).

Asymmetric Misclassification Costs

Up to this point we have been using the misclassification rate as the criterion for judging the efficacy of a classifier. However, there are circumstances where this measure is not appropriate. Sometimes the error of misclassifying a case belonging to one class is more serious than for the other class. For example, misclassifying a household as unlikely to respond to a sales offer when it belongs to the class that would respond incurs a greater cost (the opportunity cost of the forgone sale) than the converse error. In the former case, you are missing out on a sale worth perhaps tens or hundreds of dollars. In the latter, you are incurring the costs of mailing a letter to someone who will not purchase. In such a scenario, using the misclassification rate as a criterion can be misleading.

Note that we are assuming that the cost (or benefit) of making correct classifications is zero. At first glance, this may seem incomplete. After all, the benefit (negative cost) of classifying a buyer correctly as a buyer would seem substantial. In other circumstances (e.g., scoring our classification algorithm to fresh data to implement our decisions), it will be appropriate to consider the actual net dollar impact of each possible classification (or misclassification). Here, however, we are attempting to assess the value of a classifier in terms of classification error, so it greatly simplifies matters if we can capture all cost/benefit information in the misclassification cells. So, instead of recording the benefit of classifying a respondent household correctly, we record the cost of failing to classify it as a respondent household. It amounts to the same thing and our goal becomes the minimization of costs, whether the costs are actual costs or missed benefits (opportunity costs).

Consider the situation where the sales offer is mailed to a random sample of people for the purpose of constructing a good classifier. Suppose that the offer is accepted by 1% of those households. For these data, if a classifier simply classifies every household as a nonresponder, it will have an error rate of only 1%, but the classifier will be useless in practice. A classifier that misclassifies 2% of buying households as nonbuyers and 20% of the nonbuyers as buyers would have a higher error rate but would be better if the profit from a sale is substantially higher than the cost of sending out an offer. In these situations, if we have estimates of the cost of both types of misclassification, we can use the classification matrix to compute the expected cost of misclassification for each case in the validation data. This enables us to compare different classifiers using overall expected costs (or profits) as the criterion.

Suppose that we are considering sending an offer to 1000 more people, 1% of whom respond (1), on average. Naively classifying everyone as a 0 has an error rate of only 1%. Using a data mining routine, suppose that we can produce these classifications:

	Predict Class 0	Predict Class 1
Actual 0	970	20
Actual 1	2	8

These classifications have an error rate of $100 \times (20 + 2)/1000 = 2.2\%$—higher than the naive rate.

Now suppose that the profit from a 1 is $10 and the cost of sending the offer is $1. Classifying everyone as a 0 still has a misclassification rate of only 1% but yields a profit of $0. Using the data mining routine, despite the higher misclassification rate, yields a profit of $60.

The matrix of profit is as follows (nothing is sent to the predicted 0's, so there are no costs or sales in that column):

Profit	Predict Class 0	Predict Class 1
Actual 0	0	$-\$20$
Actual 1	0	$\$80$

Looked at purely in terms of costs, when everyone is classified as a 0, there are no costs of sending the offer; the only costs are the opportunity costs of failing to make sales to the ten 1's = $100. The cost (actual costs of sending the offer, plus the opportunity costs of missed sales) of using the data mining routine to select people to send the offer to is only $48, as follows:

Costs	Predict Class 0	Predict Class 1
Actual 0	0	$\$20$
Actual 1	$\$20$	$\$8$

However, this does not improve the actual classifications. A better method is to change the classification rules, hence the misclassification rates), as discussed in the preceding section, to reflect the asymmetric costs.

A popular performance measure that includes costs is the *average misclassification cost*, which measures the average cost of misclassification per classified observation. Denote by q_0 the cost of misclassifying a class 0 observation (as belonging to class 1) and by q_1 the cost of misclassifying a class 1 observation (as belonging to class 0). The average misclassification cost is

$$\frac{q_0 n_{0,1} + q_1 n_{1,0}}{n}$$

Thus we are looking for a classifier that minimizes this quantity. This can be computed, for instance, for different cutoff values.

It turns out that the optimal parameters are affected by the misclassification costs only through the ratio of these costs. This can be seen if we write the foregoing measure slightly differently:

$$\frac{q_0 n_{0,1} + q_1 n_{1,0}}{n} = \frac{n_{0,1}}{n_{0,0} + n_{0,1}} \frac{n_{0,0} + n_{0,1}}{n} q_0 + \frac{n_{1,0}}{n_{1,0} + n_{1,1}} \frac{n_{1,0} + n_{1,1}}{n} q_1$$

Minimizing this expression is equivalent to minimizing the same expression divided by a constant. If we divide by q_0, it can be seen clearly that the minimization depends only on q_1/q_0 and not on their individual values. This is very practical, since in many cases it is difficult to assess the cost associated with misclassifying a 0 member and that associated with misclassifying a 1 member, but estimating the ratio is easier.

This expression is a reasonable estimate of future misclassification cost if the proportions of classes 0 and 1 in the sample data are similar to the proportions of classes 0 and 1 that are expected in the future. If, instead of a random sample, we draw a sample such that one class is oversampled (as described in the next section), the sample proportions of 0's and 1's will be distorted compared to the future or population. We can then correct the average misclassification cost measure for the distorted sample proportions by incorporating estimates of the true proportions (from external data or domain knowledge), denoted by $p(C_0)$ and $p(C_1)$, into the formula

$$\frac{n_{0,1}}{n_{0,0} + n_{0,1}} p(C_0) q_0 + \frac{n_{1,0}}{n_{1,0} + n_{1,1}} p(C_1) q_1$$

Using the same logic as above, it can be shown that optimizing this quantity depends on the costs only through their ratio (q_1/q_0) and on the prior probabilities only through their ratio $[p(C_0)/p(C_1)]$. This is why software packages that incorporate costs and prior probabilities might prompt the user for ratios rather than actual costs and probabilities.

ASYMMETRIC MISCLASSIFICATION COSTS IN JMP

In JMP, asymmetric misclassification costs are assigned as a *Column Property* for the target variable using the *Profit Matrix*, as shown in Figure 5.9 (in the data table, right-click on the column and select *Column Info*, then click on *Column Properties*). The profit matrix assigns weights to predicted outcomes and calculates expected costs and profits for the predicted class when a probability formula for a model is saved to the data table. The Profit Matrix is also available in some modeling platforms in JMP 12.

Figure 5.9 shows the setup of the Profit Matrix for the following situation: The profit from a 1 is $10 and the cost of sending the offer is $1. *Note:* The actual and predicted labels are reversed in the *Profit Matrix* in JMP 12. These labels have been changed in JMP 13. In addition, the cutoff for classification can be set in JMP 13 using the *Profit Matrix*.

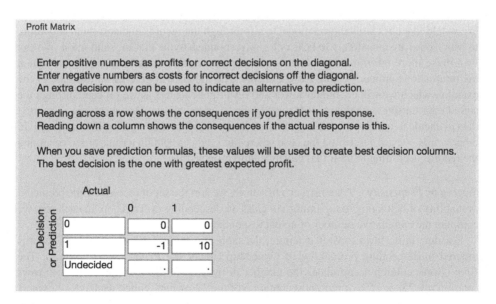

FIGURE 5.9   Specifying Misclassification Costs in JMP using the Profit Matrix Column Property.

Generalization to More Than Two Classes

All the comments made above about two-class classifiers extend readily to classification into more than two classes. Let us suppose that we have m classes $C_0, C_1, C_2, \dots, C_{m-1}$. The classification matrix has m rows and m columns. The misclassification cost associated with the diagonal cells is, of course, always zero. Incorporating prior probabilities of the various classes (where now we have m such numbers) is still done in the same manner.

However, evaluating misclassification costs using the formulas in the previous section becomes much more complicated. For an m-class case we have $m(m-1)$ types of misclassifications. While the *Profit Matrix* in JMP can accommodate misclassification costs for situations involving more than two classes, it is often difficult to elicit or obtain all the misclassification costs associated with more than two classes.

5.4   JUDGING RANKING PERFORMANCE

We now turn to the predictive goal of detecting, among a set of new records, the ones most likely to belong to a class of interest. Recall that this differs from the goal of predicting class membership for each record.

Lift Curves

Lift curves (also called lift charts, gains curves, or gains charts) are available in JMP for models involving categorical outcomes. The lift curve helps us determine how effectively we can "skim the cream" by selecting a relatively small number of cases and getting a relatively large portion of the responders.

Let us continue with the case in which a particular class is relatively rare and of much more interest than the other class: tax cheats, debt defaulters, or responders to a mailing. We would like our classification model to sift through the records and sort them according to which ones are most likely to be tax cheats, responders to the mailing, and so on. We can then make more informed decisions. For example, we can decide how many, and which tax returns to examine, looking for tax cheats. The model will give us an estimate of the extent to which we will encounter fewer and fewer noncheaters as we proceed through the sorted data starting with the records most likely to be tax cheats. Or we can use the sorted data to decide to which potential customers a limited-budget mailing should be targeted. In other words, we are describing the case where our goal is to obtain a rank-ordering among the records according to their class membership propensities.

Sorting by Propensity   To construct a lift curve, we sort the set of records by the predicted probability of belonging to the important class, in descending order. Then, in each row we compute the cumulative number of actual responders.

For our riding lawn mower example, the table in Figure 5.10 shows the 24 records ordered in descending probability of ownership (under the *Prob[owner]* column). The *Cum Owner* column accumulates the number of owners. In this example, the top 6 rows (portion = 0.25, or 25% of our data) contain 6 of the 12 owners. Since we have 12 owners and 12 non-owners, we would expect to have only 3 of the owners in the top 25% of our data by chance alone (since 12 x 0.25 = 3). The ratio of these two numbers, actual owners versus expected owners, is called *lift*. In this case, the lift at the top 25% of our data is

	Ownership	Prob[owner]	Cum Owner	Portion	Lift
1	owner	0.9960	1	0.04	2.00
2	owner	0.9900	2	0.08	2.00
3	owner	0.9857	3	0.13	2.00
4	owner	0.9799	4	0.17	2.00
5	owner	0.9679	5	0.21	2.00
6	owner	0.8882	6	0.25	2.00
7	owner	0.8791	7	0.29	2.00
8	owner	0.7699	8	0.33	2.00
9	nonowner	0.7504	8	0.38	1.78
10	owner	0.7074	9	0.42	1.80
11	owner	0.6989	10	0.46	1.82
12	nonowner	0.6066	10	0.50	1.67
13	nonowner	0.5133	10	0.54	1.54
14	owner	0.4951	11	0.58	1.57
15	nonowner	0.2650	11	0.63	1.47
16	nonowner	0.2087	11	0.67	1.38
17	owner	0.1518	12	0.71	1.41
18	nonowner	0.0879	12	0.75	1.33
19	nonowner	0.0202	12	0.79	1.26
20	nonowner	0.0139	12	0.83	1.20
21	nonowner	0.0125	12	0.88	1.14
22	nonowner	0.0069	12	0.92	1.09
23	nonowner	0.0042	12	0.96	1.04
24	nonowner	0.0005	12	1.00	1.00

FIGURE 5.10   Illustration of lift calculations for the riding mower example ("Prob[owner]" = propensities).

6/3 = 2.0, as shown in Figure 5.10. The lift curve plots the lift against the percentage (or portion) of the data for each of the classes, as shown in Figure 5.11. The curve for *owner* is the thicker (blue) line. (To change the thickness of the curve for the target class in a Lift Curve, right-click on the curve and select *Customize*. Then, select *Level Curves* and enter the desired line width).

Note: The columns Cum Owner, Portion, and Lift shown in Figure 5.10 are computed behind the scenes in JMP to create lift curves within the modeling platforms—you won't see these calculations (they were created here using the *Formula Editor* for illustration).

Interpreting the Lift Curve   What is considered good or bad performance? The ideal ranking performance would place all the owners (or 1's) at the beginning (the actual 1's would have the highest propensities and be at the top of the table), and all the nonowners (the 0's) at the end. In contrast, a useless model would be one that randomly assigns propensities (shuffling the 1's and 0's randomly in the Ownership (or actual class) column). Such behavior would increase the cumulative number of 1's, on average, by $\frac{\#1's}{n}$ in each row.

If we did not have a model but simply selected cases at random, our lift will hover at around 1.0. As a result a lift of 1.0 provides a benchmark against which we can evaluate

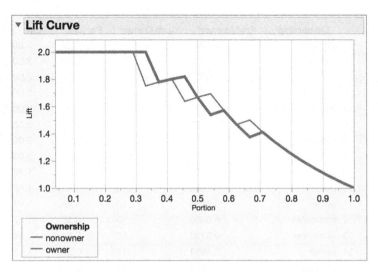

FIGURE 5.11   Lift curve for the riding mower example.

the ranking performance of the model. A more ideal situation is a lift curve that starts with a very high lift for the 1's (in a small portion of the data), then has a steep downward slope leveling out at a lift of 1.0. In the riding mower example, although our model is not perfect, it seems to perform much better than the random benchmark.

How do we read a lift curve? For a given portion of our data (x-axis), the lift value on the y-axis tells us how much better we are doing compared to random assignment. For example, looking at Figure 5.11, if we use our model to choose the top 20% of a new sample (portion = 0.20), we would catch twice as many owners than choosing a random sample of the same size. Similarly, if we use the model to choose 50% of a new sample (portion = 0.5) we would catch around 1.7 more owners compared to using a random 50%. (We can see in Figure 5.10 that for portion = 0.2 the lift is 2 and for portion = 0.5 the lift is 1.67.) The lift will vary with the portion of data we choose to act on. A good classifier will give us a high lift when we act on only a small portion of the data.

Beyond Two Classes

In JMP, lift curves can also be used with a multi-class classifier. A lift curve will be provided for each class. Lift curves are not available for prediction models with continuous responses.

Lift Curves Incorporating Costs and Benefits

When the benefits and costs of correct and incorrect classification are known or can be estimated, the lift curve is still a useful presentation and decision tool (see Figure 5.12). As before, a classifier is needed that assigns to each record a propensity that it belongs to a particular class. The procedure is as follows:

1. Set the misclassification costs using the *Profit Matrix*, as shown in Figure 5.9.
2. Built your model.

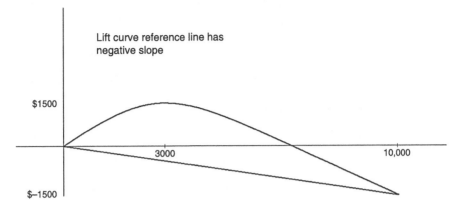

FIGURE 5.12 Hypothetical example, Lift curve incorporating costs.

3. Save the *Probability Formula* for the model to the data table. This will also calculate the *Profit* for the different classes and the *Expected Profit* for the outcome.

4. Sort the records in descending order of predicted probability of success, where *success* = belonging to the class of interest (right-click on the columns and select *Sort > Descending*).

5. Create a cumulative expected profit column. To do this, right-click on the *Expected Profit* column header, select *New Formula Column > Row > Cumulative Sum*.

6. Create a row number column. To do this, right-click on any column header, select *New Formula Column > Row > Row*.

7. Use the *Graph Builder* to manually create the lift curve, with the *Cumulative Expected Profit* column on the *y*-axis and *Row* on the *x*-axis. Use the Line graph icon (above the graph) to connect the points.

8. Use the *Line* tool on the toolbar to manually add a reference line connecting the first point (row = 1) to the last point (row = N, where *y* = the total net profit).

Note: It is entirely possible for the reference line that incorporates costs and benefits to have a negative slope if the net value for the entire dataset is negative. A hypothetical example is shown in in Figure 5.12. If the cost of mailing to a person is $0.65, the value of a responder is $25, and the overall response rate is 2%, the expected net value of mailing to a list of 10,000 is $(0.02 \times \$25 \times 10,000) - (\$0.65 \times 10,000) = \$5000 - \$6500 = -\$1500$. Hence the *y*-value at the far right of the lift curve in Figure 5.12 ($x = 10,000$) is -1500, and the slope of the reference line from the origin will be negative. The optimal point will be where the lift curve is at a maximum (i.e., mailing to about 3000 people).

5.5 OVERSAMPLING

As we saw briefly in Chapter 2, when classes are present in very unequal proportions, simple random sampling may produce too few of the rare class to yield useful information about what distinguishes them from the dominant class. In such cases stratified sampling is often used to oversample the cases from the more rare class and improve the performance of classifiers.

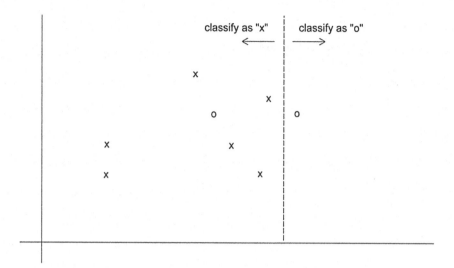

FIGURE 5.13   Classification assuming equal costs of misclassification.

It is often the case that the more rare events are the more interesting or important ones: responders to a mailing, those who commit fraud, defaulters on debt, and the like. This same stratified sampling procedure is sometimes called *weighted sampling* or *undersampling*, the latter referring to the fact that the more plentiful class is undersampled, relative to the rare class. We will stick to the term *oversampling*.

> In all discussions of *oversampling*, we assume the common situation in which there are two classes, one of much greater interest than the other. Data with more than two classes do not lend themselves to this procedure.

Consider the data in Figure 5.13, where × represents nonresponders, and ○, responders. The two axes correspond to two predictors. The dashed vertical line does the best job of classification under the assumption of equal costs: it results in just one misclassification (one ○ is misclassified as an ×). If we incorporate more realistic misclassification costs—let us say that failing to catch a ○ is five times as costly as failing to catch an ×—the costs of misclassification jump to 5. In such a case a horizontal line as shown in Figure 5.14, does a better job: it results in misclassification costs of just 2.

Oversampling is one way of incorporating these costs into the training process. In Figure 5.15 we can see that classification algorithms would automatically determine the appropriate classification line if four additional ○'s were present at each existing ○. We can achieve appropriate results either by taking five times as many ○'s as we would get from simple random sampling (by sampling with replacement if necessary), or by replicating the existing ○'s fourfold.

Oversampling without replacement in accord with the ratio of costs (the first option above) is the optimal solution, but it may not always be practical. There may not be an adequate number of responders to assure that there will be enough of them to fit a model if they constitute only a small proportion of the total. Also it is often the case that our interest

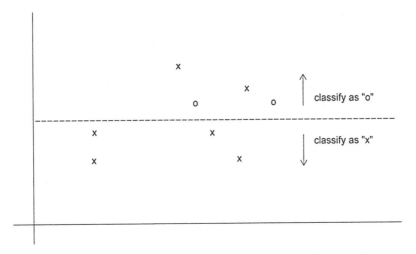

FIGURE 5.14   Classification assuming unequal costs of misclassification.

in discovering responders is known to be much greater than our interest in discovering nonresponders, but the exact ratio of costs is difficult to determine. When faced with very low response rates in a classification problem, practitioners often sample equal numbers of responders and nonresponders as a relatively effective and convenient approach. Whatever approach is used, when it comes time to assess and predict model performance, we will need to adjust for the oversampling in one of two ways:

1. Score the model to a validation set that has been selected without oversampling (i.e., via simple random sampling).
2. Score the model to an oversampled validation set, and reweight the results to remove the effects of oversampling.

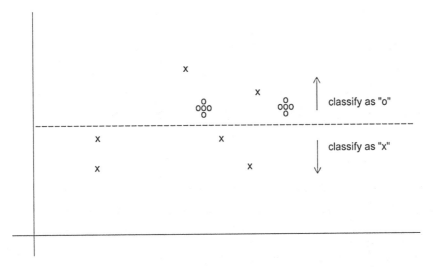

FIGURE 5.15   Classification using oversampling to account for unequal costs.

The first method is more straightforward and easier to implement. We describe how to oversample and how to evaluate performance for each of the two methods.

When classifying data with very low response rates, practitioners typically:

- Train models on data that are 50% responder, 50% nonresponder.
- Validate the models with an unweighted (simple random) sample from the original data.

Oversampling the Training Set

How is weighted sampling done? One common procedure, where responders are sufficiently scarce that you will want to use all of them, follows:

Step 1. The response and nonresponse data are separated into two distinct sets, or *strata*.

Step 2. Records are randomly selected for the training set from each stratum. Typically, one might select half the (scarce) responders for the training set, then an equal number of nonresponders.

Step 3. Remaining responders are put in the validation set.

Step 4. Nonresponders are randomly selected for the validation set in sufficient numbers to maintain the original ratio of responders to nonresponders.

Step 5. If a test set is required, it can be taken randomly from the validation set.

STRATIFIED SAMPLING AND OVERSAMPLING IN JMP

In JMP, stratified sampling can be accomplished using the *Random, Stratify* option from the *Tables > Subset* platform. Oversampling can be accomplished using the *Stratified Split Balanced* add-in, available from the JMP User Community (community.jmp.com, search for *Stratified Data Partitioning*). This add-in allows you to oversample the training data or both the training and the validation data.

Evaluating Model Performance Using a Nonoversampled Validation Set

Although the oversampled data can be used to train models, they are often not suitable for evaluating model performance, since the number of responders will (of course) be exaggerated. The most straightforward way of gaining an unbiased estimate of model performance is to apply the model to regular data (i.e., data not oversampled). To recap: Train the model on oversampled data, but validate it with regular data.

Evaluating Model Performance If Only Oversampled Validation Set Exists

In some cases very low response rates may make it more practical to use oversampled data not only for the training data but also for the validation data. This might happen, for example, if an analyst is given a dataset for exploration and prototyping that is already oversampled to boost the proportion with the rare response of interest (perhaps because it is more convenient to transfer and work with a smaller dataset). In such cases it is still possible to assess how well the model will do with real data, but this requires the oversampled validation set to be reweighted, in order to restore the class of observations that were underrepresented in the sampling process. This adjustment should be made to the classification matrix and to the lift curve in order to derive good accuracy measures. These adjustments are described next.

Adjusting the Confusion Matrix for Oversampling.   Suppose the response rate in the data as a whole is 2%, and that the data were oversampled, yielding a sample in which the response rate is 25 times higher (50% responders). The relationship is as follows:

Responders: 2% of the whole data; 50% of the sample

Nonresponders: 98% of the whole data, 50% of the sample

Each responder in the whole data is worth 25 responders in the sample (50/2). Each nonresponder in the whole data is worth 0.5102 nonresponders in the sample (50/98). We call these values *oversampling weights*.

Assume that the validation classification matrix looks like this:

Classification Matrix, Oversampled Data (validation)

	Predicted 0	Predicted 1	Total
Actual 0	390	110	500
Actual 1	80	420	500
Total	470	530	1000

At this point the misclassification rate appears to be $(80 + 110)/1000 = 19\%$, and the model ends up classifying 53% of the records as 1's. However, this reflects the performance on a sample where 50% are responders.

To estimate predictive performance when this model is used to score the original population (with 2% responders), we need to undo the effects of the oversampling. The actual number of responders must be divided by 25, and the actual number of nonresponders divided by 0.5102.

The revised classification matrix is as follows:

Classification Matrix, Reweighted

	Predicted 0	Predicted 1	Total
Actual 0	$390/0.5102 = 764.4$	$110/0.5102 = 215.6$	980
Actual 1	$80/25 = 3.2$	$420/25 = 16.8$	20
Total			1000

The adjusted misclassification rate is $(3.2 + 215.6)/1,000 = 21.9\%$. The model ends up classifying $(215.6 + 16.8)/1000 = 23.24\%$ of the records as 1's, when we assume 2% responders.

In JMP, the misclassification rate can be adjusted using a *Weight* column. This is a column in the data table created using the *Formula Editor* (see Figure 5.16). To create a classification matrix using these weights, use the *Fit Y by X* platform, with the actual class as the *Y, Response*, the most likely class as the *X, Factor*, and the weight column in the *Weight* field.

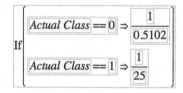

FIGURE 5.16   Formula to adjust for oversampling.

Adjusting the Lift Curve for Oversampling.   The lift curve is likely to be a more useful measure in low-response situations, where our interest lies not so much in classifying all the records correctly as in finding a model that guides us toward those records most likely to contain the response of interest (under the assumption that scarce resources preclude examining or contacting all the records). Typically, our interest in such a case is in maximizing value or minimizing cost.

The procedure presented earlier for creating a lift curve incorporating costs will be used, with one difference: A new column, *Adjusted Expected Profit*, will be used instead of the *Expected Profit* column. To create this column use the *Formula Editor*, and multiply the expected profit column by a column with the sampling weights (created as shown in Figure 5.16).

APPLYING SAMPLING WEIGHTS IN JMP

In JMP, sampling weights can be saved when creating a stratified random sample (using the *Tables > Subset* menu), or a column of sampling weights can be created using the *Formula Editor* (see Figure 5.16).

Sampling weights can be applied when fitting models by using the *Weight* field in the model launch dialogs. The resulting misclassification rates, ROC curves, lift curves, and model statistics will be adjusted based on these weights. Sampling weights can also be applied after the fact (after fitting a model and saving the prediction formula to the data table) to adjust the classification matrix for oversampling.

PROBLEMS

5.1 A data mining routine has been applied to a transaction dataset and has classified 88 records as fraudulent (30 correctly so) and 952 as nonfraudulent (920 correctly so). Construct the classification matrix and calculate the error rate.

5.2 Suppose that this routine has an adjustable cutoff (threshold) mechanism by which you can alter the proportion of records classified as fraudulent. Describe how moving the cutoff up or down would affect

a. the classification error rate for records that are truly fraudulent;

b. the classification error rate for records that are truly nonfraudulent.

5.3 Consider Figure 5.17 in which a lift curve for the transaction data model is applied to new data.

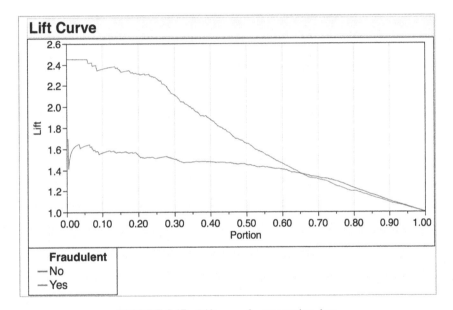

FIGURE 5.17   Lift curve for transaction data.

a. Interpret the meaning of the top curve (for Fraudulent = Yes), at portion = 0.2.

b. Explain how you might use this information in practice.

c. Another analyst comments that you could improve the accuracy of the model by classifying everything as nonfraudulent. If you do that, what is the error rate?

d. Comment on the usefulness, in this situation, of these two metrics of model performance (error rate and lift).

5.4 A large number of insurance records are to be examined to develop a model for predicting fraudulent claims. Of the claims in the historical database, 1% were judged to be fraudulent. A sample is taken to develop a model, and oversampling is used to provide a balanced sample in light of the very low response rate. When applied to this sample ($n = 800$), the model ends up correctly classifying 310 frauds, and 270 nonfrauds. It missed 90 frauds, and classified 130 records incorrectly as frauds when they were not.

a. Produce the classification matrix for the sample as it stands.

b. Find the adjusted misclassification rate (adjusting for the oversampling).

c. What percentage of new records would you expect to be classified as fraudulent?

PART IV

PREDICTION AND CLASSIFICATION METHODS

6

MULTIPLE LINEAR REGRESSION

In this chapter we introduce linear regression models for the purpose of prediction. We discuss the differences between fitting and using regression models for the purpose of inference (as in classical statistics) and for prediction. A predictive goal calls for evaluating model performance on a validation set, and for using predictive metrics. We then raise the challenges of using many predictors and describe variable selection algorithms that are often implemented in linear regression procedures.

6.1 INTRODUCTION

The most popular model for making predictions is the *multiple linear regression model* encountered in most introductory statistics courses and textbooks. This model is used to fit a relationship between a quantitative *dependent variable Y* (also called the *outcome, target, or response variable*) and a set of *predictors* $X_1, X_2, \ldots, X_p$ (also referred to as *independent variables, input variables, regressors*, or *covariates*). The assumption is that the following function approximates the relationship between the input and outcome variables:

$$Y = \beta_0 + \beta_1 x_1 + \beta_2 x_2 + \cdots + \beta_p x_p + \epsilon \tag{6.1}$$

where $\beta_0, \ldots, \beta_p$ are *coefficients* and ϵ is the *noise* or *unexplained* part. Data are then used to estimate the coefficients and to quantify the noise. In predictive modeling, the data are also used to evaluate model performance.

Regression modeling means not only estimating the coefficients but also choosing which input variables to include and in what form. For example, a numerical input can be included as-is, or in logarithmic form ($\log(X)$), or in a binned form (e.g., age group). Choosing the right form depends on domain knowledge, data availability, and needed predictive power.

Multiple linear regression is applicable to numerous predictive modeling situations. Examples are predicting customer activity on credit cards from their demographics and historical activity patterns, predicting the time to failure of equipment based on utilization

Data Mining for Business Analytics: Concepts, Techniques, and Applications with JMP Pro®, First Edition.
Galit Shmueli, Peter C. Bruce, Mia L. Stephens, and Nitin R. Patel.
© 2017 John Wiley & Sons, Inc. Published 2017 by John Wiley & Sons, Inc.

and environmental conditions, predicting expenditures on vacation travel based on historical frequent flyer data, predicting staffing requirements at help desks based on historical data and product and sales information, predicting sales from cross selling of products from historical information, and predicting the impact of discounts on sales in retail outlets.

6.2 EXPLANATORY VERSUS PREDICTIVE MODELING

Before introducing the use of linear regression for prediction, we must clarify an important distinction that often escapes those with earlier familiarity with linear regression from courses in statistics. In particular, there are the two popular but different objectives behind fitting a regression model:

1. Explaining or quantifying the average effect of inputs on an output (explanatory or descriptive task, respectively).
2. Predicting the outcome value for new records, given their input values (predictive task).

The classical statistical approach is focused on the first objective. In that scenario, the data are treated as a random sample from a larger population of interest. The regression model estimated from this sample is an attempt to capture the *average* relationship in the larger population. This model is then used in decision making to generate statements such as "a unit increase in service speed (X_1) is associated with an average increase of 5 points in customer satisfaction (Y), all other factors $(X_2, X_3, \ldots, X_p)$ being equal." If X_1 is known to *cause* Y, then such a statement indicates actionable policy changes—this is called explanatory modeling. When the causal structure is unknown, then this model quantifies the degree of *association* between the inputs and output, and the approach is called descriptive modeling.

In predictive analytics, however, the focus is typically on the second goal: predicting new individual observations. Here we are not interested in the coefficients themselves, nor in the "average record," but rather in the predictions that this model can generate for new records. In this scenario, the model is used for micro decision making at the record level. In our previous example we would use the regression model to predict customer satisfaction for each new customer of interest.

Both explanatory and predictive modeling involve using a dataset to fit a model (i.e., to estimate coefficients), checking model validity, assessing its performance, and comparing to other models. However, the modeling steps and performance assessment differ in the two cases, usually leading to different final models. Therefore the choice of model is closely tied to whether the goal is explanatory or predictive.

In explanatory and descriptive modeling, where the focus is on modeling the average record, we try to fit the best model to the data in an attempt to learn about the underlying relationship in the population. In contrast, in predictive modeling (data mining), the goal is to find a regression model that best predicts new individual records. A regression model that fits the existing data too well is not likely to perform well with new data. Hence we look for a model that has the highest predictive power by evaluating it on a holdout set and using predictive metrics (see Chapter 5).

Let us summarize the main differences in using a linear regression in the two scenarios:

1. A good explanatory model is one that fits the data closely, whereas a good predictive model is one that predicts new cases accurately. Choices of input variables and their form can therefore differ.

2. In explanatory models, the entire dataset is used for estimating the best-fit model, to maximize the amount of information that we have about the hypothesized relationship in the population. When the goal is to predict outcomes of new individual cases, the data are typically split into a training set and a validation set. The training set is used to estimate the model, and the validation or *holdout set* is used to assess this model's predictive performance on new, unobserved data.

3. Performance measures for explanatory models measure how close the data fit the model (how well the model approximates the data) and how strong the average relationship is, whereas in predictive models performance is measured by predictive accuracy (how well the model predicts new individual cases).

For these reasons it is extremely important to know the goal of the analysis before beginning the modeling process. A good predictive model can have a looser fit to the data on which it is based, and a good explanatory model can have low prediction accuracy. In the remainder of this chapter we focus on predictive models, because these are more popular in data mining and because most statistics textbooks focus on explanatory modeling.

6.3   ESTIMATING THE REGRESSION EQUATION AND PREDICTION

Once we determine the input variables to include and their form, we estimate the coefficients of the regression formula from the data using a method called *ordinary least squares* (OLS). This method finds values $\hat{\beta}_0, \hat{\beta}_1, \hat{\beta}_2, \ldots, \hat{\beta}_p$ that minimize the sum of squared deviations between the actual values (Y) and their predicted values based on that model ($\hat{Y}$).

To predict the value of the output variable for a record with input values $x_1, x_2, \ldots, x_p$, we use the equation

$$\hat{Y} = \hat{\beta}_0 + \hat{\beta}_1 x_1 + \hat{\beta}_2 x_2 + \cdots + \hat{\beta}_p x_p. \tag{6.2}$$

Predictions based on this equation are the best predictions possible in the sense that they will be unbiased (equal to the true values on average) and will have the smallest average squared error compared to any unbiased estimates *if* we can safely make the following assumptions:

1. The noise ϵ (or equivalently, Y) follows a normal distribution.
2. The choice of variables and their form is correct (*linearity*).
3. The cases are independent of each other.
4. The variability in Y values for a given set of predictors is the same regardless of the values of the predictors (*homoskedasticity*).

An important and interesting fact for the predictive goal is that *even if we drop the first assumption and allow the noise to follow an arbitrary distribution, these estimates are*

TABLE 6.1 Variables in the Toyota Corolla Example

Variable	Description
Price	Offer price in euros
Age	Age in months as of August 2004
Kilometers	Accumulated kilometers on odometer
Fuel type	Fuel type (*Petrol*, *Diesel*, *CNG*)
Horse Power	Horsepower
Metallic	Metallic color? (Yes = 1, No = 0)
Automatic	Automatic (Yes = 1, No = 0)
CC	Cylinder volume in cubic centimeters
Doors	Number of doors
Quart tax	Quarterly road tax in euros
Weight	Weight in kilograms

very good for prediction, in the sense that among all linear models, as defined by equation (6.1), the model using the least squares estimates, $\hat{\beta}_0, \hat{\beta}_1, \hat{\beta}_2, \ldots, \hat{\beta}_p$, will have the smallest average squared errors. The assumption of a normal distribution is required in explanatory modeling, where it is used for constructing confidence intervals and statistical tests for the model coefficients.

Even if the other assumptions are violated, it is still possible that the resulting predictions are sufficiently accurate and precise for predictive purposes. The key is to evaluate predictive performance of the model, which is the main priority. Satisfying assumptions is of secondary interest and residual analysis can give clues to potential improved models to examine.

Example: Predicting the Price of Used Toyota Corolla Automobiles

A large Toyota car dealership offers purchasers of new Toyota cars the option to buy their used car as part of a trade-in. In particular, a new promotion promises to pay high prices for used Toyota Corolla cars for purchasers of a new car. The dealer then sells the used cars for a small profit. To ensure a reasonable profit, the dealer needs to be able to predict the price that the dealership will get for the used cars. For that reason, data were collected on all previous sales of used Toyota Corollas at the dealership. The data include the sales price and other information on the car, such as its age, mileage, fuel type, and engine size. A description of each of the key variables in the dataset is given in Table 6.1. Several other variables, which won't be used in this example, have been grouped, hidden, and excluded in JMP. A sample of this dataset is shown in Figure 6.1.

The total number of records in the dataset is 1000 cars (we use the first 1000 cars from the dataset `ToyotoCorolla.jmp`). These data are found in the file `Toyota-Corolla1000.jmp`. We create a validation column using the *Make Validation Column* option (from *Cols > Modeling Utilities*), assigning 60% to training set and 40% to the validation set. This validation column, which is used to assign records to the training, validation, or (if used) test partition, has been saved in the data table (it is the last column in Figure 6.1). To fit a multiple linear regression model in JMP Pro, we use *Analyze > Fit*

	Price	Age	Mileage	Fuel Type	Horse Power	Metalic	Automatic	CC	Doors	QuartTax	Weight	Validation
1	13500	23	46986	Diesel	90	1	0	2000	3	210	1165	Training
2	13750	23	72937	Diesel	90	1	0	2000	3	210	1165	Validation
3	13950	24	41711	Diesel	90	1	0	2000	3	210	1165	Validation
4	14950	26	48000	Diesel	90	0	0	2000	3	210	1165	Training
5	13750	30	38500	Diesel	90	0	0	2000	3	210	1170	Training
6	12950	32	61000	Diesel	90	0	0	2000	3	210	1170	Training
7	16900	27	94612	Diesel	90	1	0	2000	3	210	1245	Validation
8	18600	30	75889	Diesel	90	1	0	2000	3	210	1245	Validation
9	21500	27	19700	Petrol	192	0	0	1800	3	100	1185	Training
10	12950	23	71138	Diesel	69	0	0	1900	3	185	1105	Training
11	20950	25	31461	Petrol	192	0	0	1800	3	100	1185	Validation
12	19950	22	43610	Petrol	192	0	0	1800	3	100	1185	Training
13	19600	25	32189	Petrol	192	0	0	1800	3	100	1185	Validation
14	21500	31	23000	Petrol	192	1	0	1800	3	100	1185	Validation
15	22500	32	34131	Petrol	192	1	0	1800	3	100	1185	Validation

FIGURE 6.1   Prices and attributes for sample of 15 cars.

Model, with Price as the *Y, Response* variable, Validation in the *Validation* field, and the other variables as model effects. This fits a model using only the training data.

Figure 6.2 shows the parameter estimates (i.e., the coefficients) from the *Fit Least Squares* analysis window. The regression coefficients are used to predict prices of individual used Toyota Corolla cars based on their age, mileage, and so on. Notice that the Fuel Type predictor has three categories (Petrol, Diesel, and CNG) and we therefore have two indicator variables in the model: Fuel Type[CNG] and Fuel Type[Diesel]. Behind the scenes, JMP has created indicator variables for Fuel Type, and has included only the first two levels (alphabetically) in the model. The indicator for Petrol is a perfect linear combination of the Diesel and CNG, and as a result Petrol is completely redundant to the other two and is not included in the model.[1]

Parameter Estimates

Term	Estimate	Std Error	t Ratio	Prob>|t|
Intercept	−1412.934	1571.885	−0.90	0.3691
Age	−134.1376	4.774744	−28.09	<.0001*
Mileage	−0.019905	0.002369	−8.40	<.0001*
Fuel Type[CNG]	−933.3714	325.1658	−2.87	0.0042*
Fuel Type[Diesel]	−804.1305	265.9956	−3.02	0.0026*
Horse Power	33.955119	5.375333	6.32	<.0001*
Metalic[0]	19.024549	60.16051	0.32	0.7519
Automatic	224.9384	269.0697	0.84	0.4035
CC	0.0209207	0.095982	0.22	0.8275
Doors	−3.003272	61.79519	−0.05	0.9613
QuartTax	22.903512	2.485834	9.21	<.0001*
Weight	12.938552	1.512499	8.55	<.0001*

FIGURE 6.2   Estimated coefficients for regression model of price vs. car attributes.

[1] Inclusion of this redundant variable would cause typical regression software to fail due to a *multicollinearity* error (see Section 4.5). In JMP, when there are such linear dependencies among model effects, a *Singularity Details Report* is provided. This report (when shown) provides a summary of the redundant terms in the model.

CODING OF CATEGORICAL VARIABLES IN REGRESSION

In multiple regression, categorical predictors are added directly to the model and JMP applies indicator coding for these variables. The user need not dummy code categorical predictors prior to building the model. Note that JMP uses $-1/1$ effect coding in *Fit Model* rather than the typical 0/1 indicator or dummy coding.

With this $-1/1$ coding, we interpret each coefficient of a categorical variable as the difference between the average response for that category and the overall average (holding everything else constant). For example, the coefficient for Fuel Type[CNG] means that the average price of cars that use CNG is \$933.37 lower than the average car price. To display coefficients using the traditional 0/1 indicator coding, select *Estimates > Indicator Parameterization Estimates* from the top red triangle.

If you'd prefer to use dummy-coded indicator variables instead, use the *Make Indicator Columns* utility under the *Cols* menu to create these variables. The modeling types for the indicator variables should be changed to continuous, and when fitting the model, you'll need to leave one indicator variable out of the model.

Since we used a validation column, validation statistics (RSquare and RASE) are also provided in the Fit Least Squares window under the *Crossvalidation* outline (see Figure 6.3). As expected, RASE is lower for the training set than the validation set, although RSquare for the training set is lower in this example. See Chapter 5 for details on validation statistics.

To show predicted prices for cars in the dataset, we save the prediction formula to the data table (use *Save Columns > Prediction Formula* from the red triangle). We also save residuals to the data table (using *Save Columns > Residuals*). Figure 6.4 shows predicted prices (in the *Pred Formula Price* column) for the first 15 cars, using the estimated model. To view the prediction formula, right-click on the column header and select *Formula* (see Figure 6.5).

We create a histogram, boxplot, and summary statistics of the residuals for the validation set (see Figure 6.6) for additional measures of predictive accuracy using *Distribution*, with *Validation* in the *By* field. Note that the average error is \$111. The boxplot and *Quantiles* boxplot of the residuals (Figure 6.6) shows that 50% of the errors are approximately between ±\$850. This might be small relative to the car price but should be taken into account when considering the profit. Another observation of interest is the large positive residuals (underpredictions), which may or may not be a concern, depending on the application. Measures such as RASE (or RMSE) are used to assess the predictive performance of a model and to compare models. We discuss such measures in the next section.

This example also illustrates the point about the relaxation of the normality assumption. A histogram or probability plot (a *Normal Quantile Plot* in JMP) of prices shows a right-

▼ **Crossvalidation**

Source	RSquare	RASE	Freq
Training Set	0.8613	1349.9	600
Validation Set	0.8694	1410.3	400

FIGURE 6.3 Crossvalidation statistics for price.

	Price	Validation	Pred Formula Price	Residual Price
1	13500	Training	16715.408642	−3215.408642
2	13750	Validation	16198.841086	−2448.841086
3	13950	Validation	16686.272524	−2736.272524
4	14950	Training	16330.860719	−1380.860719
5	13750	Training	16048.105239	−2298.105239
6	12950	Training	15331.956322	−2381.956322
7	16900	Validation	16265.923137	634.07686298
8	18600	Validation	16236.200913	2363.7990866
9	21500	Training	20500.303797	999.69620285
10	12950	Training	14208.649693	−1258.649693
11	20950	Validation	20534.470476	415.52952362
12	19950	Training	20695.051438	−745.0514383
13	19600	Validation	20519.979274	−919.9792744
14	21500	Validation	19860.016097	1639.9839034
15	22500	Validation	19504.310392	2995.6896079

FIGURE 6.4   Predicted prices and residuals (errors) for the first 15 cars in the dataset.

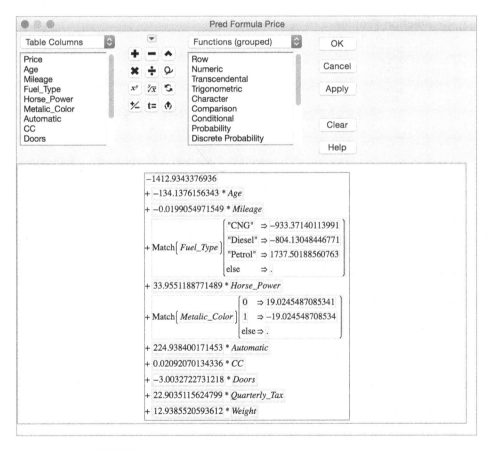

FIGURE 6.5   Prediction formula for price, saved to the data table.

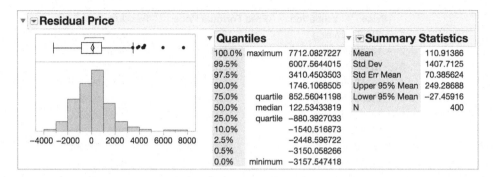

FIGURE 6.6   Histogram, boxplot and summary statistics for model residuals (based on validation set).

skewed distribution. In a descriptive/explanatory modeling case where the goal is to obtain a good fit to the data, the output variable would be transformed (e.g., by taking a logarithm) to achieve a more "normal" variable. Although the fit of such a model to the training data is expected to be better, it will not necessarily yield a significant predictive improvement.

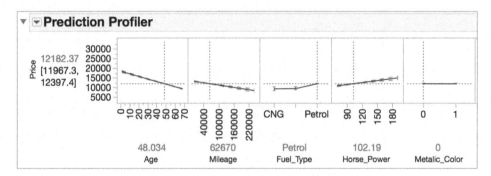

FIGURE 6.7   Prediction profiler for price model.

ADDITIONAL OPTIONS FOR REGRESSION MODELS IN JMP

Many options are for exploring results, diagnosing potential problems with the model, and saving results to the data table are available under the red triangles within the Fit Least Squares analysis window.

   The regression model can be explored interactively using the *Prediction Profiler* (or, the *Profiler*). This graph is a red triangle option under *Factor Profiling*. To predict Price based on given values of the predictors, simply drag the vertical red lines for the predictors to the desired values. The predicted value for price, along with a confidence interval for the prediction, displays on the left of the profiler. The Profiler for the Price model is partially displayed in Figure 6.7.

   The default output can be changed by using the JMP *Preferences*, found under the JMP or File menu. In Preferences, select *Platforms*, and then *Fit Least Squares* to view and change the default output for the platform.

6.4 VARIABLE SELECTION IN LINEAR REGRESSION

Reducing the Number of Predictors

A frequent problem in data mining is that of using a regression equation to predict the value of a dependent variable when we have many variables available to choose as predictors in our model. Given the high speed of modern algorithms for multiple linear regression calculations, it is tempting in such a situation to take a kitchen-sink approach: Why bother to select a subset? Just use all the variables in the model.

Another consideration favoring the inclusions of numerous variables is the hope that a previously hidden relationship will emerge. For example, a company found that customers who had purchased anti-scuff protectors for chair and table legs had lower credit risks. However, there are several reasons for exercising caution before throwing all possible variables into a model.

- It may be expensive or not feasible to collect a full complement of predictors for future predictions.
- We may be able to measure fewer predictors more accurately (e.g., in surveys).
- The more predictors there are, the higher the chance of missing values in the data. If we delete or impute cases with missing values, multiple predictors will lead to a higher rate of case deletion or imputation.
- *Parsimony* is an important property of good models. We obtain more insight into the influence of predictors in models with few parameters.
- Estimates of regression coefficients are likely to be unstable, due to *multicollinearity* in models with many variables. (Multicollinearity is the presence of two or more predictors sharing the same linear relationship with the outcome variable.) Regression coefficients are more stable for parsimonious models. One very rough rule of thumb is to have a number of cases n larger than $5(p + 2)$, where p is the number of predictors.
- It can be shown that using predictors that are uncorrelated with the dependent variable increases the variance of predictions.
- It can be shown that dropping predictors that are actually correlated with the dependent variable can increase the average error (bias) of predictions.

The last two points mean that there is a trade-off between too few and too many predictors. In general, accepting some bias can reduce the variance in predictions. This *bias–variance trade-off* is particularly important for large numbers of predictors, since in that case it is very likely that there are variables in the model that have small coefficients relative to the standard deviation of the noise and also exhibit at least moderate correlation with other variables. Dropping such variables will improve the predictions, as it will reduce the prediction variance. This type of bias–variance trade-off is a basic aspect of most data mining procedures for prediction and classification. In light of this, methods for reducing the number of predictors p to a smaller set are often used.

How to Reduce the Number of Predictors

The first step in trying to reduce the number of predictors should always be to use domain knowledge. It is important to understand what the various predictors are measuring and why they are relevant for predicting the response. With this knowledge the set of predictors should be reduced to a sensible set that reflects the problem at hand. Some practical reasons for predictor elimination are expense of collecting this information in the future, inaccuracy, high correlation with another predictor, many missing values, or simply irrelevance. Also helpful in examining potential predictors are summary statistics and graphs, such as frequency and correlation tables, predictor-specific summary statistics and plots, and missing value counts.

Manual Variable Selection

We may want to manually explore the effect of removing or including a particular predictor or set of predictors. The *Effect Summary* table, which is provided at the top of the *Fit Least Squares* analysis window in JMP (as of Version 12), provides an interactive summary of the terms in the model. The terms are sorted from the most significant to the least (as shown in Figure 6.8.) This table allows you to remove terms from the model or add new terms, including interaction or quadratic effects. As you change terms in the model, all of the statistical output, including the parameter estimates and validation statistics, automatically updates. This is useful to compare validation statistics for different models directly in the *Fit Least Squares* analysis window.

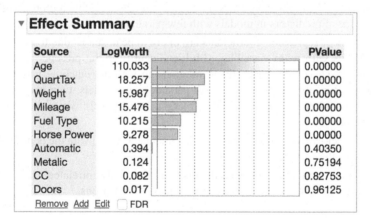

FIGURE 6.8   Effect summary table for price.

Automated Variable Selection

A more formal approach to reducing the number of predictors in the model is to use a built-in algorithm for model reduction. We'll discuss two general methods in this chapter. The first is an exhaustive search for the "best" subset of predictors by fitting regression models with all the possible combinations of predictors. In JMP, this is called *All Possible Models*. The second is to search through a partial set of models using algorithms that consider variables one at a time. These approaches are housed in the *Stepwise* platform, which is available from the *Fit Model* dialog window (under *Personality*). The initial Stepwise Fit window for the Toyota Corolla data is shown in Figure 6.9.

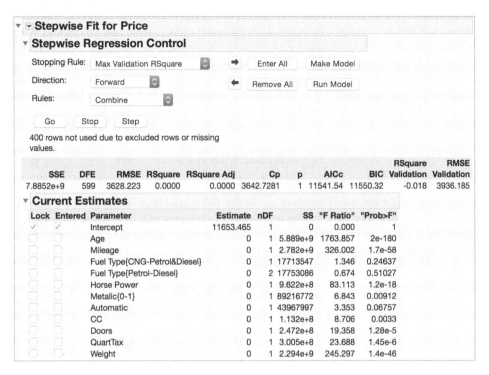

										RSquare	RMSE
SSE	**DFE**	**RMSE**	**RSquare**	**RSquare Adj**		**Cp**	**p**	**AICc**	**BIC**	**Validation**	**Validation**
7.8852e+9	599	3628.223	0.0000	0.0000	3642.7281		1	11541.54	11550.32	-0.018	3936.185

▼ **Current Estimates**

Lock	Entered	Parameter	Estimate	nDF		SS	"F Ratio"	"Prob>F"
✓	✓	Intercept	11653.465	1		0	0.000	1
☐	☐	Age		0	1	5.889e+9	1763.857	2e-180
☐	☐	Mileage		0	1	2.782e+9	326.002	1.7e-58
☐	☐	Fuel Type{CNG-Petrol&Diesel}		0	1	17713547	1.346	0.24637
☐	☐	Fuel Type{Petrol-Diesel}		0	2	17753086	0.674	0.51027
☐	☐	Horse Power		0	1	9.622e+8	83.113	1.2e-18
☐	☐	Metalic{0-1}		0	1	89216772	6.843	0.00912
☐	☐	Automatic		0	1	43967997	3.353	0.06757
☐	☐	CC		0	1	1.132e+8	8.706	0.0033
☐	☐	Doors		0	1	2.472e+8	19.358	1.28e-5
☐	☐	QuartTax		0	1	3.005e+8	23.688	1.45e-6
☐	☐	Weight		0	1	2.294e+9	245.297	1.4e-46

FIGURE 6.9   Selection of the Stepwise personality from the Fit Model dialog window.

CODING OF CATEGORICAL VARIABLES IN STEPWISE REGRESSION

In the *Stepwise* platform, JMP applies a different type of coding for categorical variables. In this platform, $k - 1$ new indicator variables are created for each k-level categorical variable. When the response is continuous, these new variables represent hierarchical groups of levels, where the levels are grouped in way that maximizes the difference in the average response. This difference is measured in terms of the sum of squares between groups (SSB). In the Toyota Corolla example, Fuel Type has three levels. JMP has created two indicator variables (see Figure 6.9). The first variable, Fuel Type{CNG-Petrol&Diesel}, is an indicator variable for CNG versus Petrol and Diesel. The second variable, Fuel Type{Petrol-Diesel}, is an indicator variable for Petrol versus Diesel.

Exhaustive Search (All Possible Models)   The idea here is to evaluate all subsets. Since the number of subsets for even moderate values of p is very large, after the algorithm creates the subsets and runs all the models, we need some way to examine the most promising subsets and to select from them. Criteria for evaluating and comparing models are based on metrics computed from the training data.

One popular criterion that is often used for subset selection is known as *Mallow's C_p* (see formula below). This criterion assumes that the full model (with all predictors) is unbiased, although it may have predictors that if dropped would reduce prediction variability. With this assumption we can show that if a subset model is unbiased, the average C_p value equals the number of parameters $p + 1$ (= number of predictors + 1), the size of the subset. So a reasonable approach to identifying subset models with small bias is to examine those with values of C_p that are near $p + 1$. C_p is also an estimate of the error[2] for predictions at the x-values observed in the training set. Thus good models are those that have values of C_p near $p + 1$ and that have small p (i.e., are of small size). C_p is computed from the formula

$$C_p = \frac{SSE}{\hat{\sigma}^2_{\text{full}}} + 2(p + 1) - n \qquad (6.3)$$

where $\hat{\sigma}^2_{\text{full}}$ is the estimated value of σ^2 in the full model that includes all predictors. It is important to remember that the usefulness of this approach depends heavily on the reliability of the estimate of σ^2 for the full model. This requires that the training set contain a large number of observations relative to the number of predictors.

Another popular criterion is the *adjusted R^2*, which is defined as

$$R^2_{\text{adj}} = 1 - \frac{n - 1}{n - p - 1}(1 - R^2)$$

where R^2 is the proportion of explained variability in the model (in a model with a single predictor, this is the squared correlation). Like R^2, higher values of adjusted R^2 indicate better fit. Unlike R^2, which does not account for the number of predictors used, adjusted R^2 uses a penalty on the number of predictors. This avoids the artificial increase in R^2 that can result from simply increasing the number of predictors but not the amount of information. It can be shown that using R^2_{adj} to choose a subset is equivalent to picking the subset that minimizes $\hat{\sigma}^2$.

A challenge in predictive modeling is to create a model that represents the underlying relationship between the predictors and the response. A model may be too simplistic and not include important parameters (the model is *underfit*), or may be overly complex, describing random noise rather than the true relationship (the model is *overfit*). Two criteria for balancing underfitting and overfitting are the *Akaike Information Criterion* (AIC) and the *Bayesian Information Criterion* (BIC).[3] These criteria measure the information lost by fitting a given model. Therefore models with smaller AIC and BIC values are considered better.

AIC and BIC measure the goodness of fit (i.e., quality) of a model but also include a penalty that is a function of the number of parameters in the model. As such, they can be used to compare various models for the same dataset. It can be shown that Mallows's C_p is equivalent to AIC in the case of multiple linear regression. BIC applies a larger penalty than AIC does, and may lead to smaller models.

---

[2]In particular, it is the sum of the MSE standardized by dividing by σ^2.

[3]AIC and BIC are especially effective with large datasets, when p-values are not as useful. In Stepwise, JMP reports AIC_c, the bias-corrected version of AIC, which is good for modeling with small datasets.

Finally, a useful point to note is that for a fixed size of subset, R^2, R^2_{adj}, C_p, and AIC_c all select the same subset. In fact, there is no difference between them in the order of merit they ascribe to subsets of a fixed size. This is good to know if comparing models with the same number of predictors, but often we want to compare models with different numbers of predictors.

The *All Possible Models* option is available under the red triangle in the Stepwise platform. The resulting window allows you to specify the maximum number of terms in the model, and the best number of models of each size to display. For the Toyota Corolla price data (with the 11 predictors) we ask for the best two models for a single predictor, two predictors, and so on up to 11 predictors. The results of this search are shown in Figure 6.10. The C_p indicates a model with 7 to 11 predictors is good. It can be seen that both RMSE and AIC_c decrease until around 7 predictors, and then stabilize. Both RMSE and AIC_c point to a 7 predictor model. The dominant predictor in all models is the age of the car, with horsepower and mileage (kilometers) playing important roles as well.

WORKING WITH THE ALL POSSIBLE MODELS OUTPUT

By default, the best model of each size will be selected (highlighted) and several statistics will display for each model, as shown in Figure 6.10. To display Mallow's C_p, right-click on the table and select *Columns > Cp*. You can sort on any statistic in the output by right-clicking on the table and selecting *Sort by Column*. This is particularly useful when working with a large number of models. To select a model and view model fit statistics and estimates, use the radio button at the right of the table. R^2_{adj}, C_p, $AICc$, and a number of other fit statistics are reported for the selected model.

Popular Subset Selection Algorithms   The second method of finding the best model relies on an iterative search through the possible predictors. The end product is one best subset of predictors. This approach is computationally cheaper than the exhaustive search, but it has the potential of missing "good" combinations of predictors. In JMP Pro, different stopping rules are available to assist in finding the best model. These include *P-value Threshold*, *Max Validation RSquare*, *Minimum AIC_c*, and *Minimum BIC*. In this section we limit our discussion to the first two stopping rules.[4]

Three popular iterative search algorithms are *forward selection, Backward elimination*, and *Stepwise (mixed)* regression. In *Forward selection* we start with no predictors and then add predictors one by one. Each predictor added is the one that is the most significant of the predictors not already in the model. The algorithm stops when the stopping rule is satisfied. When a validation column has been selected (in JMP Pro), the default stopping rule is *Max Validation RSquare*. When forward selection is used, and the model is built on the training data, the algorithm will select the model with the maximum RSquare for the validation set. The *Step History* report shows the order in which terms were added, the fit statistics, and the model that was selected as the best according to the stopping rule.

The main disadvantage of forward selection is that the algorithm will miss pairs or groups of predictors that perform very well together but perform poorly as single predictors. This

---

[4]For information on *Minimum AICc* and *Minimum BIC*, refer to the book *Fitting Linear Models* in the *JMP Help* (under *Books*).

▼ All Possible Models

Ordered up to best 2 models up to 11 terms per model.

Number	RSquare	RMSE	AICc	BIC	Cp	Model
1	0.7468	1827.18	10719.4	10732.5	477.7124	Age
1	0.3528	2921.26	11282.5	11295.6	2148.5336	Mileage
2	0.7879	1673.79	10615.2	10632.7	305.5056	Age,Horse Power
2	0.7742	1726.81	10652.6	10670.1	363.3937	Age,Weight
3	0.8176	1553.43	10526.7	10548.6	181.4955	Age,Horse Power,Weight
3	0.8162	1559.52	10531.4	10553.3	187.5723	Age,Mileage,Weight
4	0.8405	1454.09	10448.4	10474.7	86.5939	Age,Mileage,Horse Power,Weight
4	0.8353	1477.39	10467.5	10493.7	108.4461	Age,Mileage,Horse Power,QuarTax
5	0.8509	1406.81	10409.8	10440.4	44.2411	Age,Mileage,Fuel Type{Petrol-Diesel},QuarTax,Weight
5	0.8492	1414.78	10416.6	10447.1	51.4216	Age,Mileage,Horse Power,QuarTax,Weight
6	0.8592	1368.31	10377.5	10412.5	11.1074	Age,Mileage,Fuel Type{Petrol-Diesel},Horse Power,QuarTax,Weight
6	0.8516	1404.97	10409.3	10444.2	43.5229	Age,Mileage,Fuel Type{CNG-Petrol&Diesel},Fuel Type{Petrol-Diesel},QuarTax,Weight
7	0.8611	1360.07	10371.3	10410.6	4.9381	Age,Mileage,Fuel Type{Petrol-Diesel},Horse Power,Automatic,QuarTax,Weight
7	0.8593	1368.89	10379.1	10418.4	12.6023	Age,Mileage,Fuel Type{CNG-Petrol&Diesel},Fuel Type{Petrol-Diesel},Horse Power,QuarTax,Weight
8	0.8613	1360.31	10372.6	10416.2	6.1495	Age,Mileage,Fuel Type{CNG-Petrol&Diesel},Fuel Type{Petrol-Diesel},Horse Power,Automatic,QuarTax,Weight
8	0.8612	1361.07	10373.3	10416.9	6.8044	Age,Mileage,Fuel Type{CNG-Petrol&Diesel},Fuel Type{Petrol-Diesel},Horse Power,CC,QuarTax,Weight
9	0.8613	1361.34	10374.6	10422.5	8.0492	Age,Mileage,Fuel Type{CNG-Petrol&Diesel},Fuel Type{Petrol-Diesel},Horse Power,Metalic{0-1},Automatic,QuarTax,Weight
9	0.8613	1361.41	10374.6	10422.5	8.1066	Age,Mileage,Fuel Type{CNG-Petrol&Diesel},Fuel Type{Petrol-Diesel},Horse Power,Automatic,CC,QuarTax,Weight
10	0.8613	1362.45	10376.6	10428.8	10.0024	Age,Mileage,Fuel Type{CNG-Petrol&Diesel},Fuel Type{Petrol-Diesel},Horse Power,Metalic{0-1},Automatic,CC,QuarTax,Weight
10	0.8613	1362.50	10376.6	10428.9	10.0475	Age,Mileage,Fuel Type{CNG-Petrol&Diesel},Fuel Type{Petrol-Diesel},Horse Power,Metalic{0-1},Automatic,Doors,QuarTax,Weight
11	0.8613	1363.60	10378.7	10435.2	12.0000	Age,Mileage,Fuel Type{CNG-Petrol&Diesel},Fuel Type{Petrol-Diesel},Horse Power,Metalic{0-1},Automatic,CC,Doors,QuarTax,Weight

FIGURE 6.10 Exhaustive search (all possible models) result for reducing predictors in Toyota Corolla price example.

is similar to interviewing job candidates for a team project one by one, thereby missing groups of candidates who may perform superiorly together but poorly on their own.

In *Backward elimination* we start with all predictors in the model and then at each step eliminate the least useful predictor (according to statistical significance). The algorithm stops when the stopping rule has been satisfied.

Mixed stepwise regression is like *Forward selection* except that at each step we consider dropping predictors that are not statistically significant, as we do in *Backward elimination*. So, with *Mixed stepwise*, we both add and remove terms from the model. The stopping rule used with *Mixed stepwise* is *P-value Threshold*, and terms are added and then removed according to the *p*-values specified under *Prob to Enter* and *Prob to Leave*.

WHEN USING A STOPPING ALGORITHM IN JMP

- Select the *Stopping Rule*.
- Select the *Direction*. If using *Backward*, click the *Enter All* button to enter all terms into the model. Note that *Mixed* is only available with the *P-value Threshold* stopping rule.
- If your model includes categorical variables with more than two levels, JMP will recode the variables into groups that drive the greatest separation in the response.
- Use *Step* to step through one variable at a time, or *Go* to run the algorithm until the stopping rule is satisfied.
- Once you've selected a model, use the *Run Model* button to launch the *Fit Least Squares analysis* window.

Note: When using Stepwise with categorical predictors with more than two levels, change the *Rules* from *Combine* to *Whole Effects*. This adds the *whole effect* for the predictor to the model and makes model estimates easier to interpret.

For the Toyota Corolla price example, forward selection using the default *Max Validation RSquare* stopping rule yields a 5 parameter model (4 terms plus the intercept). This model retains only Age, Mileage, Horsepower, and Weight (see Figure 6.11). In comparison, *Backward elimination* starts with the full model and then drops predictors one by one. Running this algorithm, again with the default *Max Validation RSquare* rule, yields the same model as the best 8 predictor model from All Possible Models (as shown in Figure 6.12). For this model, Metallic, CC, and Doors are dropped from the model.

Changing the stopping rule can also produce different results. In Figure 6.13 we see the results of *Backward elimination* using the *P-value Threshold* stopping rule. The *p*-value to remove a term (*Prob to Leave*) is set at 0.15. With this stopping rule, terms are removed from the model one at a time until only terms with *p*-values less than the cutoff (0.15 here) remain. *Forward selection* with the *P-value Threshold* works in much the same manner. Terms are added to the model one at a time based on the *p*-value. Only terms below the threshold *Prob to Enter* are entered.

SSE	DFE	RMSE	RSquare	RSquare Adj	Cp	p	AICc	BIC	RSquare Validation	RMSE Validation
1.2581e+9	595	1454.0948	0.8405	0.8394	86.59387	5	10448.41	10474.65	0.8774	1366.245

▼ **Current Estimates**

Lock	Entered	Parameter	Estimate	nDF	SS	"F Ratio"	"Prob>F"
✓	✓	Intercept	−682.78425	1	0	0.000	1
☐	☑	Age	−135.30498	1	1.54e+9	728.464	2e-105
☐	☑	Mileage	−0.0208229	1	1.802e+8	85.216	4.6e-19
☐	☐	Fuel Type{CNG-Petrol&Diesel}	0	1	2997672	1.419	0.23409
☐	☐	Fuel Type{Petrol-Diesel}	0	2	4522031	1.070	0.34381
☐	☑	Horse Power	41.1579977	1	1.915e+8	90.560	4.4e-20
☐	☐	Metalic{0-1}	0	1	689586.8	0.326	0.56838
☐	☐	Automatic	0	1	1238284	0.585	0.44457
☐	☐	CC	0	1	369366	0.174	0.67634
☐	☐	Doors	0	1	323881.4	0.153	0.69586
☐	☐	QuartTax	0	1	69118432	34.532	6.99e-9
☐	☑	Weight	14.9151065	1	3.448e+8	163.066	3.6e-33

FIGURE 6.11   Forward selection results in Toyota Corolla price example, using the *Max Validation RSquare* stopping rule.

SSE	DFE	RMSE	RSquare	RSquare Adj	Cp	p	AICc	BIC	RSquare Validation	RMSE Validation
1.0936e+9	591	1360.3081	0.8613	0.8594	6.149462	9	10372.59	10416.19	0.8696	1408.942

▼ **Current Estimates**

Lock	Entered	Parameter	Estimate	nDF	SS	"F Ratio"	"Prob>F"
✓	✓	Intercept	−1648.0331	1	0	0.000	1
☐	☑	Age	−134.0259	1	1.476e+9	797.639	1e-111
☐	☑	Mileage	−0.0198961	1	1.32e+8	71.333	2.3e-16
☐	☑	Fuel Type{CNG-Petrol&Diesel}	−706.77093	2	92715153	25.052	3.6e-11
☐	☑	Fuel Type{Petrol-Diesel}	1265.20038	1	74463635	40.241	4.5e-10
☐	☑	Horse Power	34.052001	1	75996454	41.069	3e-10
☐	☐	Metalic{0-1}	0	1	186375.9	0.101	0.75126
☐	☑	Automatic	234.715517	1	1466437	0.792	0.37371
☐	☐	CC	0	1	79726.28	0.043	0.83577
☐	☐	Doors	0	1	10099.67	0.005	0.94118
☐	☑	QuartTax	22.933752	1	1.588e+8	85.823	3.6e-19
☐	☑	Weight	12.9338836	1	1.443e+8	77.960	1.2e-17

FIGURE 6.12   Backward elimination results using the *Max Validation RSquare* stopping rule.

In this case, *Backward elimination*, *Forward selection* and *Mixed stepwise* with the *P-value Threshold* (0.15 for both entering an leaving) all yield the same model, removing Metallic, CC, and Doors, and also removing Automatic. The R^2_{adj}, AIC_c, and other measures indicate that this is exactly the same subset as the best 7 term model suggested by All Possible Models. In other words, these algorithms correctly identify Doors, CC, Metallic, and Automatic as the least useful predictors.

Comparing the reduced models with the full model, the model with the lowest RMSE on the training data is the full model, while the model with the highest R^2_{adj} is the 8 parameter model. The 5 parameter model (from *Forward Selection*, with *Max Validation RSquare*) has both the highest RSquare and the lowest RMSE of the the models explored. However, this model has the largest C_p and AIC_c of the reduced models, indicating the the model is underfit. This example shows that the search algorithms yield fairly good solutions, but we

	SSE	DFE	RMSE	RSquare	RSquare Adj	Cp	p	AICc	BIC	RSquare Validation	RMSE Validation
	1.0951e+9	592	1360.0696	0.8611	0.8595	4.9381209	8	10371.33	10410.59	0.8693	1410.57

▼ **Current Estimates**

Lock	Entered	Parameter	Estimate	nDF	SS	"F Ratio"	"Prob>F"
✓	✓	Intercept	−1834.7531	1	0	0.000	1
☐	☑	Age	−133.7316	1	1.477e+9	798.292	8e-112
☐	☑	Mileage	−0.0199964	1	1.336e+8	72.245	1.5e-16
☐	☑	Fuel_Type{CNG-Petrol&Diesel}	−693.5013	2	93869232	25.373	2.7e-11
☐	☑	Fuel_Type{Petrol-Diesel}	1278.61028	1	76486799	41.349	2.6e-10
☐	☑	Horse_Power	33.9189915	1	75463588	40.796	3.4e-10
☐	☐	Metalic_Color{0-1}	0	1	169580.7	0.092	0.76234
☐	☐	Automatic	0	1	1466437	0.792	0.37371
☐	☐	CC	0	1	248577.5	0.134	0.71426
☐	☐	Doors	0	1	20437.61	0.011	0.91639
☐	☑	Quarterly_Tax	22.9071811	1	1.585e+8	85.667	3.9e-19
☐	☑	Weight	13.1274675	1	1.52e+8	82.149	1.8e-18

FIGURE 6.13 *Backward elimination*, *Forward selection*, and *Mixed stepwise* results using the *P-value Threshold* stopping rule.

need to carefully determine the number of predictors to retain. It also shows the merits of running a few searches and using the combined results to determine the subset to choose. There is a popular (but false) notion that *Mixed stepwise* regression is superior to *Backward elimination* and *Forward selection* because of its ability to add and to drop predictors. This example shows clearly that it is not always so.

Finally, additional ways to reduce the dimension of the data, prior to building regression models, are by using principal components (Chapter 4) and regression trees (Chapter 9).

OTHER REGRESSION PROCEDURES IN JMP PRO—GENERALIZED REGRESSION

Multiple linear regression and forward stepwise regression are also available in JMP Pro from the *Generalized Regression* personality in *Fit Model* (see Figure 6.14). Generalized regression allows for the selection of a variety of different response distributions (the default is Normal), provides a number of built-in options for model exploration, selection and validation, and also provides procedures for *penalized* or *regularized* regression, including *Lasso*, *Ridge Regression*, and *Elastic Net*. These penalized procedures are commonly used when predictors are highly correlated.

For more information on Generalized Regression, see the book *Specialized Models* under the *JMP Help*.

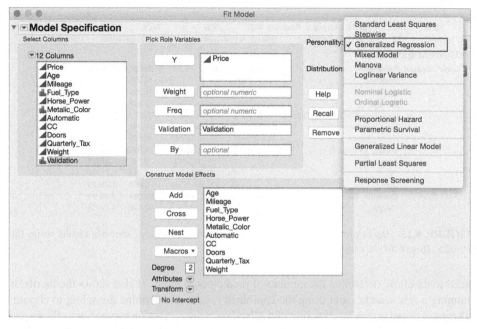

FIGURE 6.14   *Generalized Regression* option in the Fit Model.

PROBLEMS

6.1   Predicting Boston Housing Prices. The file `BostonHousing.jmp` contains information collected by the US Bureau of the Census concerning housing in the area of Boston, Massachusetts. The dataset includes information on 506 census housing tracts in the Boston area in the 1970s. The goal is to predict the median house price in new tracts based on information such as crime rate, pollution, and number of rooms. The dataset contains 12 predictors, and the response is the median house price (MEDV). Table 6.2 describes each of the predictors and the response.

TABLE 6.2   Description of Variables for Boston Housing Example

CRIM	Per capita crime rate by town
ZN	Proportion of residential land zoned for lots over 25,000 ft^2
INDUS	Proportion of nonretail business acres per town
CHAS	Charles River dummy variable (= 1 if tract bounds river; = 0 otherwise)
NOX	Nitric oxide concentration (parts per 10 million)
RM	Average number of rooms per dwelling
AGE	Proportion of owner-occupied units built prior to 1940
DIS	Weighted distances to five Boston employment centers
RAD	Index of accessibility to radial highways
TAX	Full-value property-tax rate per $10,000
PTRATIO	Pupil/teacher ratio by town
LSTAT	% Lower status of the population
MEDV	Median value of owner-occupied homes in $1000s

a. Why should the data be partitioned into training and validation sets? What will the training set be used for? What will the validation set be used for?

b. Fit a multiple linear regression model to the median house price (MEDV) as a function of CRIM, CHAS, and RM. Write the equation for predicting the median house price from the predictors in the model.

c. Using the estimated regression model, what median house price is predicted for a tract in the Boston area that does not bound the Charles River, has a crime rate of 0.1, and where the average number of rooms per house is 6? What is the prediction error?

d. Consider the 12 predictors:

   i. Which predictors are likely to be measuring the same thing among the entire set of predictors? Discuss the relationships among INDUS, NOX, and TAX.

   ii. Compute the correlation table for the numerical predictors and search for highly correlated pairs. These have potential redundancy and can cause multicollinearity. Choose which ones to remove based on this table.

   iii. Use an exhaustive search (All Possible Models) to reduce the remaining predictors as follows: First, choose the top three models. Then run each of these models and compare their predictive accuracy for the validation set. Compare RMSE, C_p, AIC_c, and Validation RSquare. Finally, describe the best model.

6.2 **Predicting Software Reselling Profits.** Tayko Software is a software catalog firm that sells games and educational software. It started out as a software manufacturer and then added third-party titles to its offerings. It recently revised its collection of items in a new catalog, which it mailed out to its customers. This mailing yielded 1000 purchases. Based on these data, Tayko wants to devise a model for predicting the spending amount that a purchasing customer will yield. The file `TaykoPur-chasesOnly.jmp` contains information on 1000 purchases. Table 6.3 describes the variables to be used in the problem (the file contains additional variables).

TABLE 6.3   Description of Variables for Tayko Software Example

FREQ	Number of transactions in the preceding year
LAST_UPDATE	Number of days since last update to customer record
WEB	Whether customer purchased by Web order at least once
GENDER	Female = 0, male = 1
ADDRESS_RES	Whether it is a residential address
ADDRESS_US	Whether it is a US address
SPENDING (response)	Amount spent by customer in test mailing (in dollars)

a. Explore the spending amount by creating a tabular summary of the categorical variables and computing the average and standard deviation of spending in each category.

b. Explore the relationship between spending and each of the two continuous predictors by creating two scatterplots (SPENDING vs. FREQ, and SPENDING vs. LAST_UPDATE). Does there seem to be a linear relationship?

c. To fit a predictive model for SPENDING:

    i. Partition the 1000 records into training and validation sets. (Note that the dataset has a validation column—create your own validation column for this exercise.)

    ii. Run a multiple linear regression model for SPENDING vs. all six predictors. Give the estimated predictive equation.

    iii. Based on this model, what type of purchaser is most likely to spend a large amount of money?

    iv. If we used backward elimination to reduce the number of predictors, which predictor would be dropped first from the model?

    v. Show how the prediction and the prediction error are computed for the first purchase in the validation set.

    vi. Evaluate the predictive accuracy of the model by examining its performance on the validation set.

    vii. Create a histogram of the model residuals. Do they appear to follow a normal distribution? How does this affect the predictive performance of the model?

6.3 **Predicting Airfare on New Routes.** The following problem takes place in the United States in the late 1990s, when many major US cities were facing issues with airport congestion, partly as a result of the 1978 deregulation of airlines. Both fares and routes were freed from regulation, and low-fare carriers such as Southwest began competing on existing routes and starting nonstop service on routes that it previously lacked. Building completely new airports is generally not feasible, but sometimes decommissioned military bases or smaller municipal airports can be reconfigured as regional or larger commercial airports. There are numerous players and interests involved in the issue (airlines, city, state and federal authorities, civic groups, the military, airport operators), and an aviation consulting firm is seeking advisory contracts with these players. The firm needs predictive models to support its consulting service. One thing the firm might want to be able to predict is fares, in the event a new airport is brought into service. The firm starts with the file `Airfares.jmp`, which contains real data that were collected between Q3-1996 and Q2-1997. The variables in these data are listed in Table 6.4, and are believed to be important in predicting FARE. Some airport-to-airport data are available, but most data are at the city-to-city level. One question that will be of interest in the analysis is the effect that the presence or absence of Southwest (SW) has on FARE.

a. Explore the numerical predictors and response (FARE) by creating histograms, a correlation table and a scatterplot matrix. Examining potential relationships between FARE and those predictors. What seems to be the best single predictor of FARE?

b. Explore the categorical predictors (excluding the first four) by computing the percentage of flights in each category. Create a tabular summary with the average fare in each category. Which categorical predictor seems best for predicting FARE?

c. Find a model for predicting the average fare on a new route:

    i. Partition the data into training and validation sets. The model will be fit to the training data and evaluated on the validation set.

    ii. Use mixed stepwise regression to reduce the number of predictors. You can ignore the first four predictors (S_CODE, S_CITY, E_CODE, E_CITY). Report the estimated model selected.

iii. Repeat (ii) using All Possible Models instead of *Mixed stepwise* regression. Compare the resulting best model to the one you obtained in (ii) in terms of the predictors that are in the model.

iv. Compare the predictive accuracy of both models (ii) and (iii) using measures such as RMSE, C_p, AIC_c, and Validation RSquare.

v. Using model (iii), predict the average fare on a route with the following characteristics: COUPON = 1.202, NEW = 3, VACATION = No, SW = No, HI = 4442.141, S_INCOME = $28,760, E_INCOME = $27,664, S_POP = 4,557,004, E_POP = 3,195,503, SLOT = Free, GATE = Free, PAX = 12,782, DISTANCE = 1976 miles.

vi. Using model (iii), predict the reduction in average fare on the route in (v) if Southwest decides to cover this route.

vii. In reality, which of the factors will not be available for predicting the average fare from a new airport (i.e., before flights start operating on those routes)? Which ones can be estimated? How?

viii. Select a model that includes only factors that are available before flights begin to operate on the new route. Use an exhaustive search (All Possible Models) to find such a model.

ix. Use model (viii) to predict the average fare on a route with characteristics COUPON = 1.202, NEW = 3, VACATION = No, SW = No, HI = 4442.141, S_INCOME = $28,760, E_INCOME = $27,664, S_ POP = 4,557,004, E_POP = 3,195,503, SLOT = Free, GATE = Free, PAX = 12,782, DISTANCE = 1976 miles.

TABLE 6.4   Description of Variables for Airfare Example

S_CODE	Starting airport's code
S_CITY	Starting city
E_CODE	Ending airport's code
E_CITY	Ending city
COUPON	Average number of coupons (a one-coupon flight is a nonstop flight, a two-coupon flight is a one-stop flight, etc.) for that route
NEW	Number of new carriers entering that route between Q3-96 and Q2-97
VACATION	Whether (Yes) or not (No) a vacation route
SW	Whether (Yes) or not (No) Southwest Airlines serves that route
HI	Herfindahl index: measure of market concentration
S_INCOME	Starting city's average personal income
E_INCOME	Ending city's average personal income
S_POP	Starting city's population
E_POP	Ending city's population
SLOT	Whether or not either endpoint airport is slot controlled (this is a measure of airport congestion)
GATE	Whether or not either endpoint airport has gate constraints (this is another measure of airport congestion)
DISTANCE	Distance between two endpoint airports in miles
PAX	Number of passengers on that route during period of data collection
FARE	Average fare on that route

    x. Compare the predictive accuracy of this model with model (iii). Is this model good enough, or is it worthwhile reevaluating the model once flights begin on the new route?

  d. In competitive industries, a new entrant with a novel business plan can have a disruptive effect on existing firms. If a new entrant's business model is sustainable, other players are forced to respond by changing their business practices. If the goal of the analysis was to evaluate the effect of Southwest Airlines' presence on the airline industry rather than predicting fares on new routes, how would the analysis be different? Describe technical and conceptual aspects.

6.4 **Predicting Prices of Used Cars.** The file `ToyotaCorolla.jmp` contains data on used cars (Toyota Corolla) on sale during late summer of 2004 in the Netherlands. It has 1436 records containing details on 38 attributes, including Price, Age, Kilometers, HP, and other specifications. The goal is to predict the price of a used Toyota Corolla based on its specifications. (The example in Section 6.3 is a subset of this dataset.)

    Data preprocessing. Split the data into training (50%), validation (30%), and test (20%) datasets. Run a multiple linear regression with the output variable Price and input variables: Age_08_04, KM, Fuel_Type, HP, Automatic, Doors, Quarterly_Tax, Mfg_Guarantee, Guarantee_Period, Airco, Automatic_Airco, CD_Player, Powered_Windows, Sport_Model, and Tow_Bar.

  a. What appear to be the three or four most important car specifications for predicting the car's price? Use the *Prediction Profiler* to explore the relationship between Price and these variables.

  b. Use the *Stepwise* platform to develop a reduced predictive model for Price. What are the predictors in your reduced model?

  c. Using metrics you consider useful, assess the performance of the reduced model in predicting prices.

7

k-NEAREST NEIGHBORS (*k*-NN)

In this chapter we describe the *k*-nearest neighbors algorithm that can be used for classification (of a categorical outcome) or prediction (of a numerical outcome). To classify or
predict a new record, the method relies on finding "similar" records in the training data.
These "neighbors" are then used to derive a classification or prediction for the new record
by voting (for classification) or averaging (for prediction). We explain how similarity is
determined, how the number of neighbors is chosen, and how a classification or prediction
is computed. *k*-NN is a highly automated data-driven method. We discuss the advantages
and weaknesses of the *k*-NN method in terms of performance and practical considerations
such as computational time.

7.1   THE *k*-NN CLASSIFIER (CATEGORICAL OUTCOME)

The idea in *k*-nearest neighbors methods is to identify *k* records in the training dataset that
are similar to a new record that we wish to classify. We then use these similar (neighboring)
records to classify the new record into a class, assigning the new record to the predominant
class among these neighbors. Denote by $(x_1, x_2, \ldots, x_p)$ the values of the predictors for this
new record. We look for records in our training data that are similar or "near" the record
to be classified in the predictor space (i.e., records that have values close to $x_1, x_2, \ldots, x_p$).
Then, based on the classes to which those proximate records belong, we assign a class to
the record that we want to classify.

Determining Neighbors

The *k*-nearest neighbors algorithm is a classification method that does not make assumptions
about the form of the relationship between the class membership (Y) and the predictors
$X_1, X_2, \ldots, X_p$. This is a nonparametric method because it does not involve estimation
of parameters in an assumed function form, such as the linear form assumed in linear

Data Mining for Business Analytics: Concepts, Techniques, and Applications with JMP Pro®, First Edition.
Galit Shmueli, Peter C. Bruce, Mia L. Stephens, and Nitin R. Patel.
© 2017 John Wiley & Sons, Inc. Published 2017 by John Wiley & Sons, Inc.

regression (Chapter 6). Instead, this method draws information from similarities among the predictor values of the records in the dataset.

A central question is how to measure the distance between records based on their predictor values. The most popular measure of distance is the Euclidean distance. The Euclidean distance between two records $(x_1, x_2, \ldots, x_p)$ and $(u_1, u_2, \ldots, u_p)$ is

$$\sqrt{(x_1 - u_1)^2 + (x_2 - u_2)^2 + \cdots + (x_p - u_p)^2} \tag{7.1}$$

You will find a host of other distance metrics in Chapters 12 and 14 for both numerical and categorical variables. However, the *k*-NN algorithm relies on many distance computations (between each record to be predicted and every record in the training set), and therefore the Euclidean distance, which is computationally cheap, is the most popular in *k*-NN.

To equalize the scales that the various predictors may have, note that in most cases predictors should first be standardized before computing a Euclidean distance.

Classification Rule

After computing the distances between the record to be classified and existing records, we need a rule to assign a class to the record to be classified, based on the classes of its neighbors. The simplest case is $k = 1$, where we look for the record that is closest (the nearest neighbor) and classify the new record as belonging to the same class as its closest neighbor. It is a remarkable fact that this simple, intuitive idea of using a single nearest neighbor to classify records can be very powerful when we have a large number of records in our training set. In turns out that the misclassification error of the 1-nearest neighbor scheme has a misclassification rate that is no more than twice the error when we know exactly the probability density functions for each class.

The idea of the **1-nearest neighbor** can be extended to $k > 1$ neighbors as follows:

1. Find the nearest *k*-neighbors to the record to be classified.
2. Use a majority decision rule to classify the record, where the record is classified as a member of the majority class of the *k*-neighbors.

Example: Riding Mowers

A riding mower manufacturer would like to find a way of classifying families in a city into those likely to purchase a riding mower and those not likely to buy one. A pilot random sample is undertaken of 12 owners and 12 nonowners in the city. The data are shown in Table 7.1. We first partition the data into training data (18 households) and validation data (6 households). Obviously, this dataset is too small for partitioning, which can result in unstable results, but we continue with this for illustration purposes. A scatterplot for the training set is shown in Figure 7.1 (data are in `RidingMowerskNN.jmp`).

Now consider a new household with $60,000 income and lot size 20,000 ft^2 (also shown in Figure 7.1, marked with an x). Among the households in the training set, the one closest to the new household (in Euclidean distance after normalizing income and lot size) is household 4, with $61,500 income and lot size 20,800 ft^2. If we use a 1-NN classifier, we would classify the new household as an owner, like household 4. If we use $k = 3$, the three nearest households are 4, 9, and 14. The first two are owners of riding mowers, and the

TABLE 7.1   Lot Size, Income, and Ownership of a Riding Mower for 24 Households

Household Number	Income ($000s)	Lot Size (000s ft^2)	Ownership of Riding Mower
1	60.0	18.4	Owner
2	85.5	16.8	Owner
3	4.8	21.6	Owner
4	61.5	20.8	Owner
5	87.0	23.6	Owner
6	110.1	19.2	Owner
7	108.0	17.6	Owner
8	82.8	22.4	Owner
9	69.0	20.0	Owner
10	93.0	20.8	Owner
11	51.0	22.0	Owner
12	81.0	20.0	Owner
13	75.0	19.6	Nonowner
14	52.8	20.8	Nonowner
15	64.8	17.2	Nonowner
16	43.2	20.4	Nonowner
17	84.0	17.6	Nonowner
18	49.2	17.6	Nonowner
19	59.4	16.0	Nonowner
20	66.0	18.4	Nonowner
21	47.4	16.4	Nonowner
22	33.0	18.8	Nonowner
23	51.0	14.0	Nonowner
24	63.0	14.8	Nonowner

last is a nonowner. The majority vote is therefore *owner*, and the new household would be classified as an owner.

Choosing *k*

The advantage of choosing $k > 1$ is that higher values of k provide smoothing that reduces the risk of overfitting due to noise in the training data. Generally speaking, if k is too low, we may be fitting to the noise in the data. However, if k is too high, we will miss out on the method's ability to capture the local structure in the data, one of its main advantages. In the extreme, $k = n =$ the number of records in the training dataset. In that case we simply assign all records to the majority class in the training data, regardless of the values of $(x_1, x_2, \ldots, x_p)$, which coincides with the naive rule! This is clearly a case of oversmoothing in the absence of useful information in the predictors about the class membership. In other words, we want to balance between overfitting to the predictor information and ignoring this information completely. A balanced choice depends greatly on the nature of the data. The more complex and irregular the structure of the data, the lower is the optimum value of k. Typically, values of k fall in the range 1 to 20. Often an odd number is chosen, to avoid ties.

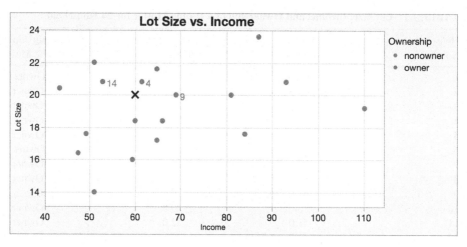

FIGURE 7.1   Scatterplot of *Lot Size* vs. *Income* for the 18 households in the training set and the new household to be classified.

K NEAREST NEIGHBORS IN JMP PRO

In JMP 12 Pro, *K Nearest Neighbors* is an option under *Analyze > Modeling* in the Partition platform, found under the drop down box for *Method*. The dialog is specified as shown in Figure 7.2. After clicking OK, a value is specified for the maximum number of nearest neighbors, *K*, and a model is computed for each value of *K* between 1 and K. The platform can be used both for categorical responses with two or more levels and for continuous responses, and prediction formulas can be saved to the data table to make predictions for new records.

Here are a few use notes:

- In the *Partition* platform (in JMP 12), the Validation column must be coded as numeric.
- Each continuous predictor is scaled by its standard deviation (so standardization of continuous predictors, in advance, is not required).
- Indicator coding is applied for each categorical predictor (so dummy coding of categorical predictors, in advance, is not required).
- Missing values for a continuous predictor are replaced by the mean of the predictor, while missing values for a categorical predictor are replaced by a value of zero.

Note: In JMP 13, the *Analyze* menu structure has changed.

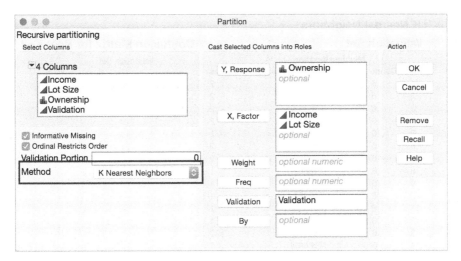

FIGURE 7.2   Selection of the *K Nearest Neighbor* option from the *Partition* platform in JMP.

So how is k chosen? Answer: We choose the k that has the best classification performance. We use the training data to classify the records in the validation data, then compute error rates for various choices of k. For our example, on the one hand, if we choose $k = 1$, we will classify in a way that is very sensitive to the local characteristics of the training data. On the other hand, if we choose a large value of k, such as $k = 18$, we would simply predict the most frequent class in the dataset in all cases. This is a very stable prediction, but it completely ignores the information in the predictors. To find a balance, we examine the misclassification rate (of the validation set) for different choices of k between 1 and 18. This is shown in Figure 7.3. JMP chooses $k = 8$, which minimizes the misclassification rate in the validation set.[1] Since the validation set is used to determine the optimal value of k, ideally we would want a third test set to evaluate the performance of the method on data that it did not see.

Once k is chosen, we can save the prediction formula to the data table to generate classifications for new records. An example is shown in Figure 7.4, where 8 neighbors are used to classify a new household, which was added as a new row in the data table (row 25). This household was classified as an owner.

The Cutoff Value for Classification

k-NN uses a majority decision rule to classify a new record, where the record is classified as a member of the majority class of the k-neighbors. The definition of "majority" is directly linked to the notion of a cutoff value applied to the class membership probabilities. Let

---

[1]Partitioning such a small dataset into training and validation sets is unwise in practice, as results will heavily rely on the particular partition. For instance, if you use a different partitioning, you might obtain a different "optimal" k. We use this example for illustration only.

K Nearest Neighbors

Validation Set

K	Count	Misclassification Rate	Misclassifications
1	6	0.33333	2
2	6	0.33333	2
3	6	0.33333	2
4	6	0.33333	2
5	6	0.33333	2
6	6	0.33333	2
7	6	0.33333	2
8	6	0.16667	1
9	6	0.16667	1
10	6	0.16667	1
11	6	0.33333	2
12	6	0.16667	1
13	6	0.33333	2
14	6	0.16667	1
15	6	0.33333	2
16	6	0.50000	3
17	6	0.33333	2
18	6	0.50000	3

Confusion Matrix for Best K=8

Validation Set

Actual	Predicted	
Ownership	nonowner	owner
nonowner	2	1
owner	0	3

FIGURE 7.3   *a* Misclassification rate of validation set for various choices of *k*, and *b* the confusion matrix for *k* = *8*.

	Income	Lot Size	Ownership	Validation	Predicted Formula Ownership 8
18	49.2	17.6	nonowner	0	nonowner
19	59.4	16	nonowner	0	nonowner
20	66	18.4	nonowner	0	nonowner
21	47.4	16.4	nonowner	0	nonowner
22	33	18.8	nonowner	1	nonowner
23	51	14	nonowner	0	nonowner
24	63	14.8	nonowner	1	nonowner
25	60	20		•	owner

FIGURE 7.4   Classifying a new household using the "best *k*"= 8.

us consider a binary outcome case. For a new record, the proportion of class 1 members among its neighbors is an estimate of its propensity (probability) of belonging to class 1. In the riding mowers example with *k* = 3, we found that the three nearest neighbors to the new household (with income = \$60,000 and lot size = 20,000 ft^2) are households 4, 9, and 14. Since 4 and 9 are owners and 14 is a nonowner, we can estimate for the new household a probability of 2/3 of being an owner (and 1/3 for being a nonowner). Using a simple majority rule is equivalent to setting the cutoff value to 0.5.

***k*-NN PREDICTIONS AND PREDICTION FORMULAS IN JMP PRO**

Finding the nearest neighbor for every record in a dataset is time-consuming, particularly in large datasets. As a result predictions aren't made until the prediction formula for a *k*-NN model is saved to the data table. The prediction formula contains all of the training data, so distances for each observation in the training set can be computed and predictions for new data can be made in more a timely manner. The fitted probabilities for categorical responses in *k*-NN models are not saved to the data table. As a result the ability to change the cutoff for classification in *k*-NN is not available (in JMP 12 Pro).

k-NN with More Than Two Classes

The *k*-NN classifier can easily be applied to an outcome with m classes, where $m > 2$. The "majority rule" means that a new record is classified as a member of the majority class of its *k*-neighbors.

7.2 *K*-NN FOR A NUMERICAL RESPONSE

The idea of *k*-NN can readily be extended to predicting a continuous value (as is our aim with multiple linear regression models). The first step of determining neighbors by computing distances remains unchanged. The second step, where a majority vote of the neighbors is used to determine class, is modified such that we take the average response value of the *k*-nearest neighbors to determine the prediction. Often this average is a weighted average, with the weight decreasing with increasing distance from the point at which the prediction is required.

Another modification is in the error metric used for determining the "best *k*." Rather than the overall error rate used in classification, RMSE and SSE are used to determine the best *k* in when the response is continuous (see Chapter 5).

PANDORA

Pandora is an Internet music radio service that allows users to build customized "stations" that play music similar to a song or artist that they have specified. Pandora uses a KNN style clustering/classification process called the Music Genome Project to locate new songs or artists that are close to the user-specified song or artist.

Pandora was the brainchild of Tim Westergren, who worked as a musician and a nanny when he graduated from Stanford in the 1980s. Together with Nolan Gasser, who was studying medieval music, he developed a "matching engine" by entering data about a song's characteristics into a spreadsheet. The first result was surprising—a Beatles song matched to a Bee Gees song, but they built a company around the concept. The early days were hard—Westergren racked up over $300,000 in personal debt, maxed out 11 credit cards, and ended up in the hospital once due to stress-induced heart palpitations. A venture capitalist finally invested funds in 2004 to rescue the firm, and as of 2013 it is listed on the NY Stock Exchange.

In simplified terms, the process works roughly as follows for songs:

1. Pandora has established hundreds of variables on which a song can be measured on a scale from 0 to 5. Four such variables from the beginning of the list are
   - Acid Rock Qualities
   - Accordion Playing
   - Acousti-Lectric Sonority
   - Acousti-Synthetic Sonority

2. Pandora pays musicians to analyze tens of thousands of songs, and rate each song on each of these attributes. Each song will then be represented by a row vector of values between 0 and 5. For example, for Led Zeppelin's Kashmir:

   Kashmir 4 0 3 3 ... (high on acid rock attributes, no accordion, etc.)

   This step represents a costly investment, and lies at the heart of Pandora's value. That is to say, these variables have been tested and selected because they accurately reflect the essence of a song, and provide a basis for defining highly individualized preferences.

3. The online user specifies a song that s/he likes (the song must be in Pandora's database).

4. Pandora then calculates the statistical distance[2] between the user's song, and the songs in its database. It selects a song that is close to the user-specified song and plays it.

5. The user then has the option of saying "I like this song," "I don't like this song," or saying nothing.

6. If "like" is chosen, the original song, plus the new song are merged into a 2-song cluster[3] that is represented by a single vector, comprised of means of the variables in the original two song vectors.

7. If "dislike" is chosen, the vector of the song that is not liked is stored for future reference. (If the user does not express an opinion about the song, in our simplified example here the new song is not used for further comparisons.)

8. Pandora looks in its database for a new song, one whose statistical distance is close to the "like" song cluster[4], and not too close to the "dislike" song. Depending on the user's reaction, this new song might be added to the "like" cluster or "dislike" cluster.

Over time, Pandora develops the ability to deliver songs that match a particular taste of a particular user. A single user might build up multiple stations around different song clusters. Clearly, this is a less limiting approach than selecting music in terms of which "genre" it belongs to.

While the process described above is a bit more complex than the basic "classification of new data" process described in this chapter, the fundamental process—classifying a record according to its proximity to other records—is the same at its core. Note the role of domain knowledge in this machine learning process—the variables have been tested and selected by the project leaders, and the measurements have been made by human experts.

Further reading: See www.pandora.com, Wikipedia's article on the Music Genome Project, and Joyce John's article "Pandora and the Music Genome Project," in the September 2006 *Scientific Computing* [23 (10): 14, 40–41]

---

[2]See Chapter 12.5 for an explanation of statistical distance.

[3]See Chapter 14 for more on clusters.

[4]See Case 18.6 "Segmenting Consumers of Bath Soap" for an exercise involving the identification of clusters, which are then used for classification purposes.

7.3 ADVANTAGES AND SHORTCOMINGS OF *K*-NN ALGORITHMS

The main advantage of *k*-NN methods is their simplicity and lack of parametric assumptions. In the presence of a large enough training set, these methods perform surprisingly well, especially when each class is characterized by multiple combinations of predictor values. For instance, in real-estate databases there are likely to be multiple combinations of {home type, number of rooms, neighborhood, asking price, etc.} that characterize homes that sell quickly as opposed to those that remain for a long period on the market.

There are three difficulties with the practical exploitation of the power of the *k*-NN approach. First, although no time is required to estimate parameters from the training data, as we discussed earlier, the time to find the nearest neighbors in a large training set can be prohibitive. A number of ideas have been implemented to overcome this difficulty. The main ideas are as follows:

- Reduce the time taken to compute distances by working in a reduced dimension using dimension reduction techniques such as principal components analysis (Chapter 4).
- Use sophisticated data structures such as search trees to speed up identification of the nearest neighbor. This approach often settles for an "almost nearest" neighbor to improve speed. An example is using bucketing, where the records are grouped into buckets so that records within each bucket are close to each other. For a to-be-predicted record, buckets are ordered by their distance to the record. Starting from the nearest bucket, the distance to each of the records within the bucket is measured.

The algorithm stops when the distance to a bucket is larger than the distance to the closest record thus far.

Second, the number of records required in the training set to qualify as large increases exponentially with the number of predictors p. This is because the expected distance to the nearest neighbor goes up dramatically with p unless the size of the training set increases exponentially with p. This phenomenon is known as the *curse of dimensionality*, a fundamental issue pertinent to all classification, prediction, and clustering techniques. This is why we often seek to reduce the number of predictors through methods such as selecting subsets of the predictors for our model or by combining them using methods such as principal components analysis, singular value decomposition, and factor analysis (see Chapter 4).

Third, k-NN is a "lazy learner": the time-consuming computation is deferred to the time of prediction. For every record to be predicted, we compute its distances from the entire set of training records only at the time of prediction. This behavior prohibits using this algorithm for real-time prediction of a large number of records simultaneously.

PROBLEMS

7.1   **Calculating Distance with Categorical Predictors.** This exercise with a tiny dataset illustrates the calculation of Euclidean distance and the creation of binary dummies. (note that creating dummies is not required in JMP, but it is useful to understand how they are derived and used).

The online education company Statistics.com segments its customers and prospects into three main categories: IT professionals (IT), statisticians (Stat), and other (Other). It also tracks, for each customer, the number of years since first contact (years). Consider the following customers; information about whether they have taken a course or not (the outcome to be predicted) is included:

Customer 1: Stat, 1 year, did not take course
Customer 2: Other, 1.1 year, took course

a. Consider now the following new prospect:
Prospect 1: IT, 1 year
Using the information given above on the two customers and one prospect, create one dataset for all three with the categorical predictor variable transformed into 2 dummy variables, and a similar dataset with the categorical predictor variable transformed into 3 dummy variables.

b. For each derived dataset, calculate the Euclidean distance between the prospect and each of the other two customers.

c. Using k-NN with $k = 1$, classify the prospect as taking or not taking a course using each of the two derived datasets. Does it make a difference whether you use 2 or 3 dummies?

7.2   **Personal Loan Acceptance.** Universal Bank is a relatively young bank growing rapidly in terms of overall customer acquisition. The majority of these customers are liability customers (depositors) with varying sizes of relationship with the bank. The customer base of asset customers (borrowers) is quite small, and the bank is interested in expanding this base rapidly to bring in more loan business. In particular, it wants to explore ways of converting its liability customers to personal loan customers (while retaining them as depositors).

A campaign that the bank ran last year for liability customers showed a healthy conversion rate of over 9% success. This has encouraged the retail marketing department to devise smarter campaigns with better target marketing. The goal is to use k-NN to predict whether a new customer will accept a loan offer. This will serve as the basis for the design of a new campaign.

The file `UniversalBank.jmp` contains data on 5000 customers. The data include customer demographic information (age, income, etc.), the customer's relationship with the bank (mortgage, securities account, etc.), and the customer response to the last personal loan campaign (*Personal Loan*). Among these 5000 customers, only 480 (= 9.6%) accepted the personal loan that was offered to them in the earlier campaign.

Partition the data into training (60%) and validation (40%) sets.

a. Consider the following customer:

Age = 40, Experience = 10, Income = 84, Family = 2, CCAvg = 2, Education = 2, Mortgage = 0, Securities Account = 0, CD Account = 0, Online = 1, and Credit Card = 1. Perform a k-NN classification with all predictors except ID and ZIP code using k = 10. How would this customer be classified? (*Note:* This analysis may take a few minutes.)

b. What is a choice of k that balances between overfitting and ignoring the predictor information?

c. Show the classification matrix for the validation data that results from using the best k.

d. Consider the following customer: Age = 40, Experience = 10, Income = 84, Family = 2, CCAvg = 2, Education = 2, Mortgage = 0, Securities Account = 0, CD Account = 0, Online = 1 and Credit Card = 1. Classify the customer using the best k.

e. Repartition the data, this time into training, validation, and test sets (50% : 30% : 20%). Apply the k-NN method with the k chosen above. Compare the classification matrix of the test set with that of the training and validation sets. Comment on the differences and their reason.

7.3   **Predicting Housing Median Prices.** The file `BostonHousing.jmp` contains information on over 500 census tracts in Boston, where for each tract multiple variables are recorded. The last column (CAT.MEDV) was derived from MEDV, such that it obtains the value 1 if MEDV>30 and 0 otherwise. Consider the goal of predicting the median value (MEDV) of a tract, given the information in the first 12 columns.   Partition the data into training (60%) and validation (40%) sets.

a. Perform a k-NN prediction with all 12 predictors (ignore the CAT.MEDV column), trying values of k from 1 to 5. Make sure to normalize the data (click "normalize input data"). What is the best k chosen? What does it mean?

b. Predict the MEDV for a tract with the following information, using the best k:

CRIM	ZN	INDUS	CHAS	NOX	RM	AGE	DIS	RAD	TAX	PTRATIO	LSTAT
0.2	0	7	0	0.538	6	62	4.7	4	307	21	10

(Save the prediction formula to the data table, add a new row, and enter these values.)

c. Why is the validation data error overly optimistic compared to the error rate when applying this *k*-NN predictor to new data?

d. Perform another *k*-NN prediction with all 12 predictors, but this time normalize (standardize) the continuous predictors first (select all columns in the data table, right-click, and select *New Formula Column > Distributional > Standardize*). Again, try values of k from 1 to 5. What is the best k chosen? Describe any differences you observe. What is the predicted value for the tract above?

e. If the purpose is to predict MEDV for several thousands of new tracts, what would be the disadvantage of using *k*-NN prediction? List the operations that the algorithm goes through in order to produce each prediction.

8

THE NAIVE BAYES CLASSIFIER

In this chapter we introduce the naive Bayes classifier, which can be applied to data with categorical predictors. We review the concept of conditional probabilities, then present the complete, or exact, Bayesian classifier. We next see how it is impractical in most cases, and learn how to modify it and instead use the "naive Bayes" classifier, which is more generally applicable.

8.1 INTRODUCTION

The naive Bayes method (and, indeed, an entire branch of statistics) is named after the Reverend Thomas Bayes (1702–1761). To understand the naive Bayes classifier, we first look at the complete, or exact, Bayesian classifier. The basic principle is simple.

Naive Bayes Method

For each new record to be classified:

1. Find all of the training records with the same predictor profile (i.e., records having the same predictor values).
2. Determine what classes the records belong to and which class is most prevalent.
3. Assign that class to the new record.

Alternatively (or in addition), it may be desirable to adjust the method so that it answers the question: "What is the propensity of belonging to the class of interest?" instead of "Which class is the most probable?" Obtaining class probabilities allows using a sliding cutoff to classify a record as belonging to class i, even if i is not the most probable class for that record. This approach is useful when there is a specific class of interest that we want to identify and we are willing to "overidentify" records as belonging to this class.

Data Mining for Business Analytics: Concepts, Techniques, and Applications with JMP Pro®, First Edition.
Galit Shmueli, Peter C. Bruce, Mia L. Stephens, and Nitin R. Patel.

(See Chapter 5 for more details on the use of cutoffs for classification and on asymmetric misclassification costs.)

Cutoff Probability Method

For each new record to be classified:

1. Establish a cutoff probability for the class of interest (C_1) above which we consider that a record belongs to that class.
2. Find all the records with the same predictor profile as the new record (i.e., records having the same predictor values).
3. Determine the probability that these records belong to the class of interest.
4. If this probability is above the cutoff probability, assign the new record to the class of interest.

Conditional Probability

Both procedures incorporate the concept of *conditional probability*, or the probability of event A given that event B has occurred, which is denoted $P(A|B)$. We will be looking at the probability of the record belonging to class i given that its predictor values are $x_1, x_2, \ldots, x_p$. In general, for a response with m classes $C_1, C_2, \ldots, C_m$, and the predictor values $x_1, x_2, \ldots, x_p$, we want to compute

$$P(C_i|x_1, \ldots, x_p) \tag{8.1}$$

To classify a record, we compute its probability of belonging to each of the classes in this way, then either classify the record to the class that has the highest probability or use the cutoff probability to decide whether it should be assigned to the class of interest.

From this definition, we see that the Bayesian classifier works only with categorical predictors. If we use a set of numerical predictors, then it is highly unlikely that multiple records will have identical values on these numerical predictors. Therefore numerical predictors must be binned and converted to categorical predictors. *The Bayesian classifier is the only classification or prediction method presented in this book that is especially suited for (and limited to) categorical predictor variables.*

Example 1: Predicting Fraudulent Financial Reporting

An accounting firm has many large companies as customers. Each customer submits an annual financial report to the firm, which is then audited by the accounting firm. For simplicity, we will designate the outcome of the audit as "fraudulent" or "truthful," referring to the accounting firm's assessment of the customer's financial report. The accounting firm has a strong incentive to be accurate in identifying fraudulent reports—passing a fraudulent report as truthful could have legal consequences.

The accounting firm notes that in addition to all the financial records, it has information on whether the customer has had prior legal trouble (criminal or civil charges of any nature filed against it). This information has not been used in previous audits, but the accounting firm is wondering whether it could be used in the future to identify reports that merit more

TABLE 8.1 Tabular Summary of Financial Reporting Data

	Prior Legal ($X = 1$)	No Prior Legal ($X = 0$)	Total
Fraudulent (C_1)	50	50	100
Truthful (C_2)	180	720	900
Total	230	770	1000

intensive review. Specifically, it wants to know whether having had prior legal trouble is predictive of fraudulent reporting.

Each customer is a record, and the response of interest has two classes into which a company can be classified: C_1 = fraudulent and C_2 = truthful. The predictor variable—"prior legal trouble"—has two values: 0 (no prior legal trouble) and 1 (prior legal trouble).

The accounting firm has data on 1000 companies that it has investigated in the past. For each company it has information on whether the financial report was judged fraudulent or truthful and whether the company had prior legal trouble. The counts in the training set are shown in Table 8.1. The raw data are available in `Financial Reporting Raw.jmp`.

8.2 APPLYING THE FULL (EXACT) BAYESIAN CLASSIFIER

Now consider the financial report from a new company, which we wish to classify as fraudulent or truthful by using these data. To do this, we compute the probabilities, as above, of belonging to each of the two classes.

If the new company had had prior legal trouble, the probability of belonging to the fraudulent class would be P(fraudulent | prior legal) = 50/230 (there were 230 companies with prior legal trouble in the training set, and 50 of them had fraudulent financial reports). The probability of belonging to the other class, "truthful," is, of course, the remainder = 180/230. (These numbers can be verified using the raw data and the *Distribution* platform, with Outcome as *Y, Columns* and filtering on Prior Legal Trouble using the *Local Data Filter*.)

Using the "Assign to the Most Probable Class" Method

If the new company had prior legal trouble, we assign it to the "truthful" class. Similar calculations for the case of no prior legal trouble are left as an exercise to the reader. In this example, using the rule "assign to the most probable class," all records are assigned to the "truthful" class. This is the same result as the naive rule of "assign all records to the majority class."

Using the Cutoff Probability Method

In this example we are more interested in identifying the fraudulent reports—these are the ones that can land the auditor in jail. We recognize that in order to identify the fraudulent reports, some truthful reports will be misidentified as fraudulent, and the overall classification accuracy may decline. Our approach is therefore to establish a cutoff value for the probability of being fraudulent, and classify all records above that value as fraudulent. The Bayesian formula for the calculation of this probability that a record belongs to class C_i is as follows:

$$P(C_i | x_1, \ldots, x_p) = \frac{P(x_1, \ldots, x_p | C_i) P(C_i)}{P(x_1, \ldots, x_p | C_1) P(C_1) + \cdots + P(x_1, \ldots, x_p | C_m) P(C_m)}$$

(8.2)

In this example (where frauds are more rare), if the cutoff were established at 0.20, we would classify a new company with prior legal trouble as fraudulent because P(fraudulent | prior legal) = 50/230 = 0.22. The user can treat this cutoff as a "slider" to be adjusted to optimize performance, like other parameters in any classification model.

Practical Difficulty with the Complete (Exact) Bayes Procedure

The approach outlined above amounts to finding all the records in the sample that have the same predictor values as the new record. This was easy in the small example presented above, where there was just one predictor.

As the number of predictors grows (even to a modest number like 20), many of the records to be classified will be without exact matches. This can be understood in the context of a model to predict voting on the basis of demographic variables. For example, even a sizable sample may not contain even a single match for a new record that is a male Hispanic with medium income from the US midwest who voted in the last election, did not vote in the prior election, has three daughters and one son, and is divorced. And this is just eight variables, a small number for most data mining exercises. The addition of just a single new variable with five equally frequent categories reduces the probability of a match by a factor of 5.

Solution: Naive Bayes

In the naive Bayes solution, we no longer restrict the probability calculation to those records that exactly match the record to be classified. Instead, we use the entire dataset.

Returning to our original basic classification procedure outlined at the beginning of the chapter, we recall the procedure for classifying a new record:

1. Find all the training records with the same predictor profile (i.e., having the same predictor values).
2. Determine what classes the records belong to and which class is most prevalent.
3. Assign that class to the new record.

The naive Bayes modification (to the basic classification procedure) is as follows:

1. For class C_1, estimate the individual conditional probabilities for each predictor $P(x_j | C_1)$—these are the probabilities that the predictor value in the record to be classified occurs in class C_1. For example, for X_1 this probability is estimated by the proportion of x_1 values among the records belonging to class C_1 in the training set.

2. Multiply the proportion of records in class C_1 by the product of these conditional probabilities (see the numerator in the equation below).

3. Repeat steps 1 and 2 for each class.

4. For each class, estimate a probability for class i by taking the value calculated in step 2 for class i and dividing it by the sum of such values for all classes (see the denominator in the equation below).

5. Assign the record to the class with the highest estimated probability for this set of predictor values.

The preceding steps lead to the naive Bayes formula for calculating the probability that a record with a given set of predictor values $x_1, \ldots, x_p$ belongs to class C_1 among m classes. The formula can be written as follows:

$$P_{nb}(C_1 \mid x_1, \ldots x_p) =$$

$$\frac{P(C_1)[P(x_1 \mid C_1)P(x_2 \mid C_1) \cdots P(x_p \mid C_1)]}{P(C_1)[P(x_1 \mid C_1)P(x_2 \mid C_1) \cdots P(x_p \mid C_1)] + \cdots + P(C_m)[P(x_1 \mid C_m)P(x_2 \mid C_m) \cdots P(x_p \mid C_m)]}$$

$$(8.3)$$

This is a somewhat formidable formula; see Example 2 for a simpler numerical version. Note that all the necessary quantities can be obtained from tabular summaries of Y instead of from each of the categorical predictors (using the *Tabulate* platform).

The Naive Bayes Assumption of Conditional Independence   In probability terms, we have made the simplifying assumption that the exact *conditional probability* of seeing a record with predictor profile $x_1, x_2, \ldots, x_p$ within a certain class, $P(x_1, x_2, \ldots, x_p | C_i)$, is well approximated by the product of the individual conditional probabilities $P(x_1 | C_i) \times P(x_2 | C_i) \cdots \times P(x_p | C_i)$. The two quantities are identical when the predictors within each class are independent.

For example, suppose that "lost money last year?" is an additional variable in the accounting fraud example. The simplifying assumption we make with naive Bayes is that, within a given class, we no longer need to look for the records characterized both by "prior legal trouble" and "lost money last year." Rather, on the assumption that the two are independent, we can simply multiply the probability of past legal trouble by the probability of a prior year loss.

In practice, however, some correlation between predictors is expected, violating the independence assumption. Despite this, the procedure works quite well—primarily because what is usually needed is not a propensity for each record that is accurate in absolute terms but just a reasonably accurate *rank ordering* of propensities. Even when the independence assumption is violated, the rank ordering of the records' propensities is typically preserved.

Note that if all we are interested in is a rank ordering, and the denominator remains the same for all classes, it is sufficient to concentrate only on the numerator. The disadvantage of this approach is that the probability values (the propensities) obtained, while ordered correctly, are not on the same scale as the exact values that the user would anticipate.

Using the Cutoff Probability Method   The procedure above is for the basic case where we seek maximum classification accuracy for all classes. In the case of the *relatively rare class of special interest*, the procedure is as follows:

1. Establish a cutoff probability for the class of interest (C_1) above which we consider a record to belong to this class.
2. For the class of interest, compute the probability that each individual predictor value in the record to be classified occurs in the training data.
3. Multiply the proportion of records in class C_1 by the product of the probabilities obtained in step 2 above.
4. Estimate the probability for the class of interest by taking the value calculated in step 3 for the class of interest and dividing it by the sum of the similar values for all classes.
5. If this value falls above the cutoff, assign the new record to the class of interest, otherwise not.
6. Adjust the cutoff value as needed.

Example 2: Predicting Fraudulent Financial Reports, Two Predictors

Let us expand the financial reports example to two predictors and, using a small subset of data, compare the complete (exact) Bayes calculations to the naive Bayes calculations.

Consider the 10 customers of the accounting firm listed in Table 8.2 (data are available in `Financial Reporting 10 Companies.jmp`). For each customer, we have information on whether it had prior legal trouble, whether it is a small or large company, and whether the financial report was found to be fraudulent or truthful. Using this information, we will calculate the conditional probability of fraud, given each of the four possible combinations {*yes*, small}, {*yes*, large}, {*no*, small}, {*no*, large}.

TABLE 8.2   Information on 10 Companies

Prior Legal Trouble	Company Size	Status
Yes	Small	Truthful
No	Small	Truthful
No	Large	Truthful
No	Large	Truthful
No	Small	Truthful
No	Small	Truthful
Yes	Small	Fraudulent
Yes	Large	Fraudulent
No	Large	Fraudulent
Yes	Large	Fraudulent

Complete (Exact) Bayes Calculations   The probabilities are computed as

$$P(\text{fraudulent}|\text{PriorLegal} = y, \text{Size} = \text{small}) = 1/2 = 0.5$$

$$P(\text{fraudulent}|\text{PriorLegal} = y, \text{Size} = \text{large}) = 2/2 = 1$$

$$P(\text{fraudulent}|\text{PriorLegal} = n, \text{Size} = \text{small}) = 0/3 = 0$$

$$P(\text{fraudulent}|\text{PriorLegal} = n, \text{Size} = \text{large}) = 1/3 = 0.33$$

Naive Bayes Calculations   Now we can compute the naive Bayes probabilities. For the conditional probability of fraudulent behaviors given {Prior Legal = *yes*, Size = small}, the numerator is the product of the proportion of {Prior Legal= *yes*} instances among the fraudulent companies and the proportion of {Size = small} instances among the fraudulent companies, times the proportion of fraudulent companies: $(3/4)(1/4)(4/10) = 0.075$. To get the actual probabilities, we must also compute the numerator for the conditional probability of *truthful* behaviors given {Prior Legal = *yes*, Size = small: $(1/6)(4/6)(6/10) = 0.067$. The denominator is then the sum of these two conditional probabilities $(0.075 + 0.067 = 0.14)$. The conditional probability of fraudulent behaviors given {Prior Legal = *yes*, Size = small} is therefore $0.075/0.14 = 0.53$. In a similar way we compute all four conditional probabilities:

$$P_{nb}(\text{fraudulent}|\text{PriorLegal} = y, \text{Size} = \text{small}) = \frac{(3/4)(1/4)(4/10)}{(3/4)(1/4)(4/10) + (1/6)(4/6)(6/10)}$$
$$= 0.53$$
$$P_{nb}(\text{fraudulent}|\text{PriorLegal} = y, \text{Size} = \text{large}) = 0.87$$
$$P_{nb}(\text{fraudulent}|\text{PriorLegal} = n, \text{Size} = \text{small}) = 0.07$$
$$P_{nb}(\text{fraudulent}|\text{PriorLegal} = n, \text{Size} = \text{large}) = 0.31$$

Note how close these naive Bayes probabilities are to the exact Bayes probabilities. Although they are not equal, both would lead to exactly the same classification for a cutoff of 0.5 (and many other values). It is often the case that the rank ordering of probabilities is even closer to the exact Bayes method than are the probabilities themselves, and for classification purposes it is the rank orderings that matter.

To perform these calculations in JMP, we use the *Naive Bayes* add-in, which is available from the *JMP User Community*. The add-in provides a dialog for selection of the response and the features (see Figure 8.1). Note that in order to use the add-in, all variables must have the *Character* Data Type and the *Nominal* Modeling Type.

The output includes a data table, a cell plot, and an interactive dialog for computing conditional probabilities. The data table provides the calculations required to compute the probabilities for each combination of outcome class and the predictor values. These conditional probabilities are used to compute the most likely outcome (using a cutoff for classification of 0.5). The cell plot provides a graphical summary of the results and the most likely outcome.

The interactive dialog allows you to select desired classes for the features, and displays the resulting conditional probabilities (using the calculations in the data table). In

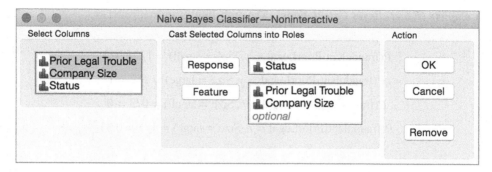

FIGURE 8.1   Dialog for selecting the outcome and features in the JMP Naive Bayes Add-in for the 10 companies example.

Figure 8.2 we see the conditional probabilities for fraudulent and truthful behaviors given {Prior Legal = *yes*, Size = small}.

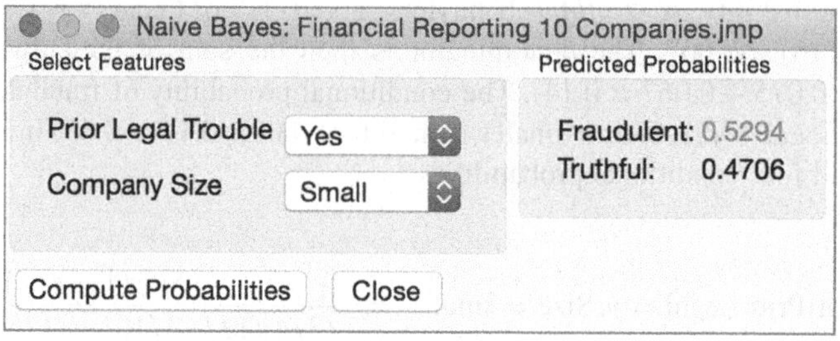

FIGURE 8.2   The JMP Naive Bayes add-in interactive dialog for displaying conditional probabilities.

USING THE JMP NAIVE BAYES ADD-IN

The *Naive Bayes* add-in is available on the JMP User Community (search for *Naive Bayes* at community.jmp.com). Note that the add-in does not provide an option for validation, and does not have built in tools for evaluating model performance (e.g., confusion matrices or lift curves). Variables must be coded as Nominal and cannot have the *Value Labels* column property. As of JMP 13, Naive Bayes will be available directly within the software.

We now consider a larger example, where flight data is used to predict delays (data are in the file `Flight Delays NB Train.jmp`).

Example 3: Predicting Delayed Flights

Predicting flight delays can be useful to a variety of organizations: airport authorities, airlines, and aviation authorities. At times, joint task forces have been formed to address the

TABLE 8.3   Description of Variables for Flight Delays Example

Day of Week	Coded as: 1 = Monday, 2 = Tuesday, ..., 7 = Sunday
Departure Time Block	Broken down into 16 intervals between 6:00 AM and 10:00 PM
Origin	Three airport codes: DCA (Reagan National), IAD (Dulles), BWI (Baltimore–Washington Int'l)
Destination	Three airport codes: JFK (Kennedy), LGA (LaGuardia), EWR (Newark)
Carrier	Eight airline codes: CO (Continental), DH (Atlantic Coast), DL (Delta), MQ (American Eagle), OH (Comair), RU (Continental Express), UA (United), and US (USAirways)
Weather	Coded as 1 if there was a weather-related delay

problem. If such an organization were tasked with providing ongoing, real-time assistance with flight delays, it would benefit from being able to predict which flights are likely to be delayed.

In this simplified illustration, we look at six predictors (see Table 8.3). The outcome of interest is whether the flight is delayed (*delayed* means arrived more than 15 minutes late). Our dataset consists of a sample of 1321 flights, taken from all flights from the Washington, DC, area into the New York City area during January 2004. For this sample, 19.4% of these flights were delayed. The data were obtained from the Bureau of Transportation Statistics (available on the web at). The goal is to accurately predict whether or not a new flight (not in this dataset), will be delayed.

Each record corresponds to a particular flight. The response is whether the flight was delayed, and thus it has two classes (*1 = Delayed* and *0 = On Time*). Information on the available predictors is listed in Table 8.3.

The table on the left in Figure 8.3 shows the percentages of delayed flights and on-time flights (referred to as the *Prior Class Probabilities*). The table on the right in Figure 8.3 shows conditional probabilities for each class, as a function of the predictor values. These conditional probabilities were computed using the *Tabulate* platform in JMP, where *Delayed* is used as the column variable, each of the predictors is used as a row variable, and *Column %* is the summary statistic.

Note that in this example all combinations of predictors and the response classes are represented in the dataset except for on-time flights (Class = 0) when the weather was bad (Weather = 1). That is, when the weather was bad, all flights in the dataset were delayed.

Using these conditional probabilities, we can classify a new flight by comparing the probability that a flight will be delayed with the probability that it will be on time. Recall that since both probabilities will have the same denominator, we can just compare the numerators. Each numerator is computed by forming the product of the conditional probabilities of the relevant predictor values, and multiplying this result by the proportion of this class in the dataset [in this case $\hat{P}(\text{delayed}) = 0.19$]. For example, to classify a Delta flight from DCA to LGA between 10 and 11 am on a Sunday with good weather, we compute the numerators:

Tabulate		
	Delayed	
	0	**1**
	Column %	**Column %**
	80.62%	19.38%

Tabulate		
	Delayed	
	0	**1**
Day of Week	**Column %**	**Column %**
1	12.86%	20.31%
2	13.99%	16.02%
3	15.21%	12.89%
4	15.96%	12.89%
5	18.12%	16.41%
6	13.15%	7.03%
7	10.70%	14.45%
Departure Time Block		
0600-0659	6.20%	3.52%
0700-0759	6.01%	5.08%
0800-0859	7.14%	5.47%
0900-0959	5.35%	2.34%
1000-1059	5.73%	1.95%
1100-1159	3.85%	1.95%
1200-1259	6.29%	5.47%
1300-1359	6.85%	5.08%
1400-1459	11.08%	15.23%
1500-1559	6.48%	8.20%
1600-1659	7.89%	7.42%
1700-1759	9.48%	15.63%
1800-1859	4.32%	3.13%
1900-1959	4.04%	8.98%
2000-2059	3.10%	1.95%
2100-2159	6.20%	8.59%
Origin		
BWI	6.85%	9.38%
DCA	63.57%	48.44%
IAD	29.58%	42.19%
Destination		
EWR	28.36%	38.67%
JFK	17.65%	18.75%
LGA	53.99%	42.58%
Carrier		
CO	3.85%	6.64%
DH	24.32%	33.98%
DL	20.00%	10.94%
MQ	11.27%	17.97%
OH	1.78%	1.17%
RU	17.09%	21.48%
UA	1.69%	0.78%
US	20.00%	7.03%
Weather		
0	100.00%	92.58%
1	0.00%	7.42%

FIGURE 8.3   Tabulation of Airport Data showing conditional probabilities for delayed and on-time flights for each class.

$$\hat{P}(\text{delayed} \mid \text{Carrier} = \text{DL}, \text{DayofWeek} = 7, \text{DepartureTime} = 1000 - 1059,$$
$$\text{Destination} = \text{LGA}, \text{Origin} = \text{DCA}, \text{Weather} = 0)$$
$$\propto (0.11)(0.14)(0.020)(0.43)(0.48)(0.93)(0.19) = 0.000011$$

$$\hat{P}(\text{ontime} \mid \text{Carrier} = \text{DL}, \text{DayofWeek} = 7, \text{DepartureTime} = 1000 - 1059,$$
$$\text{Destination} = \text{LGA}, \text{Origin} = \text{DCA}, \text{Weather} = 0)$$
$$\propto (0.2)(0.11)(0.06)(0.54)(0.64)(1)(0.81) = 0.00034$$

The symbol $\propto$ means "is proportional to," reflecting the fact that this calculation deals only with the numerator in the naive Bayes formula (8.3).

It is therefore more likely that the flight will be on time. Note that a record with such a combination of predictor values does not exist in the training set, and therefore we use naive Bayes rather than exact Bayes.

To compute the actual probabilities, we divide each of the numerators by their sum

$$\hat{P}(\text{delayed}) \mid \text{Carrier} = \text{DL}, \text{DayofWeek} = 7, \text{DepartureTime} = 1000 - 1059,$$
$$\text{Destination} = \text{LGA}, \text{Origin} = \text{DCA}, \text{Weather} = 0)$$
$$= \frac{0.000011}{0.000011 + 0.00034} = 0.03$$
$$\hat{P}(\text{ontime} \mid \text{Carrier} = \text{DL}, \text{DayofWeek} = 7, \text{DepartureTime} = 1000 - 1059,$$
$$\text{Destination} = \text{LGA}, \text{Origin} = \text{DCA}, \text{Weather} = 0)$$
$$= \frac{0.00034}{0.000011 + 0.00034} = 0.97$$

Of course, we rely on the software to compute these probabilities for any records of interest. In Figure 8.4 we use the JMP *Naive Bayes* add-in to compute these same predicted probabilities (use the *Compute Probabilies* button to updated the predicted probabilities after selecting features). The estimated probabilities and classifications for all flights are stored in the data table produced by the add-in. These can be further explored using *Tabulate*, the *Data Filter*, and other built-in tools.

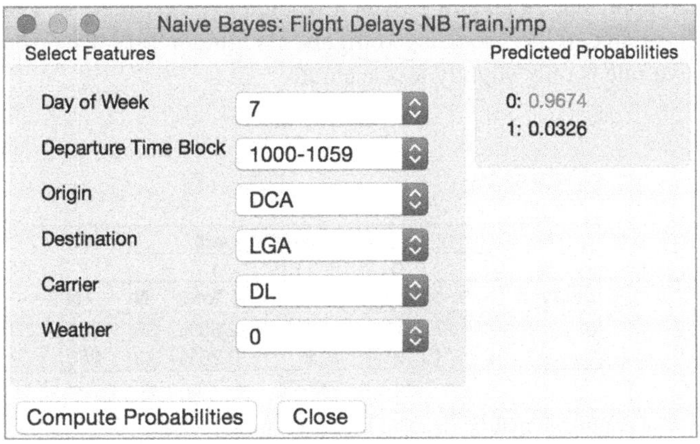

FIGURE 8.4 Predicted probabilities for delayed and ontime flights using the Naive Bayes add-in (top right).

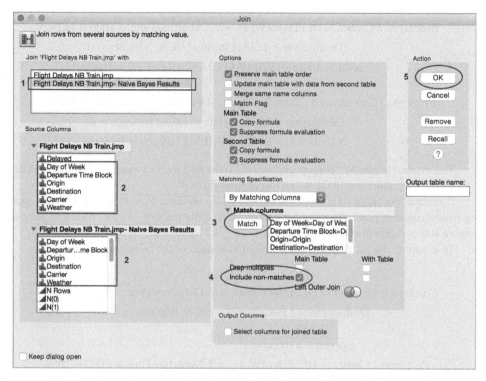

FIGURE 8.5 Using *Tables > Join* to add predicted probabilities and classifications to the original data table.

Finally, to evaluate the performance of the naive Bayes classifier for our data, we use a classification (or confusion) matrix. First, we need to add the predicted probabilities and classifications from the Naive Bayes Results table to the original data table. With the original data table (Flight Delays NB Train.jmp) active (on top of other output), we use *Tables > Join* to join the two tables, as shown in Figure 8.5.

Once the tables have been joined, we can use *Tables > Tabulate* to create a confusion matrix, as shown in Figure 8.6. We see that the overall error level is around 18.8% (213 + 35/1321). In comparison, a naive rule (which would have classified all 1321 flights as on time) would have missed the 256 delayed flights, resulting in a 19.4% error level. In other words, the naive rule is only slightly less accurate.

▼ ▼**Tabulate**

	Most Probable(Delayed)				
	0		1		
Delayed	% of Total	N	% of Total	N	N
0	77.97%	1030	2.65%	35	1065
1	16.12%	213	3.26%	43	256

FIGURE 8.6 The confusion matrix, delayed flights vs. most probably delayed flights (classified by the Naive Bayes add-in).

8.3   ADVANTAGES AND SHORTCOMINGS OF THE NAIVE BAYES CLASSIFIER

The naive Bayes classifier's beauty is in its simplicity, computational efficiency, reasonably good classification performance, and ability to handle categorical variables directly. In fact it often outperforms more sophisticated classifiers even when the underlying assumption of independent predictors is far from true. This advantage is especially pronounced when the number of predictors is very large.

There are three main issues that should be kept in mind, however. First, the naive Bayes classifier requires a very large number of records to obtain good results. Second, where a predictor category is not present in the training data, naive Bayes assumes that a new record with that category of the predictor has zero probability. This can be a problem if this rare (or unoberved) predictor value is important. For example, assume that the target variable is *bought high-value life insurance* and a predictor category is *owns yacht*. If the training data have no records with *owns yacht* = 1, for any new records where *owns yacht* = 1, naive Bayes will assign a probability of 0 to the target variable *bought high-value life insurance*. With no training records with *owns yacht* = 1, of course, no data mining technique will be able to incorporate this potentially important variable into the classification model—it will be ignored. With naive Bayes, however, the absence of this predictor actively "outvotes" any other information in the record to assign a 0 to the target value (when, in this case, it has a relatively good chance of being a 1). The presence of a large training set (and judicious binning of continuous variables, if required) helps mitigate this effect.

Finally, good performance is obtained when the goal is *classification* or *ranking* of records according to their probability of belonging to a certain class. However, when the goal is actually to *estimate* the probability of class membership (propensity), this method provides very biased results. For this reason the naive Bayes method is rarely used in credit scoring (Larsen, 2005).

One final point is that, as with all modeling algorithms, a holdout validation set should be used to evaluate the performance of the model. Since the JMP *Naive Bayes* add-in doesn't support model validation, our assessments of model performance may be overly optimistic.

SPAM FILTERING

Filtering spam in email has long been a widely familiar application of data mining. Spam filtering, which is based in large part on natural language vocabulary, is a natural fit for a naive Bayes classifier, which uses exclusively categorical variables. Most spam filters are based on this method, which works as follows:

1. Humans review a large number of emails, classify them as "spam" or "not spam," and from these select an equal (also large) number of spam emails and nonspam emails. This is the training data.

2. These emails will contain thousands of words; for each word compute the frequency with which the word occurs in the spam dataset, and the frequency with which it occurs in the nonspam dataset. Convert these frequencies into estimated probabilities (i.e., if the word "free" occurs in 500 out of 1000 spam emails, and only 100 out of 1000 nonspam emails, the probability that a spam email will contain the word "free" is 0.5, and the probability that a nonspam email will contain the word "free" is 0.1).

3. If the only word in a new message that needs to be classified as spam or not-spam is "free," we would classify the message as spam, since the Bayesian posterior probability is 0.5/(0.5+0.1) or 5/6 that, given the appearance of "free," the message is spam.

4. Of course, we will have many more words to consider. For each such word, the probabilities described in step 2 are calculated, and multiplied together, and formula 8.3 is applied to determine the naive Bayes probability of belonging to the classes. In the simple version, class membership (spam or not spam) is determined by the higher probability.

5. In a more flexible interpretation, the ratio between the "spam" and "not-spam" probabilities is treated as a score, for which the operator can establish (and change) a cutoff threshold—anything above that level is classified as spam.

6. Users have the option of building a personalized training database by classifying incoming messages as spam or not-spam, and adding them to the training database. One person's spam may be another person's substance.

It is clear that, even with the "naive" simplification, this is an enormous computational burden. Spam filters now typically operate at two levels—at servers (intercepting some spam that never makes it to your computer) and on individual computers (where you have the option of reviewing it). Spammers have also found ways to "poison" the vocabulary-based Bayesian approach, by including sequences of randomly selected irrelevant words. Since these words are randomly selected, they are unlikely to be systematically more prevalent in spam than in nonspam, and they dilute the effect of key spam terms such as "Viagra" and "free." For this reason, sophisticated spam classifiers also include variables based on elements other than vocabulary, such as the number of links in the message, the vocabulary in the subject line, determination of whether the "From:" email address is the real originator (anti-spoofing), use of HTML and images, and origination at a dynamic or static IP address (the latter are more expensive and cannot be set up quickly).

PROBLEMS

8.1   Personal Loan Acceptance. The file UniversalBank.jmp contains data on 5000 customers of Universal Bank. The data include customer demographic information (age, income, etc.), the customer's relationship with the bank (mortgage, securities account, etc.), and the customer response to the last personal loan campaign (*Personal Loan*). Among these 5000 customers, only 480 (= 9.6%) accepted the personal loan that was offered to them in the earlier campaign. In this exercise we focus on two predictors: *Online* (whether or not the customer is an active user of online banking services) and *Credit Card* (abbreviated *CC* below) (does the customer hold a credit card issued by the bank), and the outcome *Personal Loan* (abbreviated *Loan* below).

a. Check to make sure that the variables are coded as Nominal, and that none of the variables has the *Value Labels* column property (remove this column property if needed).

b. Create a summary of the data using *Tabulate*, with *Online* as a column variable, *CC* as a row variable, and *Loan* as a secondary row variable. The values inside the cells should convey the count (how many records are in that cell).

c. Consider the task of classifying a customer that owns a bank credit card and is actively using online banking services. Looking at the tabulation, what is the probability that this customer will accept the loan offer? (This is the probability of loan acceptance (*Loan* = 1) conditional on having a bank credit card (*CC* = 1) and being an active user of online banking services (*Online* = 1)).

d. Create two tabular summaries of the data. One will have *Loan* (rows) as a function of *Online* (columns) and the other will have *Loan* (rows) as a function of *CC*.

e. Compute the following quantities [$P(A|B)$ means "the probability of A given B"]:

  i. $P(CC = 1|Loan = 1)$ (the proportion of credit card holders among the loan acceptors)

  ii. $P(Online = 1|Loan = 1)$

  iii. $P(Loan = 1)$ (the proportion of loan acceptors)

  iv. $P(CC = 1|Loan = 0)$

  v. $P(Online = 1|Loan = 0)$

  vi. $P(Loan = 0)$

f. Use the quantities computed above to compute the naive Bayes probability $P(Loan = 1|CC = 1, Online = 1)$.

g. Compare this value with the one obtained from the tabulation in (b). Which is a more accurate estimate of $P(Loan = 1|CC = 1, Online = 1)$?

h. Which of the entries in this table are needed for computing $P(Loan = 1|CC = 1, Online = 1)$? In JMP, use the Naive Bayes Add-in to compute the probability that $P(Loan = 1|CC = 1, Online = 1)$. Compare this to the number you obtained in (e).

8.2 **Automobile Accidents.** The file `Accidents.jmp` contains information on 42,183 actual automobile accidents in 2001 in the United States that involved one of three levels of injury: NO INJURY, INJURY, or FATALITY. For each accident, additional information is recorded, such as day of week, weather conditions, and road type. A firm might be interested in developing a system for quickly classifying the severity of an accident based on initial reports and associated data in the system (some of which rely on GPS-assisted reporting).

Our goal here is to predict whether an accident just reported will involve an injury (MAX_SEV_IR = 1 or 2) or will not (MAX_SEV_IR = 0). For this purpose, use *Recode* to create a new variable called INJURY that takes the value "yes" if MAX_SEV_IR = 1 or 2, and otherwise "no." (See Chapter 3 for information on using *Recode*.)

a. Using the information in this dataset, if an accident has just been reported and no further information is available, what should the prediction be? (INJURY = Yes or No?) Why?

b. Create a subset of the first 12 records in the dataset and look only at the response (INJURY) and the two predictors WEATHER_R and TRAF_CON_R.

    i. Create a tabular summary that examines INJURY as a function of the two predictors for these 12 records.

    ii. Compute the exact Bayes conditional probabilities of an injury (INJURY = Yes) given the six possible combinations of the predictors. First, check to make sure that the variables are coded as Nominal, and that none of the variables has the *Value Labels* column property (remove this column property if needed).

    iii. Classify the 12 accidents using these probabilities and a cutoff of 0.5.

    iv. Compute manually the naive Bayes conditional probability of an injury given WEATHER_R = 1 and TRAF_CON_R = 1.

    v. Run a naive Bayes classifier on the 12 records and two predictors using the JMP Naive Bayes add-in. Look at the data table for probabilities and classifications for all 12 records (these display at the far right on the data table). Compare this to the exact Bayes classification. Are the resulting classifications equivalent? Is the ranking (= ordering) of observations equivalent?

c. Now return to the entire dataset.

    i. What proportion of accidents led to injuries?

    ii. Assuming that no information or initial reports about the accident itself are available at the time of prediction (only location characteristics, weather conditions, etc.), which predictors can we include in the analysis? (Hold your mouse over the column names in *Columns Panel* in the data table for descriptions of the predictors.)

    iii. Prepare the data. First, check to make sure that the relevant variables are coded as Nominal, and that none of the variables has the *Value Labels* column property. For continuous predictors (i.e., SPD_LIM), consider using *Recode* to bin the data into buckets (then apply the Nominal modeling type).

    iv. Run a naive Bayes classifier on the data set with the relevant predictors (and INJURY as the response). This may take a minute or two if you select several predictors.

    v. Using the interactive dialog, explore how the probabilities change for different conditions. Which set of conditions results in the highest probability of an injury?

    vi. What is the overall error for the validation set? (Recall that you'll need to join the Naive Bayes Results table to the `Accidents.jmp` table first, and then create a confusion matrix using *Tabulate*.)

9

CLASSIFICATION AND REGRESSION TREES

This chapter describes a flexible data-driven method that can be used for both classification (called "classification tree") and prediction (called "regression tree"). Among the data-driven methods, trees are the most transparent and easy to interpret. Trees are based on separating observations into subgroups by creating splits on predictors. These splits create logical rules that are transparent and easily understandable, such as "IF Age <55 AND Education >12, THEN the probability that class = 1 is 0.85." We discuss the two key ideas underlying trees: recursive partitioning (for constructing the tree) and pruning (for cutting the tree back).We explain that pruning is a useful strategy for avoiding overfitting, and describe alternative strategies for avoiding overfitting. As with other data-driven methods, trees require large amounts of data. However, once constructed, they are computationally cheap to deploy even on large samples. They also have other advantages such as being highly automated, robust to outliers, and able to handle missing values. In addition to prediction and classification, we describe how trees can be used for dimension reduction. Finally, we introduce "random forests" and "boosted trees," which combine results from multiple trees to improve predictive power.

9.1 INTRODUCTION

If one had to choose a classification technique that performs well across a wide range of situations without requiring much effort from the analyst while being readily understandable by the consumer of the analysis, a strong contender would be a decision tree. Many tree-based methods are available for both classification and prediction. A procedure developed by Breiman et al. (1984) is called CART (classification and regression trees). A related procedure is called C4.5.

What is a classification tree? Figure 9.1 shows a tree for classifying bank customers who receive a loan offer. Customers either accept (1) or don't accept (0) the offer. The tree

Data Mining for Business Analytics: Concepts, Techniques, and Applications with JMP Pro®, First Edition.
Galit Shmueli, Peter C. Bruce, Mia L. Stephens, and Nitin R. Patel.
© 2017 John Wiley & Sons, Inc. Published 2017 by John Wiley & Sons, Inc.

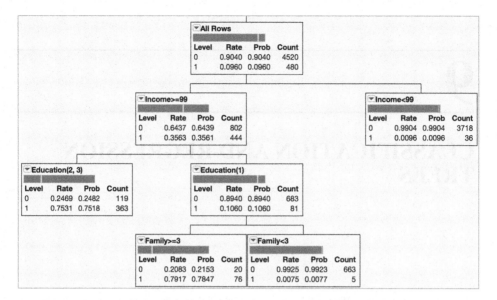

FIGURE 9.1   Classification tree after three splits.

shows the probability of accepting or not accepting, as a function of information such as income, education level, and average credit card expenditure.

One of the reasons that classification trees are very popular is that they provide easily understandable classification rules (at least if the trees are not too large). Consider the tree in the example. This tree can easily be translated into a set of rules for classifying a bank customer. For example, the bottom node under the "Family < 3" in this tree gives us the following rule:

IF($Income \geq 99$) AND ($Education = 1$) AND ($Family < 3$),
    THEN $Prob(Class = 1) = 0.0077$ (classify as don't accept).

In the following sections, we show how trees are constructed and evaluated.

9.2   CLASSIFICATION TREES

Two key ideas underlie classification trees. The first is the idea of recursive partitioning of the space of the independent variables. The second is the idea to avoid overfitting.

Recursive Partitioning

Let us denote the outcome (response) variable by y and the input (predictor) variables by $x_1, x_2, x_3, \ldots, x_p$. In classification, the outcome variable will be a categorical variable. Recursive partitioning divides up the p-dimensional space of the X variables into nonoverlapping rectangles. The X variables here are continuous, nominal, or ordinal. This division is accomplished recursively (i.e., operating on the results of prior divisions).

First, one of the variables is selected, say X_i, and a value of X_i, say s_i, is chosen to split the p-dimensional space into two parts: one part that contains all the points with $X_i < s_i$

TABLE 9.1   Lot Size, Income, and Ownership of a Riding Mower for 24 Households

Household Number	Income ($000s)	Lot Size (000s ft^2)	Ownership of Riding Mower
1	60.0	18.4	Owner
2	85.5	16.8	Owner
3	64.8	21.6	Owner
4	61.5	20.8	Owner
5	87.0	23.6	Owner
6	110.1	19.2	Owner
7	108.0	17.6	Owner
8	82.8	22.4	Owner
9	69.0	20.0	Owner
10	93.0	20.8	Owner
11	51.0	22.0	Owner
12	81.0	20.0	Owner
13	75.0	19.6	Nonowner
14	52.8	20.8	Nonowner
15	64.8	17.2	Nonowner
16	43.2	20.4	Nonowner
17	84.0	17.6	Nonowner
18	49.2	17.6	Nonowner
19	59.4	16.0	Nonowner
20	66.0	18.4	Nonowner
21	47.4	16.4	Nonowner
22	33.0	18.8	Nonowner
23	51.0	14.0	Nonowner
24	63.0	14.8	Nonowner

and the other with all the points with $X_i \geq s_i$. Then one of these two parts is divided in a similar manner by choosing a variable again (it could be X_i or another variable) and a split value for the variable. This results in three rectangular regions. This process is continued so that we get smaller and smaller rectangular regions. Here our predictor is continuous, but the same general approach is used for nominal and ordinal predictors.

We illustrate recursive partitioning with an example.

Example 1: Riding Mowers

We again use the riding mower example presented in Chapter 3. A riding mower manufacturer would like to find a way of classifying families in a city into those likely to purchase a riding mower and those not likely to buy one. A pilot random sample of 12 owners and 12 nonowners in the city is collected. The data are shown in Table 9.1 and are plotted in Figure 9.2.

If we apply the classification tree procedure to these data, the procedure will choose *Income* for the first split with a splitting value of 85.5. The Income, Lot_Size space is now divided into two rectangles, one with Income < 85.5 and the other with Income ≥ 85.5. This is illustrated in Figure 9.3.

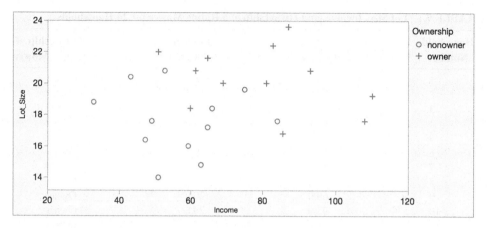

FIGURE 9.2   Scatterplot of *Lot Size* Vs. *Income* for 24 owners and nonowners of riding mowers.

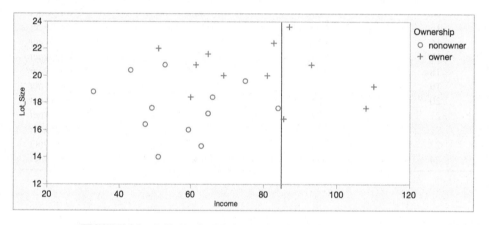

FIGURE 9.3   Splitting the 24 observations by Income value of 85.5.

Notice how the split has created two rectangles, each of which is much more homogeneous than the rectangle before the split. The rectangle on the right contains points that are all owners (five owners) and the rectangle on the left contains mostly nonowners (twelve nonowners and seven owners).

How was this particular split selected? The algorithm examined each predictor variable (in this case, Income and Lot Size) and all possible split values for each variable to find the best split. What are the possible split values for a variable? They are simply each value for the variable. There are 22 unique values for Income and 18 unique values for Lot Size. All of these split points, or *candidates*, are tested to see which split separates the probability of ownership the most. The split that drives the greatest dissimilarity in the probabilities (or proportions) for the two rectangles is selected as the split point.

Categorical Predictors

The riding mower example used numerical predictors. However, categorical predictors can also be used in the recursive partitioning context. To handle categorical predictors, the split choices for a categorical predictor are all ways in which the set of categories can be divided into two subsets. For example, a categorical variable with four categories, say $\{a, b, c, d\}$, can be split in seven ways into two subsets: $\{a\}$ and $\{b, c, d\}$, $\{b\}$ and $\{a, c, d\}$, $\{c\}$ and $\{a, b, d\}$, $\{d\}$ and $\{a, b, c\}$, $\{a, b\}$ and $\{c, d\}$, $\{a, c\}$ and $\{b, d\}$, and $\{a, d\}$ and $\{b, c\}$. When the number of categories is large, the number of splits becomes very large.[1] For ordinal predictors, JMP provides an option to restrict the ordering of the splits. So, for an ordinal variable with four categories, $\{1, 2, 3, 4\}$, the only possible splits are $\{1\}$ and $\{2, 3, 4\}$, $\{1, 2\}$ and $\{3, 4\}$, and $\{1, 2, 3\}$ and $\{4\}$.

9.3   GROWING A TREE

The initial recursive partition output for the mower data is shown in Figure 9.4. The graphic displays the overall proportion of nonowners and owners in the dataset, where the y-axis value for the horizontal line is the proportion nonowners (0.50). For each predictor, JMP tests all possible splits and measures the difference in the proportion of owners and nonowners between each pair of split groups. For each predictor, the optimal split, or *Cut Point*, is listed under *Candidates*.

The G^2 statistic, reported under Candidates, provides a measure of the differences or dissimilarity in the proportions at the cut point for each variable. The *LogWorth*, which is based on the p-value of the chi-squared statistic, is used to determine the optimal split location.[2] The bigger the LogWorth value, the better the split (Sall, 2002). In this example the predictor with the highest LogWorth is Income, and the cut point for Income is 85.5.

Growing a Tree Example

The tree after one split is shown in Figure 9.5. The graph at the top is updated to show the two split groups (Income ≥ 85.5 and Income < 85.5). The horizontal line is the proportion of nonowners in each split group. As we saw in Figure 9.3, when Income ≥ 85.5, all of the observations (5) are owners, and when Income < 85.5 12 out 17 are nonowners. The branches in the tree are called *nodes*, and these nodes are candidates for the next split. The last node in a branch is the *terminal* node.

The red triangle option *Display Options > Show Split Prob* displays both *Rate*, which is the proportion of observations in a node, and *Prob*, the predicted probabilities for the node. These numbers may differ (as they do in this example; see Figure 9.5).[3]

If Income ≥ 85.5 all the records are owners (no further splitting in that branch is possible), so the next candidate split is under Income < 85.5 (see the bottom of Figure 9.5).

---

[1] This is a difference between recursive partitioning in JMP (and also CART) and C4.5; JMP performs only binary splits, leading to binary trees, whereas C4.5 performs splits that are as large as the number of categories, leading to "bushlike" structures.

[2] G^2 is the likelihood-ratio chi-squared test statistic. Because several different splits are tested, the p-value is adjusted. The LogWorth is the -log 10(adjusted p-value).

[3] The predicted probabilites are calculated using a weighting factor to ensure that probabilities are always nonzero. This allows logs of probabilities to be calculated on validation sets (this is needed to compute the validation RSquare).

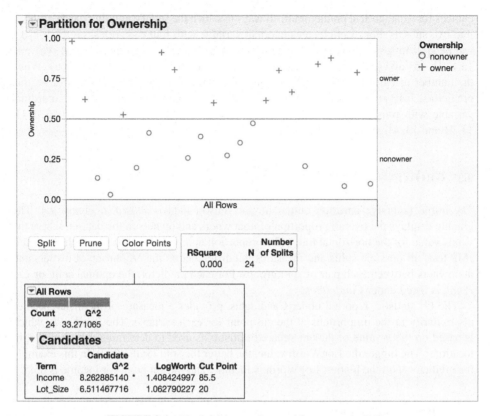

FIGURE 9.4   The initial recursive partition output in JMP.

The split variable is Lot Size, and the cut point is 20. We change the minimum split size to 1 observation (this is a top red triangle option—the default is 5), and continue to split until no further splits are permitted. Figure 9.6 shows the final tree. Each of the terminal nodes (also called leaves) contain either all owners or all nonowners, so no additional splits are possible.

To help interpret the tree, we use the *Leaf Report* (see Figure 9.7). This provides a summary of our tree, and displays the rules (the split *Prob* values) for classifying the outcome. The terminal node on the right in the tree corresponds to the last row on the leaf report: if Income < 85.5 and Lot Size < 17.6, then the probability (Ownership = owner) is 0.0752.

Classifying a New Observation

We use the estimated probabilities in the terminal nodes of the tree (or the leaf report) to classify new observations. See Chapter 5 for further discussion of the use of a cutoff value in classification, for cases where a single class is of interest.

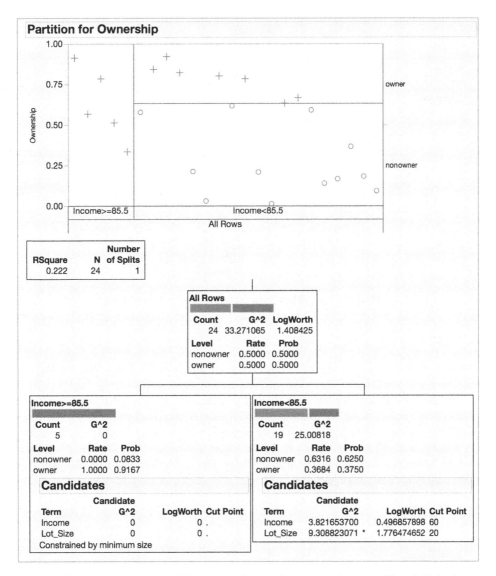

FIGURE 9.5 JMP recursive partition output after 1 split. *Display Options > Show Split Prob* was selected to show *Rates* and *Probs*.

In building a classification tree, we've built a model. We can save the prediction formula for this model to the data table (see Figure 9.8). This creates columns with the estimated probabilities for each class, along with the most likely class (the classification). When a new observation is added to the data table, the estimated probability for each class is calculated and the most likely class is provided (using a cutoff of 0.5). Note that after the prediction formula is saved, the *Prediction Profiler* in the JMP Pro *Model Comparison* platform can

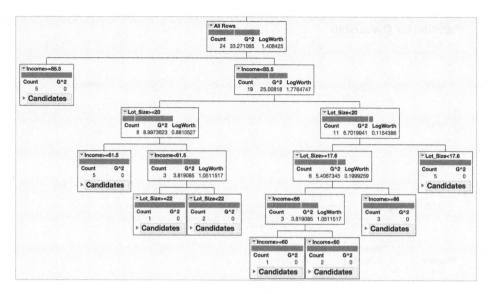

FIGURE 9.6   JMP recursive partition output for full tree.

be used to explore predicted probabilities for given values of the predictors. (In JMP 13, the Profiler is a red triangle option in the Partition output.)

In a binary classification situation (typically, with a *success* class that is relatively rare and of particular interest), we can also establish a lower cutoff to better capture those rare successes (at the cost of lumping in more nonsuccesses as successes). This can be done by creating a new formula column in the data table with a cutoff for the classification, or via the *Alternate Cutoff Confusion Matrix Add-in* (this can be installed from the JMP User Community at community.jmp.com).

▼ **Leaf Report**			
Response Prob			
Leaf Label	nonowner	.2 .4 .6 .8	owner
Income>=85.5	0.0833		0.9167
Income<85.5&Lot_Size>=20&Income>=61.5	0.0815		0.9185
Income<85.5&Lot_Size>=20&Income<61.5&Lot_Size>=22	0.2512		0.7488
Income<85.5&Lot_Size>=20&Income<61.5&Lot_Size<22	0.8342		0.1658
Income<85.5&Lot_Size<20&Lot_Size>=17.6&Income<66&Income>=60	0.2901		0.7099
Income<85.5&Lot_Size<20&Lot_Size>=17.6&Income<66&Income<60	0.8601		0.1399
Income<85.5&Lot_Size<20&Lot_Size>=17.6&Income>=66	0.8933		0.1067
Income<85.5&Lot_Size<20&Lot_Size<17.6	0.9248		0.0752

FIGURE 9.7   JMP Leaf Report for full tree.

	Income	Lot_Size	Ownership	Prob(Ownership= =nonowner)	Prob(Ownership= =owner)	Most Likely Ownership
1	60	18.4	owner	0.290122619	0.709877381	owner
2	85.5	16.8	owner	0.0833333333	0.9166666667	owner
3	64.8	21.6	owner	0.0815277778	0.9184722222	owner
4	61.5	20.8	owner	0.0815277778	0.9184722222	owner
5	87	23.6	owner	0.0833333333	0.9166666667	owner
6	110.1	19.2	owner	0.0833333333	0.9166666667	owner
7	108	17.6	owner	0.0833333333	0.9166666667	owner
8	82.8	22.4	owner	0.0815277778	0.9184722222	owner
9	69	20	owner	0.0815277778	0.9184722222	owner
10	93	20.8	owner	0.0833333333	0.9166666667	owner
11	51	22	owner	0.2512395833	0.7487604167	owner
12	81	20	owner	0.0815277778	0.9184722222	owner
13	75	19.6	nonowner	0.8933095238	0.1066904762	nonowner
14	52.8	20.8	nonowner	0.8341597222	0.1658402778	nonowner
15	64.8	17.2	nonowner	0.9248090278	0.0751909722	nonowner
16	43.2	20.4	nonowner	0.8341597222	0.1658402778	nonowner
17	84	17.6	nonowner	0.8933095238	0.1066904762	nonowner
18	49.2	17.6	nonowner	0.860081746	0.139918254	nonowner
19	59.4	16	nonowner	0.9248090278	0.0751909722	nonowner
20	66	18.4	nonowner	0.8933095238	0.1066904762	nonowner
21	47.4	16.4	nonowner	0.9248090278	0.0751909722	nonowner
22	33	18.8	nonowner	0.860081746	0.139918254	nonowner
23	51	14	nonowner	0.9248090278	0.0751909722	nonowner
24	63	14.8	nonowner	0.9248090278	0.0751909722	nonowner

FIGURE 9.8   Riding Mower Data Table with saved prediction formula and classifications from the full tree model.

FITTING CLASSIFICATION TREES IN JMP PRO

Trees are fit using *Analyze > Modeling > Partition*. Use the *Split* button to grow the tree, and the *Prune* button to prune the tree back. A few options of note:

1. To show split statistics within each node, select *Display Options > Show Split Prob* or *Show Split Count* from the top red triangle.
2. To show a high-level summary of the tree, select *Show Leaf Report* or *Small Tree View* from the top red triangle.
3. The minimum split size can be changed using a red triangle option.
4. To save the formula for the tree to the data table, select *Save Columns > Save Prediction Formula*.

Note: In JMP 13, the *Analyze* menu structure has changed.

Growing a Tree with CART

In the Introduction we discussed the CART modeling procedure. CART-type tree methods first build a full tree and then prune it down. When growing the tree, they choose the predictor for splitting based on "impurity measures." Each predictor split is ranked according to how much it reduces impurity (heterogeneity) in the resulting nodes (rectangles in our mowers example). A pure node is one that is composed of a single class (e.g., owners). We discuss CART and measures of impurity later in the chapter.

9.4 EVALUATING THE PERFORMANCE OF A CLASSIFICATION TREE

To assess the accuracy of a tree in classifying new cases, we use the tools and criteria that were discussed in Chapter 5. We start by partitioning the data into training and validation sets. The training set is used to grow the tree, and the validation set is used to determine when to stop splitting. Since the validation data are used in the construction of the model, a third set of test data is used for assessing the performance of the final tree.

Each observation in the validation (and test) data is classified according to the decision tree. These predicted classes can then be compared to the actual memberships via a classification matrix. When a particular class is of interest, a lift curve is useful for assessing the model's ability to capture those members. We use the following example to illustrate.

Example 2: Acceptance of Personal Loan

Universal Bank is a relatively young bank that is growing rapidly in terms of overall customer acquisition. The majority of these customers are liability customers with varying sizes of relationship with the bank. The customer base of asset customers is quite small, and the bank is interested in growing this base rapidly to bring in more loan business. In particular, it wants to explore ways of converting its liability (deposit) customers to personal loan customers.

A campaign the bank ran for liability customers showed a healthy conversion rate of over 9% successes. This has encouraged the retail marketing department to devise smarter campaigns with better target marketing. The goal of our analysis is to model the previous campaign's customer behavior in order to analyze what combination of factors make a customer more likely to accept a personal loan. This will serve as the basis for the design of a new campaign.

The bank's dataset includes data on 5000 customers . The data include customer demographic information (age, income, etc.), customer response to the last personal loan campaign (Personal Loan), and the customer's relationship with the bank (mortgage, securities account, etc.). Among these 5000 customers, only 480 ($= 9.6\%$) accepted the personal loan that was offered to them in the earlier campaign. Figure 9.9 contains a sample of the bank's customer database for 20 customers (the data are in `UniversalBankCT.jmp`).

First, we create a validation column (using *Cols > Modeling Utilities > Make Validation Column*), dividing the data into three random subsets: training (2500 observations), validation (1500 observations), and test (1000 observations). This validation column has been saved to the data table. We set the minimum split size to 1, and then construct a full-grown tree (clicking *Split* repeatedly until the tree stops splitting). A small tree view of the tree after 10 splits is shown in Figure 9.10.

	ID	Personal Loan	Age	Experience	Income	ZIP Code	Family	CCAvg	Education	Mortgage	Securities Account	CD Account	Online	CreditCard
1	1	No	25	1	49	91107	4	1.6	1	0	1	0	0	0
2	2	No	45	19	34	90089	3	1.5	1	0	1	0	0	0
3	3	No	39	15	11	94720	1	1	1	0	0	0	0	0
4	4	No	35	9	100	94112	1	2.7	2	0	0	0	0	0
5	5	No	35	8	45	91330	4	1	2	0	0	0	0	1
6	6	No	37	13	29	92121	4	0.4	2	155	0	0	1	0
7	7	No	53	27	72	91711	2	1.5	2	0	0	0	1	0
8	8	No	50	24	22	93943	1	0.3	3	0	0	0	0	1
9	9	No	35	10	81	90089	3	0.6	2	104	0	0	1	0
10	10	Yes	34	9	180	93023	1	8.9	3	0	0	0	0	0
11	11	No	65	39	105	94710	4	2.4	3	0	0	0	0	0
12	12	No	29	5	45	90277	3	0.1	2	0	0	0	1	0
13	13	No	48	23	114	93106	2	3.8	3	0	1	0	0	0
14	14	No	59	32	40	94920	4	2.5	2	0	0	0	1	0
15	15	No	67	41	112	91741	1	2	1	0	1	0	0	0
16	16	No	60	30	22	95054	1	1.5	3	0	0	0	1	1
17	17	Yes	38	14	130	95010	4	4.7	3	134	0	0	0	0
18	18	No	42	18	81	94305	4	2.4	1	0	0	0	0	0
19	19	Yes	46	21	193	91604	2	8.1	3	0	0	0	0	0
20	20	No	55	28	21	94720	1	0.5	2	0	1	0	0	1

FIGURE 9.9   Sample of 20 customers of Universal Bank.

A look at the top tree node reveals that the first predictor that is chosen to split the data is Income. The split value, shown in the next two nodes (the second row in the tree), is 99 ($000s).

Since the full-grown tree leads to completely pure terminal leaves, it is 100% accurate in classifying the training data. This can be seen in Figure 9.11. In contrast, the confusion matrix for the validation and test data (which were not used to construct the full-grown tree) show lower classification accuracy. The main reason is that the full-grown tree overfits the training data (to complete accuracy!). This motivates the next section, where we describe ways to avoid overfitting by either stopping the growth of the tree before it is fully grown or by pruning the full-grown tree.

9.5   AVOIDING OVERFITTING

As the riding mower example illustrated, using a full-grown tree (based on the training data) leads to overfitting of the data. As discussed in Chapter 5, overfitting will lead to poor performance on new data. If we look at the overall error at the various levels of the tree, it is expected to decrease as the number of levels grows until the point of overfitting. Of course, for the training data the overall error decreases more and more until it is zero at the maximum level of the tree. However, for new data, the overall error is expected to decrease until the point where the tree models the relationship between class and the predictors. After that, the tree starts to model the noise in the training set, and we expect the overall error for the validation set to start increasing. This is depicted in Figure 9.12. One intuitive reason for the overfitting at the high levels of the tree is that these splits are based on very small numbers of observations. In such cases, class difference is likely to be attributed to noise rather than predictor information.

Two ways to avoid exceeding this level, thereby limiting overfitting, are by setting rules to stop tree growth, or alternatively, by pruning the full-grown tree back to a level where it does not overfit. These solutions are discussed below.

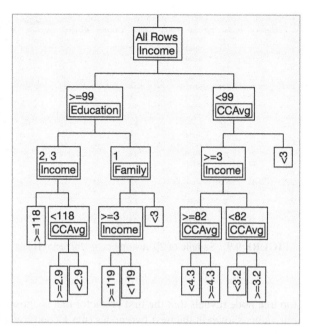

FIGURE 9.10   Small tree view of first 10 splits for the Universal Bank data using the training set (2500 observations).

Stopping Tree Growth: CHAID

One can think of different criteria for stopping the tree growth before it starts overfitting the data. Examples are tree depth (i.e., number of splits), minimum number of records in a terminal node, and minimum reduction in impurity. The problem is that it is not simple to determine what is a good stopping point using such rules.

A number of methods have been developed based on the idea of using rules to prevent the tree from growing excessively and overfitting the training data. One popular method called *CHAID* (chi-squared automatic interaction detection) is a recursive partitioning method that predates classification and regression tree (CART) procedures and is widely used in database marketing applications. It uses a well-known statistical test (the chi-square test for independence) to assess whether splitting a node improves the purity by a statistically significant amount. In particular, at each node we split on the predictor that has the strongest association with the response variable. The strength of association is measured by the *p*-value of a chi-squared test of independence. If, for the best predictor, the test does not show a significant improvement, the split is not carried out and the tree is terminated. This method is more suitable for categorical predictors, but it can be adapted to continuous predictors by binning the continuous values into categorical bins.

Growing a Full Tree and Pruning It Back

An alternative solution to stopping tree growth is pruning the full-grown tree. This is the basis of methods such as CART (developed by Breiman et al., implemented in multiple

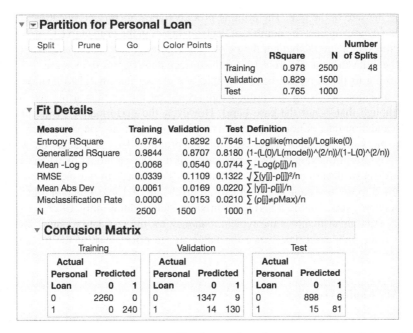

FIGURE 9.11   Universal Bank data: classification matrix and error rates for the training, validation, and test data using the full tree.

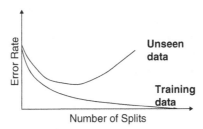

FIGURE 9.12   Error rate as a function of the number of splits for training vs. validation data: overfitting.

data mining software packages; e.g., SAS Enterprise Miner, CART, MARS, and XLMiner) and C4.5 (developed by Quinlan and implemented in, e.g., IBM SPSS Modeler). In C4.5 the training data are used both for growing and pruning the tree. In CART the innovation is to use the validation data to prune back the tree that is grown from training data. CART and CART-like procedures use validation data to prune back the tree that has deliberately been overgrown using the training data.

The idea behind pruning is to recognize that a very large tree is likely to be overfitting the training data, and that the weakest branches, which hardly reduce the error rate, should be removed. In the riding mower example, the last few splits resulted in nodes with very few points. We can see intuitively that these last splits are likely simply to be capturing noise in the training set rather than reflecting patterns that would occur in future data, such as the validation data. Pruning consists of successively lopping off the branches, thereby reducing the size of the tree. The pruning process trades off misclassification error in the validation dataset against the number of decision nodes in the pruned tree to arrive at a tree that captures the patterns, but not the noise, in the training data. We would like to find the point where the curve in Figure 9.12, for the unseen data, begins to increase.

Returning to the universal bank example, we expect that the classification accuracy of the validation set using the pruned tree would be higher than using the full-grown tree (although that is not the case here). However, the performance of the pruned tree on the validation data is not fully reflective of the performance on completely new data because the validation data were used for the pruning. This is a situation where it is particularly useful to evaluate the performance of the chosen model, whatever it may be, on a third set of data, the test set, which has not been used. In our example the pruned tree applied to the test data yields an overall error rate of 1.9% (compared to 1.27% for the validation data). This is a highly accurate model, yielding only 19 errors in the test set. This example also shows the typical tendency for the test data to show higher error rates than the validation data, since the latter are, in effect, part of the model building process.

How JMP Limits Tree Size

To prevent overfitting, JMP uses a combination of limiting the initial growth of the tree through setting the minimum split size and pruning a tree after it has grown.

When a validation column is used, JMP Pro provides automated tree building and pruning via the *Go* button. The tree is built, one split at a time, and the RSquare for the training and validation sets is tracked. If the tree reaches a point where the validation RSquare stops increasing, the software continues building the tree for 10 additional splits. If no further increase in validation RSquare is attained after these additional splits, the tree is pruned back to the point of the highest validation RSquare value.

The *Split History* report in JMP tracks the path of the RSquare measures for the training, validation and test sets. The report for the Universal Banking example is shown in Figure 9.13. While the training RSquare continues to climb after 8 splits, the validation RSquare (the middle line) actually starts to decrease. At 18 splits we continue to see no improvement in the validation RSquare, so the tree is pruned back to 8 splits. The fit statistics for the final 8-split tree are shown at the bottom of Figure 9.13. (*Split History* and *Show Fit Details* are top red triangle options.)

While the misclassification rate for the training set is higher than what we observed for the full tree, the rate for the validation set is improved (0.0193 for the full model (see Figure 9.11) versus 0.0127 for the pruned model).

9.6 CLASSIFICATION RULES FROM TREES

As described in Section 9.1, classification trees provide easily understandable *classification rules* (if the trees are not too large). Each terminal node is equivalent to a classification rule. Returning to the example above, our pruned tree has 8 splits. The leaf report provides rules for classification corresponding to the terminal nodes in our tree. The bottom row in

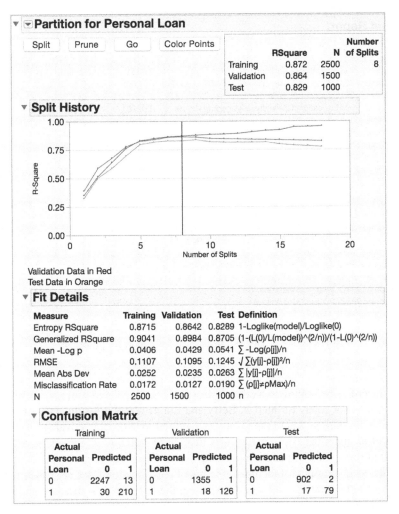

FIGURE 9.13 Fit details and split history for the training (top line on right side), validation (middle line) and test (lower line) sets for the universal bank data.

the leaf report (in Figure 9.14), which represents the right-most terminal node, gives us the rule

IF (*Income* < 99) AND (*CCAvg* < 3), THEN the probability that *Class* = 1 is 0.

However, in many cases the number of rules can be reduced by removing redundancies. For example, the rule at the top of the leaf report, corresponding to the left-most terminal node, is

IF (*Income* ≥ 99) AND (*Education* is 2 or 3) AND (*Income* ≥ 118), THEN the probability that *Class* = 1 is 0.9949.

This rule can be simplified to

IF (*Income* ≥ 118) AND (*Education* is 2 or 3), THEN the probability that *Class* = 1 is 0.9949.

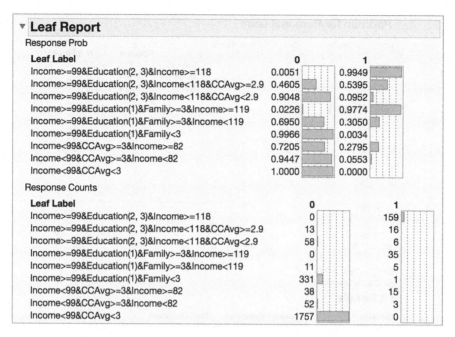

FIGURE 9.14   Universal bank data: leaf report for pruned tree.

This transparency in the process and understandability of the algorithm that leads to classifying a record as belonging to a certain class is advantageous in settings where the final classification is not solely of interest. Berry and Linoff (2000) give the example of health insurance underwriting, where the insurer is required to show that coverage denial is not based on discrimination. By showing rules that led to denial (e.g., income < $20K AND low credit history), the company can avoid law suits. Compared to the output of other classifiers, such as discriminant functions, tree-based classification rules are easily explained to managers and operating staff. As we will see, the decision tree structure and resulting model are certainly far more transparent and are easier to interpret (and explain!) than neural networks!

9.7   CLASSIFICATION TREES FOR MORE THAN TWO CLASSES

Classification trees can be used with an outcome that has more than two classes. In terms of determining optimal split or cut points, the split statistics were defined for m classes and hence can be used for any number of classes. The tree itself will have the same structure, except that its nodes include statistics for each of the m-class labels, and classifications are into one of the m-classes based on the highest estimated probability.

9.8   REGRESSION TREES

The tree method can also be used for numerical response variables. Regression trees for prediction operate in much the same fashion as classification trees. The output variable,

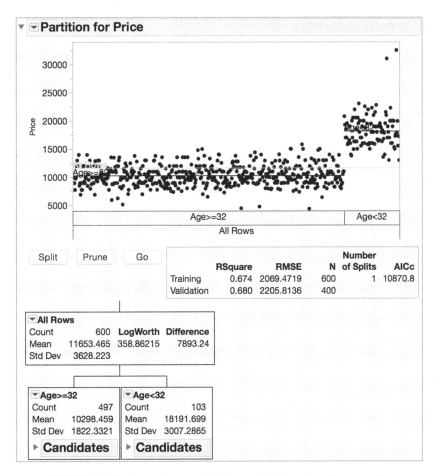

FIGURE 9.15   Regression tree for Toyota Corolla prices after one split.

Y, is a numerical variable in this case, but both the principle and the procedure are the same. Many splits are attempted, and for each split we compute a test statistic and use the LogWorth to determine the best cut point. For classification trees, the test statistic is G^2, the chi-square statistic. In regression trees, the test statistic is the *Sum of Squares*, which measures the difference in the means of the groups at the best cut point.

To illustrate a regression tree, consider the example of predicting prices of Toyota Corolla automobiles (from Chapter 6). The dataset includes information on 1000 sold Toyota Corolla cars (we use the dataset ToyotoCorolla1000.jmp). The goal is to find a predictive model of price as a function of 10 predictors (including mileage, horsepower, number of doors, etc.). A regression tree will be built using a training set of 600. The first split, shown in Figure 9.15, is on *Age*, and the cut point is 32. The difference in the means at this cut point, $7893.24, is reported in the first node. The horizontal line in the graph represents the overall mean price for each of the split groups, and the position of each point relative to the y-axis is the price for the particular observation.

The small tree and split history for the final tree are shown in Figure 9.16. The tree was pruned back, using the validation set, to 14 splits. We see that from the 10 input variables

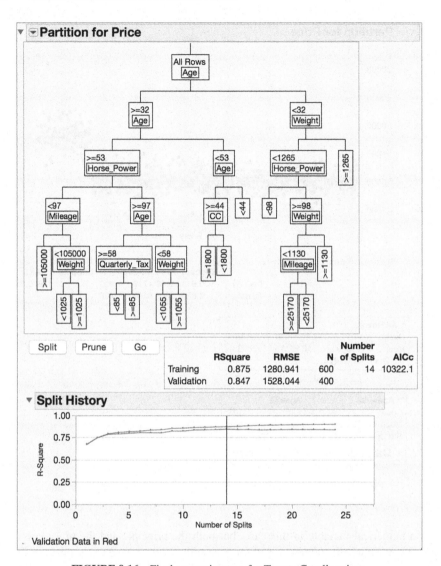

FIGURE 9.16   Final regression tree for Toyota Corolla prices.

only six predictors are useful for predicting price. This information is displayed in the *Column Contribution* report in Figure 9.17 (this is a top red triangle option). For each variable this report displays the number of splits and the total sum of squares attributed to each of the variables.

Prediction

Predicting the value of the response Y for an observation is performed in a fashion similar to the classification case: each terminal node gives a predicted price. The *Leaf Report* (bottom, in Figure 9.17) provides a summary of the regression tree, with rules for predicting the response.

▼ **Column Contributions**

Term	Number of Splits	SS		Portion
Age	4	6112962404		0.8858
Weight	4	420375738		0.0609
Horse_Power	2	191817699		0.0278
CC	1	88028798.4		0.0128
Mileage	2	58181952.8		0.0084
Quarterly_Tax	1	29384656.6		0.0043
Fuel_Type	0	0		0.0000
Metalic_Color	0	0		0.0000
Automatic	0	0		0.0000
Doors	0	0		0.0000

▼ **Leaf Report**

Leaf Label	Mean	Count
Age>=32&Age>=53&Horse_Power<97&Mileage>=105000	7975.80645	31
Age>=32&Age>=53&Horse_Power<97&Mileage<105000&Weight<1025	8686.22222	45
Age>=32&Age>=53&Horse_Power<97&Mileage<105000&Weight>=1025	9712.17949	39
Age>=32&Age>=53&Horse_Power>=97&Age>=58&Quarterly_Tax<85	8909.88889	54
Age>=32&Age>=53&Horse_Power>=97&Age>=58&Quarterly_Tax>=85	9831.97917	96
Age>=32&Age>=53&Horse_Power>=97&Age<58&Weight<1055	10136.3462	26
Age>=32&Age>=53&Horse_Power>=97&Age<58&Weight>=1055	11243.5417	24
Age>=32&Age<53&Age>=44&CC>=1800	8222.27273	11
Age>=32&Age<53&Age>=44&CC<1800	11271.3971	68
Age>=32&Age<53&Age<44	12486.7476	103
Age<32&Weight<1265&Horse_Power<98	16476.7857	42
Age<32&Weight<1265&Horse_Power>=98&Weight<1130&Mileage>=25170	16878.2143	14
Age<32&Weight<1265&Horse_Power>=98&Weight<1130&Mileage<25170	18754.6875	16
Age<32&Weight<1265&Horse_Power>=98&Weight>=1130	19794.2308	26
Age<32&Weight>=1265	26140	5

FIGURE 9.17   Column contributions and leaf report for final regression tree for Toyota Corolla prices.

For instance, to predict the price of a Toyota Corolla with $Age = 30$ and $Weight = 1300$, we use the last row in the leaf report (or the right-most terminal node in the tree) and see the predicted mean price is \$26,140. In regression trees the terminal node is determined by the average output value of the training observations in that terminal node. In the example above the value \$26,140 is the average of the 5 cars in the training set that fall in the category of Age < 32 and Weight ≥ = 1265.

> Classification trees produce a formula for making classifications, while regression trees produce formulas for making predictions. These formulas can be saved to the data table. In this example the saved formula can be used to predict prices for new observations. The JMP *Profiler* (accessed from either the *Graph* menu or from the *Model Comparison* platform) can be used to explore this formula and can also be used to make predictions. In JMP 13, the Profiler is an option in the Partition analysis window.

Evaluating Performance

As stated above, predictions are obtained by averaging the values of the responses in the nodes. We therefore have the usual definition of predictions and errors. The predictive performance of regression trees can be measured in the same way that other predictive methods are evaluated, using summary measures such as RSquare and RMSE. Additional statistics are available in the *Model Comparison* platform.

9.9   ADVANTAGES AND WEAKNESSES OF A TREE

Tree methods are good off-the-shelf classifiers and predictors. They are also useful for variable selection, with the most important predictors usually showing up at the top of the tree (and at the top of the *Column Contributions* report). Trees require relatively little effort from users in the following senses: First, there is no need for transformation of variables (any monotone transformation of the variables will give the same trees). Second, variable subset selection is automatic because it is part of the split selection. In the loan example, note that the best pruned tree has automatically selected just four variables (Income, CCAvg, Family, and Education) out of the set 12 variables available.

Unlike models that assume a particular relationship between the response and predictors (e.g., a linear relationship such as in linear regression and linear discriminant analysis), classification and regression trees are nonlinear and nonparametric. This allows for a wide range of relationships between the predictors and the response. However, this can also be a weakness: since the splits are done on single predictors rather than on combinations of predictors, the tree is likely to miss relationships between predictors, in particular linear structures like those in linear or logistic regression models. Classification trees are useful classifiers in cases where horizontal and vertical splitting of the predictor space adequately divides the classes. But consider, for instance, a dataset with two predictors and two classes, where separation between the two classes is most obviously achieved by using a diagonal line (as shown in Figure 9.18). A classification tree is therefore expected to have lower performance than methods such as discriminant analysis. One way to improve performance is to create new predictors that are derived from existing predictors (e.g., computing ratios combining values into buckets), which can capture hypothesized relationships between predictors (similar to interactions in regression models).

Another performance issue with classification trees is that they require a large dataset in order to construct a good classifier. From a computational aspect, trees can be relatively expensive to grow because of the multiple sorting involved in computing all possible splits on every variable. Pruning the data using the validation set adds further computation time. However, with modern computing and technological advancements, computation time is generally not an issue.

Although trees are useful for variable selection, one challenge is that they "favor" predictors with many potential split points. This includes categorical predictors with many categories and numerical predictors with many different values. Such predictors have a higher chance of appearing in a tree due to the sheer number of splits that are tested. Simplistic solutions include combining multiple categories into a smaller set (using *Recode*) and binning numerical predictors with many values (using a formula column). Alternatively, there are special algorithms that avoid this problem by using a different splitting criterion (e.g., QUEST classification trees; see Loh and Shih, 1997, and www.math.ccu.edu.tw/ yshih/quest.html). JMP addresses this issue by using the LogWorth for selection of split variables and cut points. The LogWorth, which is based on an adjusted p-value, takes into consideration the number of possible ways a split can occur. This adjustment minimizes the tendency to split on variables with many levels or values.

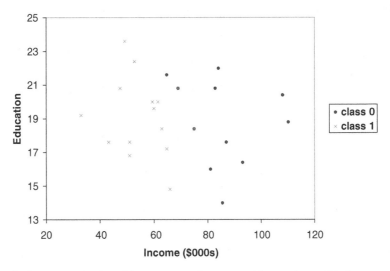

FIGURE 9.18   Scatterplot describing a two-predictor case with two classes. The best separation is achieved with a diagonal line, which classification trees cannot do.

An appealing feature of trees is that they handle missing data without having to impute values or delete observations with missing values. However, for cases where many values for a variable are missing or values are not missing randomly, this can be problematic. The *Informative Missing* option in the Partition dialog window in JMP provides missing value coding and imputation to help address this issue.

Finally, a very important practical advantage of trees is the transparent rules that they generate. Such transparency makes the results easy to communicate, and is often useful in managerial applications.

9.10   IMPROVING PREDICTION: MULTIPLE TREES

To address the shortcomings of a single tree—especially poor predictive power—researchers have developed several extensions to trees that combine results from multiple trees. These are examples of *ensembles* (see Chapter 13). Two popular multi-tree approaches are *random forests* and *boosted trees*.

Breiman and Cutler introduced *random forests*,[4] a method for improving predictive power by combining multiple classifiers or prediction algorithms.

Here is the basic idea in random forests:

1. Draw multiple random samples, with replacement, from the data (this sampling approach is called the bootstrap).
2. Fit a classification (or regression) tree to each sample (and thus obtain a forest).

---

[4]For further details on random forests see www.stat.berkeley.edu/users/breiman/RandomForests/cc_home.htm. Random forests are a special case of *bagging*.

3. Combine the predictions/classifications from the individual trees to obtain improved predictions. Use voting for classification and averaging for prediction.

The approach to random forests used in JMP Pro, called *Bootstrap Forest*, bootstraps from the data and also randomly samples predictors in the following manner:

1. Draw multiple random samples, with replacement, from the data (this sampling approach is called *bootstrapping*).
2. Randomly sample predictors.
3. Fit a classification (or regression) tree to each sample (and thus obtain a forest).
4. Average all of the trees in the forest to create the bootstrap aggregated (*bagged*) model.

Random forests cannot easily be displayed in a treelike diagram, thereby losing the interpretability that a single tree provides.[5] However, the *Column Contribution* report can be displayed for random forests, providing a measure the relative contribution of the different predictors. The contribution for a particular predictor is the sum of squares (for a continuous response) or G^2 for a categorical response. The number of splits for each predictor is also provided.

The second type of multi-tree improvement is *boosted trees*. Here a sequence of trees is fit, so that each tree concentrates on errors from the previous tree. The general steps are as follows:

1. Fit a single tree that is relatively simple, with a random selection of predictors.
2. Calculate the scaled residuals from this tree (based on misclassifications or errors).
3. Fit a new (again simple) tree to the scaled residuals from the first tree, again with a random selection of predictors.
4. Repeat steps 2 and 3 many times.
5. Sum the models for the individual trees to produce the boosted model.

Figure 9.19 shows the result of running a boosted tree on the Universal Bank example that we saw earlier (using the default settings). Compared to the performance of the single best-pruned tree (Figure 9.13), the boosted tree has slightly better performance on the test set in terms of lower overall misclassification rate.[6]

In general, the boosted tree will better performance on the validation and test sets in terms of lower overall error rate, and especially in terms of correct classification of 1's—the rare class of special interest. Where does boosting's special talent for finding 1's come from? When one class is dominant (0's constitute over 90% of the data here), basic classifiers are tempted to classify cases as belonging to the dominant class, and the 1's in this case constitute most of the misclassifications with the single best-pruned tree. The boosting algorithm concentrates on the errors (most of our misclassifications are 1's), so it is naturally going to do well in reducing the misclassification of 1's.

---

[5]The structure of bootstrap forests can be illustrated by creating a forest with very few trees (e.g., 3) with very few splits (e.g., a maximum of 5). To view these trees, select *Show Trees* from the top red triangle or save the probability formula to the data table.

[6]JMP 13 also offers stochastic gradient boosting, which randomly selects observations at each iteration.

Boosted Tree for Personal Loan

Specifications

Target Column:	Personal Loan	Number of training rows:	2500
Validation Column:	Validation	Number of validation rows:	1500
Number of Layers:	50	Number of test rows:	1000
Splits Per Tree:	3		
Learning Rate:	0.1		
Overfit Penalty:	0.0001		

Overall Statistics

Measure	Training	Validation	Test	Definition
Entropy RSquare	0.8663	0.8647	0.8322	1-Loglike(model)/Loglike(0)
Generalized RSquare	0.9000	0.8987	0.8731	$(1-(L(0)/L(model))^{(2/n)})/(1-L(0)^{(2/n)})$
Mean -Log p	0.0423	0.0428	0.0531	$\sum -Log(\rho[j])/n$
RMSE	0.1019	0.1038	0.1148	$\sqrt{\sum(y[j]-\rho[j])^2/n}$
Mean Abs Dev	0.0307	0.0292	0.0328	$\sum |y[j]-\rho[j]|/n$
Misclassification Rate	0.0128	0.0140	0.0150	$\sum(\rho[j]\neq\rho Max)/n$
N	2500	1500	1000	n

Confusion Matrix

Training				Validation				Test		
Actual				**Actual**				**Actual**		
Personal	**Predicted**			**Personal**	**Predicted**			**Personal**	**Predicted**	
Loan	**No**	**Yes**		**Loan**	**No**	**Yes**		**Loan**	**No**	**Yes**
No	2254	6		No	1354	2		No	902	2
Yes	26	214		Yes	19	125		Yes	13	83

FIGURE 9.19 Universal bank data: confusion matrix and error rates for the training, validation, and test data based on boosted tree.

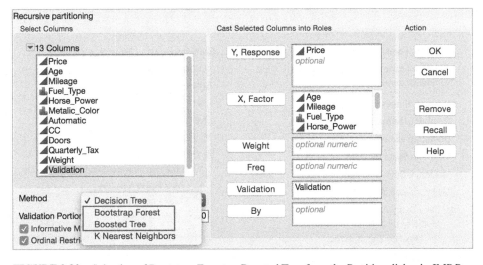

FIGURE 9.20 Selection of Bootstrap Forest or Boosted Tree from the Partition dialog in JMP Pro.

FITTING ENSEMBLE TREE MODELS IN JMP PRO

The default tree method in the *Partition* platform is *Decision Tree*. In JMP Pro, we choose *Bootstrap Forest* or *Boosted Tree* under *Method* in the Partition dialog, as shown in Figure 9.20. A variety of options for controlling tree growth and the size of the trees can be selected for each ensemble method (see Figure 9.21). *Bootstrap Forests* involve random selection of observations and predictors. So the random seed must be set before running the algorithm in order to produce repeatable results. Additional options, such as stochastic gradient boosting and the ability to set the random seed, are available in JMP Pro 13.

9.11   CART AND MEASURES OF IMPURITY

Recall that in the CART procedure, splits are formed using measures of impurity or heterogeneity. This is an intuitive approach to tree building, and we briefly revisit the concept of impurity here.

We define *pure* as containing records that belong to just one class. (Of course, this is not always possible, as there may be records that belong to different classes but have exactly the same values for every one of the predictor variables.) There are a number of ways to measure impurity. The two most popular measures are the *Gini index* and an *entropy measure*. We describe both next. Note that these measures are not available in JMP (since JMP uses a different approach for building trees).

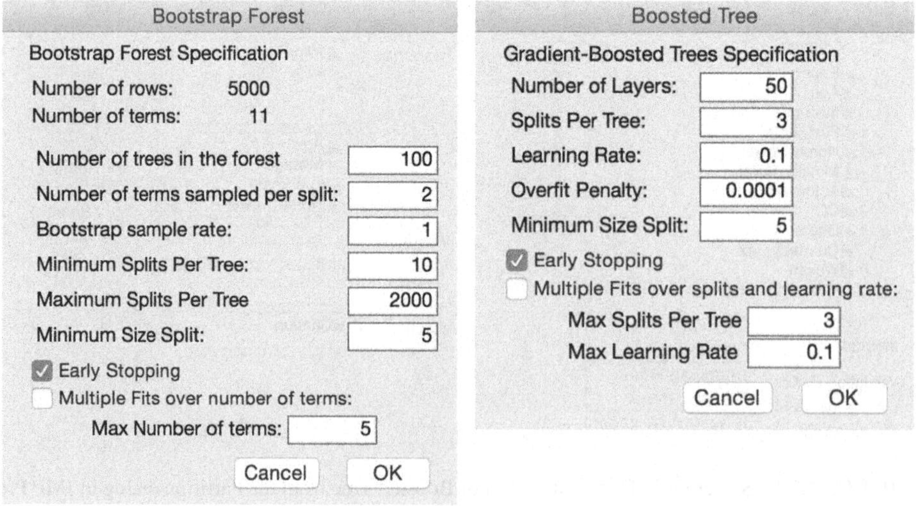

FIGURE 9.21   Options for Bootstrap Forest and Boosted Tree in JMP Pro.

Denote the m classes of the response variable by $k = 1, 2, \ldots, m$. The Gini impurity index for a rectangle A is defined by

$$I(A) = 1 - \sum_{k=1}^{m} p_k^2$$

where p_k is the proportion of observations in rectangle A that belong to class k. This measure takes values between 0 (if all the observations belong to the same class) and $(m - 1)/m$ (when all m classes are equally represented).

A second impurity measure is the entropy measure. The entropy for a rectangle A is defined by

$$\text{entropy}(A) = -\sum_{k=1}^{m} p_k \log_2(p_k)$$

This measure ranges between 0 (most pure, all observations belong to the same class) and $\log_2(m)$ (when all m classes are represented equally). In the two-class case, the entropy measure is maximized (like the Gini index) at $p_k = 0.5$.

In regression trees a typical impurity measure is the sum of the squared deviations from the mean of the terminal node. This is equivalent to the squared errors, because the mean of the terminal node is exactly the prediction. The lowest impurity possible is zero, when all values in the node are equal.

Rather than using measures of impurity to form splits in classification trees, JMP uses the G^2 likelihood-ratio statistic and the LogWorth (see the discussion in Section 9.3). For regression trees, the sum of squares (SS) values are reported in each node.

PROBLEMS

9.1 Competitive Auctions on eBay.com. The file `eBayAuctions.jmp` contains information on 1972 auctions that transacted on eBay.com during May–June 2004. The goal is to use these data to build a model that will classify auctions as competitive or noncompetitive. A *competitive auction* is defined as an auction with at least two bids placed on the item auctioned. The data include variables that describe the item (auction category), the seller (his/her eBay rating), and the auction terms that the seller selected (auction duration, opening price, currency, day-of-week of auction close). In addition, we have the price at which the auction closed. The task is to predict whether or not the auction will be competitive.

Data preprocessing. Split the data into training and validation datasets using a $60\% : 40\%$ ratio.

a. Fit a classification tree using all predictors, using the Go button. Display the leaf report, and write down the first four branches in the leaf report in terms of rules.

b. Is this model practical for predicting the outcome of a new auction?

c. Describe the interesting and uninteresting information that these rules provide.

d. Fit another classification tree, this time only with predictors that can be used for predicting the outcome of a new auction. Describe the resulting tree in terms of rules.

e. Examine the lift curve and the classification matrix for the tree. What can you say about the predictive performance of this model?

f. Plot the resulting tree on a scatterplot: Use the two axes for the two best (quantitative) predictors. Each auction will appear as a point, with coordinates corresponding to its values on those two predictors. Use different colors or symbols to separate competitive and noncompetitive auctions. Draw lines (you can use the line tool in JMP or an axis reference line) at the values that create splits. Does this splitting seem reasonable with respect to the meaning of the two predictors? Does it seem to do a good job of separating the two classes?

g. Based on this last tree (d), what can you conclude from these data about the chances of an auction obtaining at least two bids and its relationship to the auction settings set by the seller (duration, opening price, ending day, currency)? What would you recommend for a seller as the strategy that will most likely lead to a competitive auction?

9.2 **Predicting Delayed Flights.** The file `FlightDelays.jmp` contains information on all commercial flights departing the Washington, DC area and arriving at New York during January 2004. For each flight there is information on the departure and arrival airports, the distance of the route, the scheduled time and date of the flight, and so on. The variable that we are trying to predict is whether or not a flight is delayed. A delay is defined as an arrival that is at least 15 minutes later than scheduled.

Data preprocessing. Bin the scheduled departure time (CRS_DEP_TIME) into 8 bins. This will avoid treating the departure time as a continuous predictor, because it is reasonable that delays are related to rush-hour times. (Note that these data are not stored in JMP with a time format, so you'll need to explore the best way to bin this data - two options are (1) via the formula editor and (2) using the *Make Binning Formula* column utility.) Partition the data into training and validation sets.

a. Fit a classification tree to the flight delay variable using all the relevant predictors (use the binned version of the departure time) and the validation column. Do not include DEP_TIME (actual departure time) in the model because it is unknown at the time of prediction (unless we are doing our predicting of delays after the plane takes off, which is unlikely).

   i. How many splits are in the final model?

   ii. How many variables are involved in the splits?

   iii. Which variables contribute the most to the model?

   iv. Which variables were not involved in any of the splits?

   v. Express the resulting tree as a set of rules.

   vi. If you needed to fly between DCA and EWR on a Monday at 7 AM, would you be able to use this tree to predict whether the flight will be delayed? What other information would you need? Is this information available in practice? What information is redundant?

b. Fit another tree, this time using the original scheduled departure time rather than the binned version. Save the formula for this model to the data table (we'll return to this in a future exercise).

   i. Compare this tree to the original, in terms of the number of splits and the number of variables involved. What are the key differences?

ii. Does it make more sense to use the original scheduled departure time variable or the binned version? Why?

9.3 **Predicting Prices of Used Cars (Regression Trees).** The file `Toyota-Corolla.jmp` contains the data on used cars (Toyota Corolla) on sale during late summer of 2004 in The Netherlands. It has 1436 records containing details on 38 attributes, including *Price, Age, Kilometers, HP,* and other specifications. The goal is to predict the price of a used Toyota Corolla based on its specifications. (The example in Section 9.8 is a subset of this dataset).

Data preprocessing. Split the data into training (50%), validation (30%), and test (20%) datasets.

a. Run a regression tree with the output variable Price and input variables Age_08_04, KM, Fuel_Type, HP, Automatic, Doors, Quarterly_Tax, Mfg_Guarantee, Guarantee_Period, Airco, Automatic_Airco, CD_Player, Powered_Windows, Sport_Model, and Tow_Bar. Set the minimum split size to 1, and use the split button repeatedly to create a full tree (hint, use the red triangle options to hide the tree and the graph). As you split, keep an eye on RMSE and RSquare for the training, validation and test sets.

i. Describe what happens to the RSquare and RMSE for the training, validation and test sets as you continue to split the tree.

ii. How does the performance of the test set compare to the training and validation sets on these measures? Why does this occur?

iii. Based on this tree, which are the most important car specifications for predicting the car's price?

iv. Refit this model, and use the *Go* button to automatically split and prune the tree based on the validation RSquare. Save the prediction formula for this model to the data table.

v. How many splits are in the final tree?

vi. Compare RSquare and RMSE for the training, validation and test sets for the reduced model to the full model.

vii. Which model is better for making predictions? Why?

9.4 **Predicting Used Car Prices (Bootstrap Forest and Boosted Trees).** Return to the Toyota Corolla data, and refit the partition model. (Hint: Use the recall button in the partition dialog). This time, choose bootstrap forest from the dialog window. Use the default settings.

a. Compared to final reduced tree above, how does the bootstrap forest behave in terms of overall error rate on the test set? Save the prediction formula for this model to the data table.

b. Run the same model again, but this time choose boosted tree from the partition dialog. Use the default settings.

c. How does the boosted tree behave in terms of the error rate relative to the reduced model and the bootstrap forest. Save the prediction formula for this model to the data table.

d. To facilitate comparison of error rates for the different models, use the *Model Comparison* platform (under *Analyze > Modeling*). To view fit statistics for the different models, put the validation column in the *Group* field in the Model Comparison dialog.

   i. Which model performs best on the test set?

   ii. Explain why this model might have the best performance over the other models you fit.

9.5 **Predicting Flight Delays (Bootstrap Forest and Boosted Trees).** We return to the flight delays data for this exercise, and fit both a bootstrap forest and a boosted tree to the data. Use scheduled departure time (CRS_DEP_TIME) rather than the binned version for these models.

  a. Fit a bootstrap forest, with the default settings. Save the formula for this models to the data table.

    i. Look at the column contributions report. Which variables were involved in the most splits?

    ii. What is the error rate on the test set?

  b. Fit a boosted tree to the flight delays data, again with the default settings. Save the formula to the data table.

    i. Which variables were involved in the most splits? Is this similiar to what you observed with the bootstrap forest model?

    ii. What is the error rate on the test set for this model?

  c. Use the Model Comparison platform to compare these models to the final reduced model found earlier (again, put the validation column in the *Group* field in the Model Comparison dialog.

    i. Which model has the lowest overall error rate on the test set?

    ii. Explain why this model might have the best performance over the other models you fit.

10

LOGISTIC REGRESSION

In this chapter we describe the highly popular and powerful classification method called logistic regression. Like linear regression, it relies on a specific mathematical equation relating the predictors with the outcome. The user must specify the predictors to include and their form (e.g., including any interaction terms). This means that even small datasets can be used for building logistic regression classifiers, and that once the model is estimated, it is computationally fast and cheap to classify even large samples of new observations. We describe the logistic regression model formulation and its estimation from data. We also explain the concepts of "logit," "odds," and "probability" of an event that arise in the logistic model context and the relations among the three. We discuss variable importance using coefficients and statistical significance and also mention variable selection algorithms for dimension reduction. All this is illustrated on an authentic dataset of flight information where the goal is to predict flight delays. Our presentation is strictly from a data mining perspective, where classification is the goal and performance is evaluated on a separate validation set. However, because logistic regression is heavily used also in statistical analyses for purposes of inference, we give a brief review of key concepts related to coefficient interpretation, goodness-of-fit evaluation, inference, and multiclass models in the Appendixes at the end of this chapter.

10.1 INTRODUCTION

Logistic regression extends the ideas of linear regression to the situation where the dependent variable Y is categorical. We can think of a categorical variable as dividing the observations into classes. For example, if Y denotes a recommendation on holding/selling/buying a stock, we have a categorical variable with three categories. We can think of each of the stocks in the dataset (the observations) as belonging to one of three classes: the *hold* class, the *sell* class, and the *buy* class. Logistic regression can be used for classifying a new observation, where the class is unknown, into one of the classes, based on the values of

Data Mining for Business Analytics: Concepts, Techniques, and Applications with JMP Pro®, First Edition. Galit Shmueli, Peter C. Bruce, Mia L. Stephens, and Nitin R. Patel.

its predictor variables (called *classification*). It can also be used in data where the class is known to find factors distinguishing between observations in different classes in terms of their predictor variables, or "predictor profile" (called *profiling*). Logistic regression is used in applications such as:

1. Classifying customers as returning or nonreturning (classification)
2. Finding factors that differentiate between male and female top executives (profiling)
3. Predicting the approval or disapproval of a loan based on information such as credit scores (classification)

The logistic regression model is used in a variety of fields: whenever a structured model is needed to explain or predict categorical (in particular, binary) outcomes. One such application is in describing consumer choice behavior in econometrics.

LOGISTIC REGRESSION AND CONSUMER CHOICE THEORY

In the context of choice behavior, the logistic model can be shown to follow from the *random utility theory* developed by Manski (1977) as an extension of the standard economic theory of consumer behavior. In essence, the consumer theory states that when faced with a set of choices, a consumer makes the choice that has the highest utility (a numerical measure of worth with arbitrary zero and scale). It assumes that the consumer has a preference order on the list of choices that satisfies reasonable criteria such as transitivity. The preference order can depend on the person (e.g., socioeconomic characteristics) as well as attributes of the choice. The random utility model considers the utility of a choice to incorporate a random element. When we model the random element as coming from a "reasonable" distribution, we can logically derive the logistic model for predicting choice behavior.

In this chapter we focus on the use of logistic regression for classification. We deal only with a binary dependent variable having two possible classes. In the Appendixes we show how the results can be extended to the case where Y assumes more than two possible outcomes. Popular examples of binary response outcomes are success/failure, yes/no, buy/don't buy, default/don't default, and survive/die. For convenience we often code the values of a binary response Y as 0 and 1.

Note that in some cases we may choose to convert continuous data or data with multiple outcomes into binary data for the purpose of simplification, reflecting the fact that decision making may be binary (approve /don't approve the loan, make/don't make an offer). As with multiple linear regression, the independent variables $X_1, X_2, \ldots, X_k$ may be categorical or continuous variables or a mixture of these two types. While in multiple linear regression the aim is to predict the value of the continuous Y for a new observation, in logistic regression the goal is to predict which class a new observation will belong to, or simply to *classify* the observation into one of the classes. In the stock example, we would want to classify a new stock into one of the three recommendation classes: sell, hold, or buy. Or, we might want to compute for a new observation its *propensity* (= its probability) to belong to each class, and then possibly rank a set of new observations from highest to lowest propensity in order to act on those with the highest propensity.

In logistic regression we take two steps: the first step yields estimates of the *propensities* or *probabilities* of belonging to each class. In the binary case we get an estimate of $p = P(Y = 1)$, the probability of belonging to class 1 (which also tells us the probability of belonging to class 0). In the next step we use a cutoff value on these probabilities in order to classify each case into one of the classes. For example, in a binary case, a cutoff of 0.5 means that cases with an estimated probability of $P(Y = 1) \geq 0.5$ are classified as belonging to class 1, whereas cases with $P(Y = 1) < 0.5$ are classified as belonging to class 0. This cutoff need not be set at 0.5. When the event in question is a low-probability but notable event (e.g., 1 = fraudulent transaction), a lower cutoff may be used to classify more cases as belonging to class 1.

10.2   THE LOGISTIC REGRESSION MODEL

The idea behind logistic regression is straightforward: instead of using Y as the dependent variable, we use a function of it, which is called the *logit*. The logit, it turns out, can be modeled as a linear function of the predictors. Once the logit has been predicted, it can be mapped back to a probability.

To understand the logit, we take several intermediate steps. First, we look at $p = P(Y = 1)$, the probability of belonging to class 1 (as opposed to class 0). In contrast to Y, the class label, which only takes the values 0 and 1, p can take any value in the interval $[0, 1]$. However, if we express p as a linear function of the q predictors[1] in the form

$$p = \beta_0 + \beta_1 x_1 + \beta_2 x_2 + \cdots + \beta_q x_q \qquad (10.1)$$

it is not guaranteed that the right-hand side will lead to values within the interval $[0, 1]$. The fix is to use a nonlinear function of the predictors in the form

$$p = \frac{1}{1 + e^{-(\beta_0 + \beta_1 x_1 + \beta_2 x_2 + \cdots + \beta_q x_q)}} \qquad (10.2)$$

This is called the *logistic response function*. For any values of $x_1, \ldots, x_q$, the right-hand side will always lead to values in the interval $[0, 1]$. Next, we look at a different measure of belonging to a certain class, known as *odds*. The odds of belonging to class 1 ($Y = 1$) is defined as *the ratio of the probability of belonging to class 1 to the probability of belonging to class 0*:

$$\text{Odds}(Y = 1) = \frac{p}{1 - p} \qquad (10.3)$$

This metric is very popular in horse races, sports, gambling in general, epidemiology, and many other areas. Instead of talking about the *probability* of winning or contacting a disease, people talk about the *odds* of winning or contacting a disease. How are these two different? If, for example, the probability of winning is 0.5, the odds of winning are $0.5/0.5 = 1$. We can also perform the reverse calculation. That is, given the odds of an

---

[1]Unlike elsewhere in the book, where p denotes the number of predictors, in this chapter we indicate predictors by q, to avoid confusion with the probability p.

event, we can compute its probability by manipulating equation (10.3):

$$p = \frac{\text{odds}}{1 + \text{odds}} \tag{10.4}$$

Substituting (10.2) into (10.4), we can write the relationship between the odds and the predictors as

$$\text{Odds}(Y = 1) = e^{\beta_0 + \beta_1 x_1 + \beta_2 x_2 + \cdots + \beta_q x_q} \tag{10.5}$$

This last equation describes a multiplicative (proportional) relationship between the predictors and the odds. Such a relationship is interpretable in terms of percentages, for example, a unit increase in predictor x_j is associated with an average increase of $\beta_j \times 100\%$ in the odds (holding all other predictors constant).

Now, if we take a natural logarithm[2] on both sides, we get the standard formulation of a logistic model:

$$\log(\text{odds}) = \beta_0 + \beta_1 x_1 + \beta_2 x_2 + \cdots + \beta_q x_q \tag{10.6}$$

The log(odds) is called the *logit*, and it takes values from $-\infty$ (very low odds) to ∞ (very high odds).[3] A logit of 0 corresponds to even odds of 1 (probability = 0.50). Thus our final formulation of the relation between the response and the predictors uses the logit as the dependent variable and models it as a *linear function* of the q predictors.

To see the relation among the probability, odds, and logit of belonging to class 1, look at Figure 10.1, which shows the odds (top) and logit (bottom) as a function of p. Notice that the odds can take any nonnegative value, and that the logit can take any real value.

Let us examine some data to illustrate the use of logistic regression.

Example: Acceptance of Personal Loan (Universal Bank)

Recall the example described in Chapter 9, of acceptance of a personal loan by Universal Bank. The bank's dataset includes data on 5000 customers (data are in Universal-Bank.jmp). The data include customer demographic information (*Age, Income*, etc.), customer response to the last personal loan campaign (*Personal Loan*), and the customer's relationship with the bank (mortgage, securities account, etc.). Among these 5000 customers, only 480 (= 9.6%) accepted the personal loan that was offered to them in a previous campaign. The goal is to build a model that identifies customers who are most likely to accept the loan offer in future mailings.

Data Preprocessing   We start by partitioning the data randomly using a standard 60:40% ratio, into training and validation sets. We use the training set to fit a model and the validation set to assess the model's performance.

In this dataset, dummy (or indicator) variables have been created for the four two-level categorical predictors (shown on the next page). These indicator variables are coded as *Continuous* variables in JMP.

---

[2]The natural logarithm function is typically denoted ln() or log(). In this book, we use log().
[3]We use the terms *odds* and *odds* (Y = 1) interchangeably.

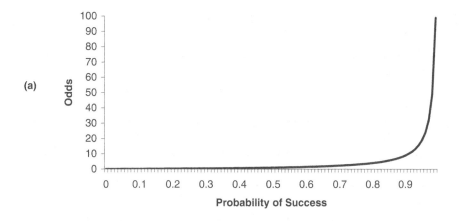

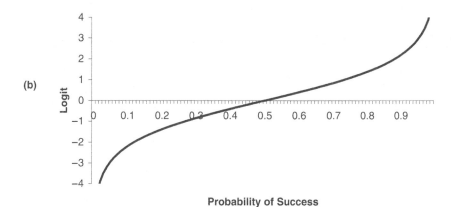

FIGURE 10.1   (*A*) Odds and (*B*) logit as a function of *P*.

The following coding has been used:

$$\text{Securities Account} = \begin{cases} 1 \text{ if customer has securities account in bank} \\ 0 \text{ otherwise} \end{cases}$$

$$\text{CD Account} = \begin{cases} 1 \text{ if customer has CD account in bank} \\ 0 \text{ otherwise} \end{cases}$$

$$\text{Online} = \begin{cases} 1 \text{ if customer uses online banking} \\ 0 \text{ otherwise} \end{cases}$$

$$\text{CreditCard} = \begin{cases} 1 \text{ if customer holds Universal Bank credit card} \\ 0 \text{ otherwise} \end{cases}$$

However, in JMP, both the response and the predictors can remain in text form (Yes, No, etc.), and it is not necessary to code categorical predictor variables into continuous indicator (dummy) variables. So the variable Education, which has three categories, can be used in its original form without dummy coding.

INDICATOR (DUMMY) VARIABLES IN JMP

By default, JMP creates indicator variables behind the scenes for all categorical predictors. The dropped category (the reference category) for each categorical variable is the last category in alphabetical order, or the highest level alphanumerically. Another point to note, is that instead of applying the typical 0/1 coding, JMP applies $-1/+1$ coding. This results in different model coefficients compared to 0/1 coding, but produces *identical* model predictions and classifications.

If desired, dummy variables can be used instead of the original categorical variables. To manually create indicator variables for categorical predictors, use the *Make Indicator Columns* feature under *Cols > Utilities*. The utility in JMP 12 produces *Nominal* indicator variables (in JMP 13 these indicator variables are coded as *Continuous*). Change the modeling types for these indicator variables to *Continuous* before fitting regression models (and, remember that you'll need to leave one indicator variable out of the model for each original categorical predictor).

For more information on indicator coding in JMP, search for *The Factor Models* in the JMP Help or the book *Fitting Linear Models* under the Help menu.

Note: In JMP 13, coefficients using the 0/1 indicator coding can be requested using a red triangle option in the logistic analysis window. (select *Indicator Parameterization Estimates*).

Model with a Single Predictor

Consider a simple logistic regression model with just one predictor. This is conceptually analogous to the simple linear regression model in which we fit a straight line to relate the outcome Y to a single predictor X.

Let us construct a simple logistic regression model for classification of customers using the single predictor Income. The equation relating the outcome variable to the predictor in terms of probabilities is

$$P(Personal\ Loan = Yes \mid Income = x) = \frac{1}{1 + e^{-(\beta_0 + \beta_1 x)}}$$

or equivalently, in terms of odds,

$$\text{Odds}(Personal\ Loan = Yes) = e^{\beta_0 + \beta_1 x} \qquad (10.7)$$

We fit a logistic regression to the training data. The estimated coefficients for the model are $b_0 = -6.3525$ and $b_1 = 0.0392$. So the fitted model is

$$P(Personal\ Loan = Yes \mid Income = x) = \frac{1}{1 + e^{6.3525 - 0.0392x}} \qquad (10.8)$$

For the single predictor case, we can plot the data and graphically display the logistic model. Examine Figure 10.2. The curve represents the probability that someone will accept

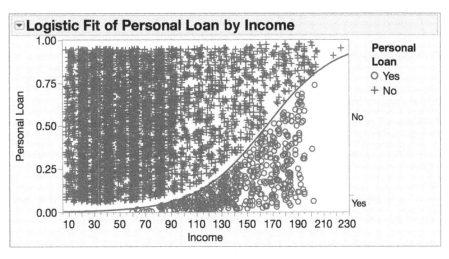

FIGURE 10.2   The logistic curve for Personal Loan as a function of Income. A row legend was added to color and mark observations in each Personal Loan group. To add a row legend, right-click over the plot and select *row legend*.

a loan as Income increases. The *y*-axis on the left represents the probability of acceptance, and the *y*-axis on the right acts as a stacked bar chart, providing a breakdown of the overall probability of Yes versus No (there were fewer accepted loans than nonaccepted). The points are plotted at the income level (on the *x*-axis), and appear either above or below the curve based on actual acceptance. Accepted loans (Yes) are plotted below the curve, and are randomly scattered relative to the *y*-axis (this is done to provide an indication of scatter relative to Income in larger datasets).

Although logistic regression can be used for prediction in the sense that we predict the *probability* of a categorical outcome, it is most often used for classification. To see the difference between the two, consider predicting the probability of a customer accepting the loan offer as opposed to classifying the customer as an acceptor/nonacceptor. From Figure 10.2 it can be seen that the loan acceptance can yield values between 0 and 1. To end up with classifications into either Yes or No (e.g., a customer either accepts the loan offer or not), we need a threshold, or cutoff value (see section on "Propensities and Cutoff for Classification" in Chapter 5). This is true in the case of multiple predictor variables as well.

In the Universal Bank example, in order to classify a new customer as an acceptor/nonacceptor of the loan offer, we use the information on his/her income by plugging it into the fitted equation in (10.8). This yields an estimated probability of accepting the loan offer. We then compare it to the cutoff value. The customer is classified as an acceptor if the probability of his/her accepting the offer is above the cutoff.[4]

---

[4]Here we compared the probability to a cutoff c (the cutoff is 0.5 by default). If we prefer to look at *odds* of accepting rather than the probability, an equivalent method is to use the equation (10.7) and compare the odds to $c/(1 - c)$. If the odds are higher than this number, the customer is classified as an acceptor. If the odds are lower, we classify the customer as a nonacceptor.

FITTING ONE PREDICTOR LOGISTIC MODELS IN JMP

Logistic regression models in JMP with one predictor are fit using *Analyze > Fit Y by X* or *Fit Model*. This displays the logistic curve shown in Figure 10.2. In the Universal Bank example, we model the probability of Personal Loan = Yes (or 1, accepted). By default, JMP will model the probability that Personal Loan = No (or 0). To model the probability of Yes rather than No, as was done in this example, use the *Value Ordering* column property, select Yes, and use the *Move Up* button to change the order of the values.
Note that in JMP 13 the target category can be selected directly in the *Fit Y by X* and *Fit Model* dialogs.

Estimating the Logistic Model from Data: Multiple Predictors

In logistic regression, the relation between Y and the β parameters is nonlinear. For this reason the β parameters are not estimated using the method of least squares (as in multiple linear regression). Instead, a method called *maximum likelihood* is used. The idea, in brief, is to find the estimates that maximize the chance of obtaining the data that we have.[5]

Algorithms to compute the coefficient estimates in logistic regression are less robust than algorithms for linear regression. Computed estimates are generally reliable for well-behaved datasets where the number of observations with outcome variable values of both 0 and 1 are large, their ratio is "not too close" to either 0 or 1, and when the number of coefficients in the logistic regression model is small relative to the sample size (e.g., no more than 10%). As with linear regression, collinearity (strong correlation among the predictors) can lead to computational difficulties. Computationally intensive algorithms have been developed recently that circumvent some of these difficulties. For technical details on the maximum likelihood estimation in logistic regression, see Hosmer and Lemeshow (2000).

To illustrate a typical output from such a procedure, look at the output in Figure 10.3 for the logistic model fitted to the training set of 3000 Universal Bank customers. The outcome variable is Personal Loan, with Yes defined as the *success* (this is equivalent to setting the outcome variable to 1 for an acceptor and 0 for a nonacceptor). Here we use all 12

---

[5]The method of maximum likelihood ensures good asymptotic (large sample) properties for the estimates. Under very general conditions, maximum likelihood estimators are (1) *consistent* (the probability of the estimator differing from the true value approaches zero with increasing sample size), (2) *asymptotically efficient* (the variance is the smallest possible among consistent estimators), and (3) *asymptotically normally distributed* (the confidence intervals and perform statistical tests are to be computed in a manner analogous to the analysis of linear multiple regression models, provided that the sample size is *large*).

Parameter Estimates

Term	Estimate	Std Error	ChiSquare	Prob>ChiSq
Intercept	−10.164069	2.4497848	17.21	<.0001*
Age	−0.044547	0.090961	0.24	0.6243
Experience	0.05658147	0.0900536	0.39	0.5298
Income	0.06576067	0.0042213	242.68	<.0001*
Family	0.57155568	0.1011896	31.90	<.0001*
CCAvg	0.18723439	0.0615372	9.26	0.0023*
Education[Undergrad]	−3.0372506	0.2432931	155.85	<.0001*
Education[Graduate]	1.55179759	0.1752704	78.39	<.0001*
Mortgage	0.00175308	0.0008038	4.76	0.0292*
Securities Account	−0.8548708	0.4186376	4.17	0.0411*
CD Account	3.46902866	0.4489309	59.71	<.0001*
Online	−0.843563	0.2283237	13.65	0.0002*
CreditCard	−0.9640741	0.2825423	11.64	0.0006*

For log odds of Yes/No

FIGURE 10.3 Logistic regression parameter estimates (coefficients) table for personal loan acceptance as a function of 12 predictors based on training data.

predictors, omitting ZIP Code (note that for each categorical predictor with m categories JMP creates $m - 1$ indicator variables).

Ignoring p-values for the coefficients, a model based on all 12 predictors would have the estimated logistic equation shown in Figure 10.4. This equation, the logit, is stored in JMP

$$
\begin{aligned}
& -10.164069254528 \\
& + -0.0445470057798 * \mathit{Age} \\
& + 0.05658146902618 * \mathit{Experience} \\
& + 0.06576067129506 * \mathit{Income} \\
& + 0.57155567973585 * \mathit{Family} \\
& + 0.18723439396087 * \mathit{CCAvg} \\
& + \text{Match}\left(\mathit{Education}\right) \begin{cases} 1 & \Rightarrow -3.0372506132999 \\ 2 & \Rightarrow 1.55179758968979 \\ 3 & \Rightarrow 1.4854530236101 \\ \text{else} & \Rightarrow . \end{cases} \\
& + 0.00175308271912 * \mathit{Mortgage} \\
& + -0.8548708298344 * \mathit{Securities\ Account} \\
& + 3.46902865946834 * \mathit{CD\ Account} \\
& + -0.8435630344762 * \mathit{Online} \\
& + -0.9640741062968 * \mathit{CreditCard}
\end{aligned}
$$

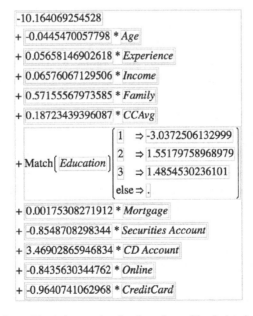

FIGURE 10.4 Estimated logistic equation for the universal bank data based on training data.

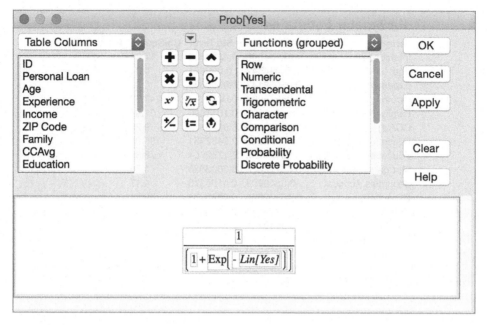

FIGURE 10.5   The equation for calculating the probability of acceptance, prob[Yes], saved to the data table as a formula column.

as *Lin[Yes]* when the probability formula is saved to the data table. Note that the coefficient for the third level of the categorical variable Education is provided in this formula.

The positive coefficients for CD Account, Graduate (Education = 2), and Professional (Education = 3) mean that holding a CD account and having graduate or professional education are associated with higher probabilities of accepting the loan offer. On the other hand, having a securities account, using online banking, and owning a Universal Bank credit card are associated with lower acceptance rates. For the continuous predictors, positive coefficients indicate that a higher value on that predictor is associated with a higher probability of accepting the loan offer (e.g., income: higher income customers tend more to accept the offer). Similarly, negative coefficients indicate that a higher value on that predictor is associated with a lower probability of accepting the loan offer (e.g., Age: older customers are less likely to accept the offer).

The probability of accepting the offer (the *propensity*) for a customer with given values of the 12 predictors can be calculated using equation 10.2.[6] In JMP, when the probability formula is saved to the data table, the formula for calculating propensities is saved as *Prob[Yes]*, as shown in Figure 10.5.

---

[6]If all q predictors are categorical, each having m_q categories, we need not compute probabilities/odds for each of the n observations. The number of different probabilities/odds is exactly $m_1 \times m_2 \times \cdots \times m_q$.

FITTING LOGISTIC MODELS IN JMP WITH MORE THAN ONE PREDIC-TOR

To run a logistic regression model with more than one predictor or with a validation column, use *Analyze > Fit Model*. The resulting model, and the relationships between the predictors and the response, can be graphically explored using the *Profiler*.

To look at the equation for the model, use *Save Probability Formula* from the top red triangle. This saves the logit (Under *Lin[1]* or *Lin[Yes]*), along with predicted probabilities for each class of the response and the predicted outcome, using a 0.50 cutoff.

10.3   EVALUATING CLASSIFICATION PERFORMANCE

The general measures of performance that were described in Chapter 5 are used to assess how well the logistic model does. Recall that there are several performance measures, the most popular being those based on the classification matrix (accuracy alone or combined with costs) and the lift curve. As in other classification methods, the goal is to find a model that accurately classifies observations to their class, using only the predictor information. A variant of this goal is to find a model that does a superior job of identifying the members of a particular class of interest (which might come at some cost to overall accuracy). Since the training data are used for selecting the model, we expect the model to perform quite well for those data, and therefore we prefer to test its performance on the validation set. Recall that the data in the validation set were not involved in the model building, and thus we can use them to test the model's ability to classify data that it has not "seen" before.

To obtain the classification matrix from a logistic regression analysis, we use the estimated equation to predict the probability of class membership (the *propensities*) for each observation in the validation set, and use the cutoff value to decide on the class assignment of these observations. We then compare these classifications to the actual class memberships of these observations. In the Universal Bank case we use the estimated model to predict the probability of offer acceptance in a validation set that contains 2000 customers (these data were not used in the modeling step). Technically, this is done (by the software) by predicting the logit using the estimated model in Figure 10.4 and then obtaining the probabilities *p* through the equation shown in Figure 10.5. We then compare these probabilities to our chosen cutoff value in order to classify each of the 2000 validation observations as acceptors or nonacceptors.

JMP Pro provides the misclassification rates for the training and validation sets automatically, and the classification (or confusion) matrix is a red triangle option. Detailed probabilities and the classification for each observation are obtained by saving the probability formula to the data table. Figure 10.6 shows a partial output of scoring the validation set (these are the shaded rows). We see that the first four customers in the validation set have a probability of accepting the offer that is lower than the cutoff of 0.5, and therefore they are classified as nonacceptors (0). The fifth customer in the validation set has a probability of acceptance estimated by the model to exceed 0.5, and he or she is therefore classified as an acceptor (Yes), which in fact is a misclassification.

	Personal Loan	Validation	Lin[Yes]	Prob[Yes]	Prob[No]	Most Likely Personal Loan
1	No	Training	−9.30521373	0.0000909405	0.9999090595	No
2	No	Validation	−10.75437659	0.0000213513	0.9999786487	No
3	No	Validation	−12.6077736	3.345893e-6	0.9999966541	No
4	No	Training	−2.009027973	0.1182582966	0.8817417034	No
5	No	Training	−5.2501519	0.005219337	0.994780663	No
6	No	Training	−5.82861105	0.0029335297	0.9970664703	No
7	No	Validation	−4.130395058	0.015822161	0.984177839	No
8	No	Validation	−8.437624603	0.0002165171	0.9997834829	No
9	No	Training	−3.113222558	0.0425651205	0.9574348795	No
10	Yes	Training	4.3908814129	0.9877618247	0.0122381753	Yes
11	No	Validation	0.2729614358	0.5678197882	0.4321802118	Yes
12	No	Training	−5.772169835	0.0031033348	0.9968966652	No
13	No	Validation	−1.019050966	0.265212302	0.734787698	No
14	No	Validation	−4.888765476	0.0074744258	0.9925255742	No
15	No	Validation	−6.409780202	0.0016426833	0.9983573167	No

FIGURE 10.6   Scoring the data: Output for the first 15 universal bank customers. Rows in validation set are shaded.

Another useful tool for assessing model classification performance is the lift curve (see Chapter 5). Figure 10.7 illustrates the lift curve obtained for the Personal Loan offer model using the validation set. The lift indicates the additional responders that you can identify, for a given portion of cases (read on the x-axis), by using the model. Taking the 10% (portion = 0.1) of the records that are ranked by the model as most probable responders yields roughly 7.7 times as many responders as would simply selecting 10% of the records at random.

Variable Selection

The next step includes searching for alternative models. As with multiple linear regression, we can build more complex models that reflect interactions among predictors by including new variables that are derived from the predictors. For example, if we hypothesize that there is an interactive effect between income and family size, we should add an interaction term of the form *Income*Family*. The choice among the set of alternative models is guided primarily by subject matter knowledge and performance on the validation data. For models that perform roughly equally well, simpler models are generally preferred over more complex models.

Note also that performance on validation data may be overly optimistic when it comes to predicting performance on data that have not been exposed to the model at all. This is because, when the validation data are used to select a final model, we are selecting based on how well the model performs with those data and therefore may be incorporating some of the random idiosyncrasies of those data into the judgment about the best model. The model still may be the best among those considered, but it will probably not do as well with the unseen data. Therefore it is useful to evaluate the chosen model on a new test set to get a sense of how well it will perform on new data. In addition one must consider practical

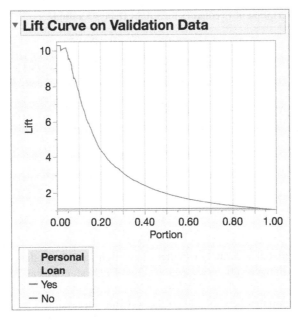

FIGURE 10.7   Lift curve of validation data for Universal Bank loan offer: logistic model ranking performance compared to random assignment.

issues such as costs of collecting variables, error-proneness, and model complexity in the selection of the final model.

As in linear regression, in logistic regression we can use automated variable selection heuristics such as mixed stepwise selection, forward selection, and backward elimination. In JMP (and JMP Pro), an exhaustive search using *All Possible Models* is not available for logistic regression models. (See Section 6.4 in Chapter 6 for details on stepwise variable selection.)

10.4   EXAMPLE OF COMPLETE ANALYSIS: PREDICTING DELAYED FLIGHTS

Predicting flight delays would be useful to a variety of organizations: airport authorities, airlines, and aviation authorities. At times, joint task forces have been formed to address the problem. Such an organization, if it were to provide ongoing real-time assistance with flight delays, would benefit from some advance notice about flights that are likely to be delayed. Data are in `Flight Delays LR.jmp`.

In this simplified illustration, we look at six predictors (see Table 10.1). The outcome of interest is whether the flight is delayed or not (*delayed* means more than 15 minutes late). Our data consist of flights from the Washington, DC, area into the New York City area during January 2004. The percent of delayed flights among these 2201 flights is 19.45%. The data were obtained from the Bureau of Transportation Statistics website (www.transtats.bts.gov).

TABLE 10.1 Description of Predictors for Flight Delays Example

Delayed	Coded as: 1 = Yes, 0 = No
Carrier	Eight airline codes: CO (Continental), DH (Atlantic Coast), DL (Delta), MQ (American Eagle), OH (Comair), RU (Continental Express), UA (United), and US (USAirways)
Day of Week	Coded as: 1 = Monday, 2 = Tuesday,..., 7 = Sunday
Departure Time	Broken down into 16 intervals
Destination	Three airport codes: JFK (Kennedy), LGA (LaGuardia), EWR (Newark)
Origin	Three airport codes: DCA (Reagan National), IAD (Dulles), BWI (Baltimore–Washington Int'l)
Weather	Coded as 1 if there was a weather-related delay

The goal is to predict accurately whether a new flight, not in this dataset, will be delayed. Our dependent variable is a binary variable called *Delayed*, coded as 1 for a delayed flight and 0 otherwise.

Other information that is available on the website, such as distance and arrival time, is irrelevant because we are looking at a certain route (distance, flight time, etc., should be approximately equal). A sample of the data for 20 flights is shown in Figure 10.8. Figures 10.9 and 10.10 show visualizations of the relationships between flight delays and different predictors or combinations of predictors. From Figure 10.9 we see that Sundays (7) and Mondays (1) have the largest proportion of delays. Delay rates also seem to differ by carrier, time of day, origin and destination airports. For weather, we see a strong distinction between delays when Weather = 1 (in that case, there is always a delay) and Weather = 0.

	Delayed	Carrier	Day of Week	Departure Time	Destination	Origin	Weather
110	0	RU	5	1300–1359	EWR	DCA	0
111	1	RU	5	1500–1559	EWR	IAD	0
112	0	RU	5	1500–1559	EWR	DCA	0
113	0	DH	6	1200–1259	LGA	IAD	0
114	0	DH	6	1700–1759	LGA	IAD	0
115	0	DH	6	0659 or earlier	LGA	IAD	0
116	0	DH	6	1000–1059	LGA	IAD	0
117	0	DH	6	0800–0859	JFK	IAD	0
118	0	DH	6	1200–1259	JFK	IAD	0
119	0	DH	6	1400–1459	JFK	IAD	0
120	0	DH	6	1600–1659	JFK	IAD	0
121	0	DH	6	1700–1759	JFK	IAD	0
122	1	DH	6	2100 or later	JFK	IAD	0
123	0	DH	6	1600–1659	JFK	IAD	0
124	0	DH	6	2100 or later	LGA	IAD	0
125	1	DL	6	1500–1559	JFK	DCA	0
126	0	DL	6	0800–0859	LGA	DCA	0
127	0	DL	6	1000–1059	LGA	DCA	0
128	0	DL	6	1200–1259	LGA	DCA	0
129	0	DL	6	1400–1459	LGA	DCA	0
130	0	DL	6	1600–1659	LGA	DCA	0

FIGURE 10.8 Sample of 20 flights.

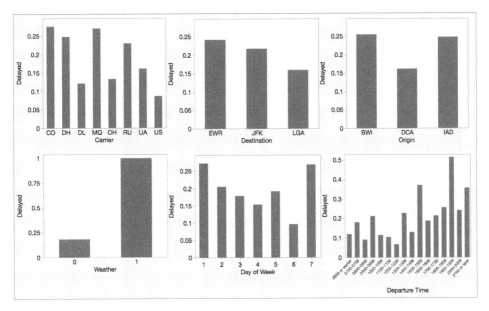

FIGURE 10.9 Proportion of delayed flights by each of the predictors. The modeling type for *Delayed* was first changed to continuous.

The heatmap in Figure 10.10 reveals some specific combinations with high delay rates, such as Sunday flights by Continental Express (Carrier RU) departing from BWI, or Sunday flights by American Eagle (MQ) departing from DCA. We can also see combinations with very low delay rates.[7]

Our main goal is to find a model that can obtain accurate classifications of new flights based on their predictor information. An alternative goal is finding a certain percentage of flights that are most/least likely to get delayed (*ranking*). And a third, different, goal is profiling flights: finding out which factors distinguish between ontime and delayed flights (not only in this sample but in the entire population of flights on this route), and for those factors we would like to quantify these effects. A logistic regression model can be used for all these goals, albeit in different ways.

Data Preprocessing

To prepare our data for modeling, we use graphical and numeric summaries, along with subject matter knowledge, to identify variables in need of recoding or reformatting. Here are some of the steps we've taken in this example:

1. To produce the bar charts and heat map in Figures 10.9 and 10.10, we first changed the modeling type for the response to continuous. We then changed the modeling type back to nominal before fitting logistic models.

---

[7]Before creating the heatmap in Figure 10.10 data were first summarized using *Tables > Summary*. To create this summary, select Delayed, and under *Statistics* select *Mean*, then select Carrier, Day of Week, and Origin, and then select *Group*.

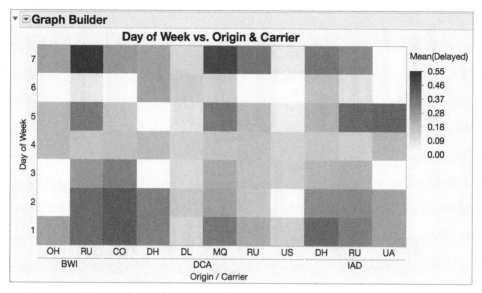

FIGURE 10.10   Heatmap of Proportion of delayed flights (darker = higher proportion) by day of week, origin, and carrier using *Graph Builder*. Data first summarized using *Tables > Summary*.

2. Since we are interested in predicting delayed flights, we need to tell JMP to model the probability of Delayed = 1. To do this, we used the *Value Ordering* column property.

3. Departure Time is derived from another column, Time of Day (not shown). This is a numeric column, but the time data aren't coded in JMP using a time format (JMP provides many date and time formats). We have binned the time data into one hour increments (using the *Formula Editor*).

4. We have grouped variables that won't be used in the analysis, and have hidden and excluded these variables (using options under the *Cols* menu).

We then partitioned the data using a 60:40% ratio into training and validation sets. We use the training set to fit a model and the validation set to assess the model's performance.

Model Fitting, Estimation and Interpretation—A Simple Model

For this example we have several categorical predictors. Instead of using indicator (dummy) variables, we rely on JMP to create indicator variables behind the scenes for each of these categorical predictors. Recall that JMP uses $-1/1$ indicator coding rather than the traditional $0/1$ coding. The resulting coefficients for these indicator variables *sum to zero*, so the coefficient for the last level is not displayed. (Note that logistic models using either $-1/1$ or $0/1$ indicator coding produce the same estimated probabilities.)

To demonstrate how JMP codes categorical predictors, we fit a model for Delayed with one predictor, Destination. In the top of Figure 10.11 we see the parameter estimates for this model. The coefficient for arrival airport JFK is estimated from the data to be 0.0822 and the coefficient for arrival at EWR is 0.2207. Since the coefficients for the categorical indicator variables in JMP sum to zero, the coefficient for arrival at LGA is $-[(0.0822) + (0.2207)] = -0.3029$. This can be seen by saving the probability formula to

▼ Parameter Estimates

Term	Estimate	Std Error	ChiSquare	Prob>ChiSq
Intercept	−1.3618315	0.0576483	558.05	<.0001*
Destination[EWR]	0.22065958	0.0778145	8.04	0.0046*
Destination[JFK]	0.08222126	0.091626	0.81	0.3695
For log odds of 1/0				

FIGURE 10.11 Estimated logistic regression model for delayed flights with one categorical predictor, destination.

the data table. The logit, saved as Lin[1], is shown in Figure 10.12. The coefficients tell us that, ignoring all other factors, flights to LGA are slightly less likely to be delayed than those to JFK or EWR.

Model Fitting, Estimation and Interpretation—The Full Model

We now fit a full model with all six predictors. The estimated model with 33 parameters (plus the intercept) is given in Figure 10.13. For each categorical variable, JMP has created indicator variables, and as we have seen, the highest level alphanumerically (or last alphabetically) is kept out of the model.

With other predictors in the model, the coefficients for the arrival airports have changed slightly. The coefficient for JFK is now −0.06799 and the coefficient for arrival at EWR is −0.0113. So the coefficient for arrival at LGA is −[(−0.06799) + (−0.0113)] = 0.0793. If we take into account the statistical significance of the coefficients, we see that in general the destination airport is not associated with the chance of delays. However, the departure airport is statistically significant.

The day of the week is also significant, with flights leaving on Monday (1) and Friday (5) having positive coefficients (higher probability of delays) and flights on Thursdays (4) and Saturdays (6) having negative coefficients (lower probability of delays). Also the coefficients appear to change over the course of the day, with lower probability of delays mostly occurring earlier in the day.

The *Profiler* (Figure 10.14) provides a visual representation of the model coefficients and the relationship between the different predictors and the probability of flight delays. We can see, for example, that the probability of a delayed flight generally increases throughout the day, peaking in the evening. We can also see that the delay rate is roughly the same for

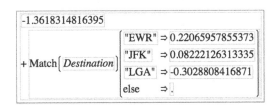

FIGURE 10.12 Logit for delayed flights with one predictor, Destination.

the different destinations, while BWI appears to have a higher delay rate than DCA and IAD.

Parameter Estimates

Term		Estimate	Std Error	ChiSquare	Prob>ChiSq
Intercept	Unstable	11.8788584	553.54245	0.00	0.9829
Carrier[CO]	Unstable	2.59702148	101.19343	0.00	0.9795
Carrier[DH]	Unstable	2.24635404	101.1929	0.00	0.9823
Carrier[DL]	Unstable	1.27877003	101.1929	0.00	0.9899
Carrier[MQ]	Unstable	2.4932625	101.19282	0.00	0.9803
Carrier[OH]	Unstable	0.68811571	101.19538	0.00	0.9946
Carrier[RU]	Unstable	2.12976488	101.19293	0.00	0.9832
Carrier[UA]	Unstable	−12.010284	708.34745	0.00	0.9865
Day of Week[1]		0.30157894	0.1811195	2.77	0.0959
Day of Week[2]		0.07546288	0.196388	0.15	0.7008
Day of Week[3]		−0.1589367	0.1944871	0.67	0.4138
Day of Week[4]		−0.3842139	0.1886445	4.15	0.0417*
Day of Week[5]		0.22013162	0.170299	1.67	0.1961
Day of Week[6]		−0.4786919	0.2401693	3.97	0.0462*
Departure Time[0659 or earlier]		−0.5474193	0.2819076	3.77	0.0522
Departure Time[0700-0759]		−0.6301708	0.4818879	1.71	0.1910
Departure Time[0800-0859]		−1.0669539	0.3485066	9.37	0.0022*
Departure Time[0900-0959]		0.39007625	0.4551093	0.73	0.3914
Departure Time[1000-1059]		−1.038575	0.4390541	5.60	0.0180*
Departure Time[1100-1159]		0.19465575	0.5541449	0.12	0.7254
Departure Time[1200-1259]		−1.5173006	0.3847497	15.55	<.0001*
Departure Time[1300-1359]		0.41106926	0.3017861	1.86	0.1732
Departure Time[1400-1459]		−0.3596341	0.272698	1.74	0.1872
Departure Time[1500-1559]		0.80398207	0.2379079	11.42	0.0007*
Departure Time[1600-1659]		−0.2156358	0.261717	0.68	0.4100
Departure Time[1700-1759]		0.11350714	0.2485986	0.21	0.6480
Departure Time[1800-1859]		0.66798459	0.2691259	6.16	0.0131*
Departure Time[1900-1959]		1.79857406	0.354208	25.78	<.0001*
Departure Time[2000-2059]		0.65792876	0.3463971	3.61	0.0575
Destination[EWR]		−0.0112986	0.2004744	0.00	0.9551
Destination[JFK]		−0.0679927	0.1568678	0.19	0.6647
Origin[BWI]		0.64053176	0.2429902	6.95	0.0084*
Origin[DCA]		−0.5265972	0.2174133	5.87	0.0154*
Weather[0]	Unstable	−14.965294	614.88871	0.00	0.9806

For log odds of 1/0

FIGURE 10.13 Estimated logistic regression model for delayed flights with all predictors, (based on the training set).

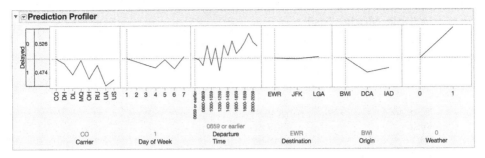

FIGURE 10.14   Prediction profiler for flight delays. The contours show how the predicted probability of delays changes for different values of the predictors.

The estimates for *Weather* and *Carrier* are marked as *unstable*. This means that the predicted probability (of delays, in this case) is 1 or near 1 for values of these factors or that there is not enough information to estimate the coefficients. All of the occurrences of *Weather* = 1 were delayed. As a result JMP is providing a warning that estimate of the coefficient for weather is unstable (note that Weather is not significant even though the coefficient is large). For *Carrier*, the warning is related to the fact that there are relatively few flights for some of the carriers. For example, OH and UA had a combined 61 flights, but only 36 of these are in the training set. In this case the *unstable* warning is a hint that data may need to be coded or grouped differently. We revisit these issues later in this example.

Model Performance

How should we measure the performance of logistic models? One possible measure is "percent of flights correctly classified." This measure, along with several other fit statistics, are provided for the training and validation sets under *Whole Model Test*. The misclassification rate for the validation set, for the whole model, is 0.1557 (see Figure 10.15). This is slightly better than the misclassification rate for the training set (0.1612).

The confusion matrix gives a sense of what type of misclassification is more frequent. From the confusion matrix at the top of Figure 10.16, it can be seen that the model does better in classifying nondelayed flights correctly and is less accurate in classifying flights that were delayed. (*Note*: The same pattern appears in the classification matrix for the training data, so it is not surprising to see it emerge for new data.) If there is an asymmetric cost structure such that one type of misclassification is more costly than the other, the cutoff value can be selected to minimize the cost (via the *Profit Matrix*). Of course, this tweaking should be carried out on the training data and assessed only using the validation data.

In most conceivable situations, it is likely that the purpose of the model will be to identify those flights most likely to be delayed so that resources can be directed toward either reducing the delay or mitigating its effects. Air traffic controllers might work to open

▼ **Whole Model Test**

Model	-LogLikelihood	DF	ChiSquare	Prob>ChiSq
Difference	131.62702	33	263.254	<.0001*
Full	533.26770			
Reduced	664.89473			

RSquare (U)	0.1980
AICc	1136.39
BIC	1310.86
Observations (or Sum Wgts)	1321

Measure	Training	Validation	Definition
Entropy RSquare	0.1980	0.0112	1-Loglike(model)/Loglike(0)
Generalized RSquare	0.2847	0.0173	$(1-(L(0)/L(model))^{(2/n)})/(1-L(0)^{(2/n)})$
Mean -Log p	0.4037	0.4705	$\sum -Log(\rho[j])/n$
RMSE	0.3531	0.3604	$\sqrt{\sum(y[j]-\rho[j])^2/n}$
Mean Abs Dev	0.2525	0.2593	$\sum |y[j]-\rho[j]|/n$
Misclassification Rate	0.1612	0.1557	$\sum (\rho[j]\neq\rho Max)/n$
N	1321	880	n

FIGURE 10.15   Fit statistics, including misclassification rates, for the flight delay data.

up additional air routes, or to allocate more controllers to a specific area for a short time. Airlines might bring on personnel to rebook passengers and to activate standby flight crews and aircraft. Hotels might allocate space for stranded travelers. In all cases the resources available are going to be limited, and resources might vary over time and from organization to organization. In this situation the most useful model would provide an ordering of flights by their probability of delay, letting the model users decide how far down that list to go in taking action. Therefore model lift is a useful measure of performance—as you move down that list of flights, ordered by their delay probability, how much better does the model do in predicting delay than would a naive model, which is simply the average delay rate for all flights? From the lift curve for the validation data (Figure 10.16), we see that our model is superior to the baseline (simple random selection of flights).

Variable Selection

From the initial exploration of predictors shown in Figures 10.9 and 10.10, the parameter estimates table in Figure 10.13, and the *Prediction Profiler* shown in Figure 10.14, it appears that several of the variables might be dropped or coded differently. Additionally we look at the number of flights in different categories to identify categories with very few or no flights—such categories are candidates for removal or merger.

First, we find that most carriers depart from a single airport (DCA). For those that depart from all three airports, the delay rates are similar regardless of airport. We use the *Effect Summary* table (at the top of the Fit Nominal Logistic window) to remove the variable Origin from the model, and find that the model performance on the validation set is not harmed. We find that the destination airport is not significant (see Figure 10.17), and decide to drop it for a practical reason: not all carriers fly to all airports (this can be confirmed with

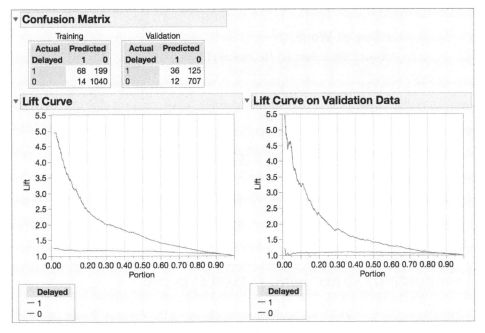

FIGURE 10.16   Confusion matrix and lift curve for the flight delay validation data

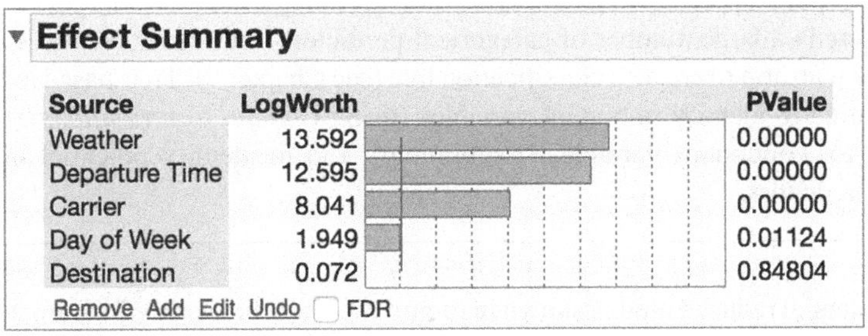

FIGURE 10.17   Effect summary table, after the removal of Origin

Tabulate). Our model would then be invalid for prediction in nonexistent combinations of carrier and destination airport.

Earlier we discussed the issue with instability of the parameter estimates for *Carrier* (see Figure 10.13). To resolve this, we try grouping the carriers into fewer categories. We also try to group *Departure Time* and *Day of Week* into fewer categories that are more distinguishable with respect to delays. From the *Prediction Profiler* in Figure 10.14, we see that groups of carriers have similar delay rates (e.g., DL, OH, UA and US) and that Sunday and Monday seem to have similar delay rates. The grouping of Sunday and Monday, using the JMP *Recode* utility, is shown in Figure 10.18.

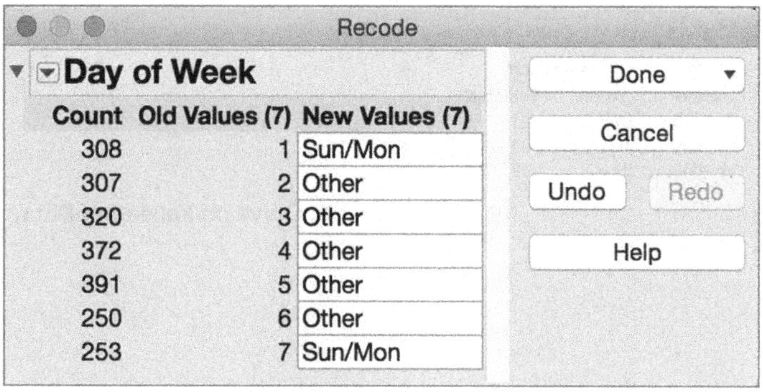

FIGURE 10.18   Using *Recode*, from the *Cols > Utilities* menu to group day of the week

---

REGROUPING AND RECODING VARIABLES IN JMP

Variable recoding in JMP can be done manually using the *Formula Editor* or using *Recode* from the column utilities menu. New recoded variables can also be created directly from Stepwise. Recall that the stepwise platform creates coded variables that maximize the difference in the response—these coded variables can be saved to the data table and used to build new models.

Another popular approach to identify potential groupings, which works particularly well when there is a large number of categorical predictors with many levels, is to use the *Partition* platform to create a classification tree (see Chapter 9). Tree-based methods can also be used to identify important variables (using *Column Contributions*), to identify cutoff's for continuous variables (for binning), and to identify potential interactions among predictors.

One potential reduced model, shown in Figure 10.19, includes only 7 parameter estimates and has the advantage of being more parsimonious. Figure 10.19 displays the training and validation fit statistics, the confusion matrices, and the lift curves. It can be seen that the misclassification rate for this model competes well with the larger model in terms of both accurate classification and lift. (Note that there are many other possible models that we might consider.)

We therefore conclude with a seven-predictor model that requires only knowledge of the carrier, the day of week, the hour of the day, and whether it is likely that there will be a delay due to weather. However, this weather variable refers to actual weather at flight time, not a forecast, and is not known in advance! If the aim is to predict in advance whether a particular flight will be delayed, a model without weather must be used. In contrast, if the goal is profiling delayed vs. nondelayed flights, we can include weather in the model to allow evaluating the impact of the other factors only when there *isn't* a weather-related delay. Since Weather = 1 corresponds to a delayed flight, we cannot explore the impact of the other factors when there *is* a weather-related delay (the probability of a delayed flight will be 1.0 under all conditions). If profiling flight delays in the event of different weather

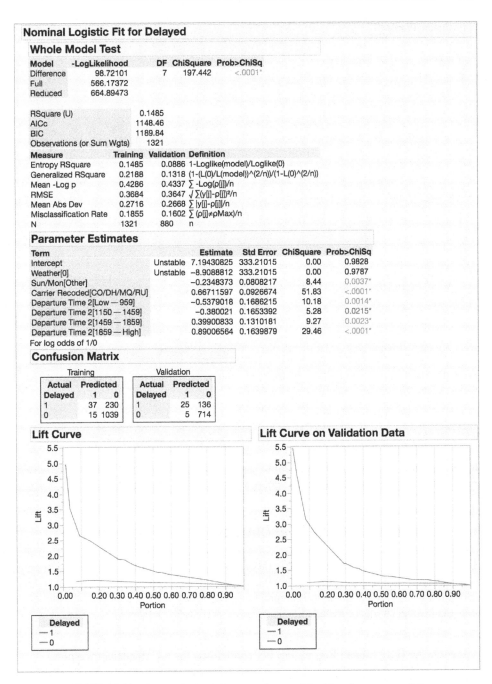

FIGURE 10.19   Output for logistic regression model with only seven predictors.

conditions is of primary concern, compiling additional information related to weather is recommended.

To conclude, based on the data from January 2004 and our final model, the highest chance of a nondelayed flight is Tuesday through Saturday (recoded as *Other*), and flying

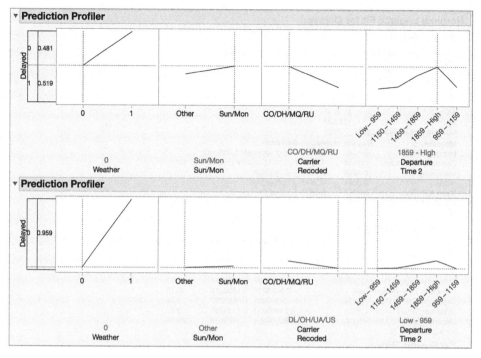

FIGURE 10.20   Prediction Profiler showing best (bottom) and worst (top) conditions for delayed flights, assuming there are no delays due to weather.

in the morning, on Delta, Comair, United, or US Airways (see the bottom *Prediction Profiler* in Figure 10.20). And clearly, good weather is advantageous!

10.5   APPENDIXES: LOGISTIC REGRESSION FOR PROFILING

The presentation of logistic regression in this chapter has been primarily from a data mining perspective, where classification or ranking are the goal and performance is evaluated by reviewing results with a validation sample. For reference, some key concepts of a classical statistical perspective, are included below.

Appendix A: Why Linear Regression Is Problematic for a Categorical Response

Now that you have seen how logistic regression works, we explain why it is considered preferable over linear regression for a binary outcome. Technically, one can apply a multiple linear regression model to this problem, treating the outcome variable Y as continuous. This is called a *Linear Probability Model*. Of course, Y must be coded numerically (e.g., 1 for customers who did accept the loan offer and 0 for customers who did not accept it). Although software will yield an output that at first glance may seem usual (e.g., Figure 10.21), a closer look will reveal several anomalies:

▼ Analysis of Variance

Source	DF	Sum of Squares	Mean Square	F Ratio
Model	3	85.67087	28.5570	494.3648
Error	2996	173.06379	0.0578	**Prob > F**
C. Total	2999	258.73467		<.0001*

▼ Parameter Estimates

Term	Estimate	Std Error	t Ratio	Prob>|t|
Intercept	−0.234629	0.013287	−17.66	<.0001*
Income	0.0031894	9.888e-5	32.26	<.0001*
Family	0.032942	0.003839	8.58	<.0001*
CD Account	0.2701636	0.017885	15.11	<.0001*

FIGURE 10.21   Output for multiple linear regression model of *Personal Loan* on three predictors.

1. Using the model to predict Y for each of the observations (or classify them) yields predictions that are not necessarily 0 or 1.

2. A look at the histogram or probability plot of the residuals reveals that the assumption that the dependent variable (or residuals) follows a normal distribution is violated. Clearly, if Y takes only the values 0 and 1, it cannot be normally distributed. In fact, a more appropriate distribution for the number of 1's in the dataset is the binomial distribution with $p = P(Y = 1)$.

3. The assumption that the variance of Y is constant across all classes is violated. Since Y follows a binomial distribution, its variance is $np(1 - p)$. This means that the variance will be higher for classes where the probability of adoption, p, is near 0.5 than where it is near 0 or 1.

The first anomaly is the main challenge when the goal is classification, especially if we are interested in propensities ($P(Y = 1)$). The second and third anomalies are relevant to profiling, where we use statistical inference that relies on standard errors.

In Figure 10.21 you will find partial output from running a multiple linear regression of Personal Loan (PL, coded as $PL = 1$ for customers who accepted the loan offer and $PL = 0$ otherwise) on three of the predictors.

The estimated model is

$$\widehat{PL} = -0.2346 + 0.0032 \text{ Income} + 0.0329 \text{ Family} + 0.27016363 \text{ CD}$$

To predict whether a new customer will accept the personal loan offer ($PL = 1$) or not ($PL = 0$), we input the information on its values for these three predictors. For example, we would predict the loan offer acceptance of a customer with an annual income of $50K with two family members who does not hold CD accounts in Universal Bank to be $-0.0995 + (0.0032)(50) + (0.0329)(2) - 0.135 = -0.0093$. Clearly, this is not a valid "loan acceptance" value. Furthermore the histogram of the residuals (Figure 10.22) reveals that the residuals are probably not normally distributed. Therefore our estimated model is based on violated assumptions.

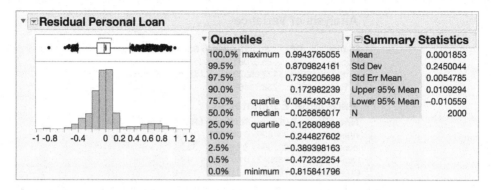

FIGURE 10.22   Histogram of residuals for the validation set from a multiple linear regression model of loan acceptance on the three predictors. This shows that the residuals do not follow the normal distribution that the model assumes.

Appendix B: Evaluating Explanatory Power

When the purpose of the analysis is profiling (i.e., identifying predictor profiles that distinguish the two classes, or explaining the differences between the classes in terms of predictor values), we are less interested in how well the model classifies new data than in how well the model fits the data it was trained on. For example, if we are interested in characterizing loan offer acceptors versus nonacceptors in terms of income, education, and so on, we want to find a model that fits the data best. We therefore mention popular measures used to assess how well the model fits the data. Clearly, we look at the training set in order to evaluate goodness-of-fit.

Overall Strength-of-Fit   As in multiple linear regression, we first evaluate the overall explanatory power of the model before looking at single predictors. We ask: Is this set of predictors better than a simple naive model for explaining the difference between classes?[8]

The *-LogLikelihood* is a statistic that measures overall goodness-of-fit of a logistic model. It is similar to the concept of sum of squared errors (SSE) in the case of least squares estimation (used in linear regression). We compare the -LogLikelihood of our model, called *Full* in JMP (top in Figure 10.23), to the -LogLikelihood of the naive model, labeled *Reduced*. If the reduction in -LogLikelihood is statistically significant (as indicated by a low p-value[9] or by a high multiple R^2), we consider our model to capture a statistically significant relationship between the output and input variables. In JMP, multiple R^2, labeled *RSquare (U)* or *Entropy RSquare*, is computed as $1 - Full/Reduced$. Given the model -LogLikelihood and the RSquare (U), we can compute the null -LogLikelihood by $Reduced = Full/(1 - RSquare(U))$.

Finally, the classification matrix and lift curve for the *training data* (Figure 10.23) give a sense of how accurately the model classifies the data. If the model fits the data well, we expect it to classify these data accurately into their actual classes.

---

[8] In a naive model, no explanatory variables exist and each observation is classified as belonging to the majority class.

[9] Twice the difference between the -LogLikelihood of a naive model and -LogLikelihood of the model at hand approximately follows a chi-squared distribution with k degrees of freedom, where k is the number of predictors in the model.

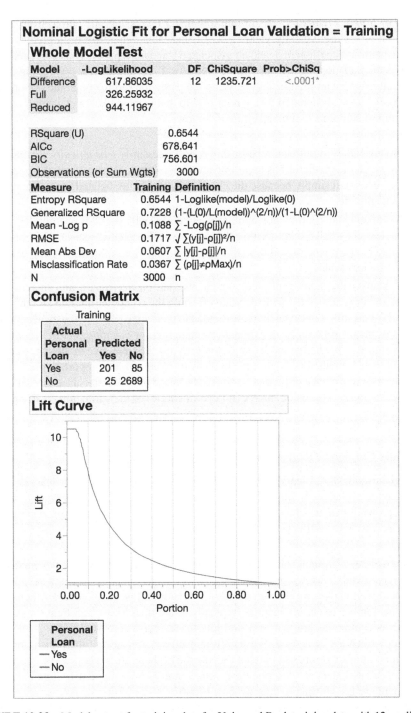

FIGURE 10.23 Model output for training data for Universal Bank training data with 12 predictors.

Impact of Single Predictors As in multiple linear regression, the output from a logistic regression procedure typically yields a coefficient table, where for each predictor X_i we have an estimated coefficient b_i and an associated standard error σ_i. The associated p-value indicates the statistical significance of the predictor X_i, with very low p-values indicating a statistically significant relationship between the predictor and the outcome (given that the other predictors are accounted for), a relationship that is most likely not a result of chance. Three important points to remember are as follows:

1. A statistically significant relationship is not necessarily a *practically significant* one, in which the predictor has great impact. If the sample is very large, the p-value will be very small simply because the chance uncertainty associated with a small sample is negligible.
2. Comparing the coefficient magnitudes, or equivalently the odds magnitudes, is meaningless unless all predictors have the same scale. Recall that each coefficient is multiplied by the predictor value, so that different predictor scales would lead to different coefficient scales.
3. A statistically significant predictor means that, *on average*, a unit increase in that predictor is associated with a certain effect on the outcome (holding all other predictors constant). It does not, however, indicate predictive power. Statistical significance is of major importance in explanatory modeling, or *profiling*, but of secondary importance in predictive modeling (*classification*). In predictive modeling, statistically significant predictors might give hints as to more and less important predictors, but the eventual choice of predictors should be based on predictive measures, such as the validation set confusion matrix (for classification) or the validation set lift curve (for ranking).

Appendix C: Logistic Regression for More Than Two Classes

The logistic model for a binary response can be extended for more than two classes. Suppose that there are m classes. Using a logistic regression model, for each observation we would have m probabilities of belonging to each of the m classes. Since the m probabilities must add up to 1, we need estimate only $m - 1$ probabilities.

Ordinal Classes Ordinal classes are classes that have a meaningful order. For example, in stock recommendations, the three classes *buy*, *hold*, and *sell* can be treated as ordered. As a simple rule, if classes can be numbered in a meaningful way, we consider them ordinal. When the number of classes is large (typically, more than 5), we can treat the dependent variable as continuous and perform multiple linear regression. When $m = 2$, the logistic model described above is used. We therefore need an extension of the logistic regression for a small number of ordinal classes ($3 \leq m \leq 5$). There are several ways to extend the binary class case. Here we describe the *proportional odds* or *cumulative logit method*. For other methods, see Hosmer and Lemeshow (2000).

For simplicity of interpretation and computation, we look at *cumulative* probabilities of class membership. For example, in the stock recommendations we have $m = 3$ classes. Let us denote them by $1 = buy$, $2 = hold$, and $3 = sell$. The probabilities that are estimated by the model are $P(Y \leq 1)$, which is the probability of a *buy* recommendation, and $P(Y \leq 2)$, which is the probability of a *buy* or *hold* recommendation. The three noncumulative probabilities of class membership can easily be recovered from the two cumulative probabilities:

$$P(Y = 1) = P(Y \leq 1)$$
$$P(Y = 2) = P(Y \leq 2) - P(Y \leq 1)$$
$$P(Y = 3) = 1 - P(Y \leq 2)$$

Next we want to model each logit as a function of the predictors. Corresponding to each of the $m - 1$ cumulative probabilities is a logit. In our example we would have

$$\text{logit(buy)} = \log \frac{P(Y \leq 1)}{1 - P(Y \leq 1)}$$
$$\text{logit (buy or hold)} = \log \frac{P(Y \leq 2)}{1 - P(Y \leq 2)}$$

Each of the logits is then modeled as a linear function of the predictors (as in the two-class case). If in the stock recommendations we have a single predictor x, we have two equations:

$$\text{logit(buy)} = \alpha_0 + \beta_1 x$$
$$\text{logit(buy or hold)} = \beta_0 + \beta_1 x$$

This means that both lines have the same slope (β_1) but different intercepts. Once the coefficients $\alpha_0, \beta_0, \beta_1$ are estimated, we can compute the class membership probabilities by rewriting the logit equations in terms of probabilities. For the three-class case, for example, we would have

$$P(Y = 1) = P(Y \leq 1) = \frac{1}{1 + e^{-(a_0 + b_1 x)}}$$
$$P(Y = 2) = P(Y \leq 2) - P(Y \leq 1) = \frac{1}{1 + e^{-(b_0 + b_1 x)}} - \frac{1}{1 + e^{-(a_0 + b_1 x)}}$$
$$P(Y = 3) = 1 - P(Y \leq 2) = 1 - \frac{1}{1 + e^{-(b_0 + b_1 x)}}$$

where a_0, b_0, and b_1 are the estimates obtained from the training set.

For each observation we now have the estimated probabilities that it belongs to each of the classes. In our example, each stock would have three probabilities: for a *buy* recommendation, a *hold* recommendation, and a *sell* recommendation. The last step is to classify the observation into one of the classes. This is done by assigning it to the class with the highest membership probability. So, if a stock had estimated probabilities $P(Y = 1) = 0.2$, $P(Y = 2) = 0.3$, and $P(Y = 3) = 0.5$, we would classify it as getting a *sell* recommendation.

Nominal Classes.   When the classes cannot be ordered and are simply different from one another, we are in the case of nominal classes. An example is the choice between several brands of cereal. A simple way to verify that the classes are nominal is when it makes sense to tag them as $A, B, C, \ldots$, and the assignment of letters to classes does not matter. For simplicity, let us assume that there are $m = 3$ brands of cereal that consumers can choose from (assuming that each consumer chooses one). Then we estimate the probabilities $P(Y = A)$, $P(Y = B)$, and $P(Y = C)$. As before, if we know two of the probabilities, the third probability is determined. We therefore use one of the classes as the reference class. Let us use C as the reference brand.

The goal, once again, is to model the class membership as a function of predictors. So in the cereals example we might want to predict which cereal will be chosen if we know the cereal's price, x.

Next we form $m - 1$ pseudologit equations that are linear in the predictors. In our example we would have

$$\text{logit}(A) = \log \frac{P(Y = A)}{P(Y = C)} = \alpha_0 + \alpha_1 x$$

$$\text{logit}(B) = \log \frac{P(Y = B)}{P(Y = C)} = \beta_0 + \beta_1 x$$

Once the four coefficients are estimated from the training set, we can estimate the class membership probabilities: [10]

$$P(Y = A) = \frac{e^{a_0 + a_1 x}}{1 + e^{a_0 + a_1 x} + e^{b_0 + b_1 x}}$$

$$P(Y = B) = \frac{e^{b_0 + b_1 x}}{1 + e^{a_0 + a_1 x} + e^{b_0 + b_1 x}}$$

$$P(Y = C) = 1 - P(Y = A) - P(Y = B)$$

where a_0, a_1, b_0, and b_1 are the coefficient estimates obtained from the training set. Finally, an observation is assigned to the class that has the highest probability.

Comparing Ordinal and Nominal Models. The estimated logistic models will differ, depending on whether the response is coded as ordinal or nominal. Lets take, for example, data on accidents from the US Bureau of Transportation Statistics. These data can be used to predict whether an accident will result in injuries or fatalities based on predictors such as alcohol involvement, time of day, and road condition. The severity of the accident has three classes: 0 = No Injury, 1 = Nonfatal Injuries, 2 = Fatalities.

Let us first consider severity as an ordinal variable, and fit a simple model with two nominal predictors, alcohol and weather. Each of these predictors is coded as No (not involved in the accident) or Yes (involved in the accident). Next we change the modeling

---

[10]From the two logit equations we see that

$$P(Y = A) = P(Y = C)e^{\alpha_0 + \alpha_1 x}$$
$$P(Y = B) = P(Y = C)e^{\beta_0 + \beta_1 x}$$

Since $P(Y = A) + P(Y = B) + P(Y = C) = 1$, we get

$$P(Y = C) = 1 - P(Y = C)e^{\alpha_0 + \alpha_1 x} - P(Y = C)e^{\beta_0 + \beta_1 x}$$
$$= \frac{1}{e^{\alpha_0 + \alpha_1 x} + e^{\beta_0 + \beta_1 x}}.$$

By plugging this form into the two equations above it, we also obtain the membership probabilities in classes A and B.

type for severity to nominal, and fit a model with the same predictors. The reference severity level, in both models, is severity = 2.

The parameter estimates for the resulting models are shown in Figures 10.24 and 10.25. The ordinal logistic model has estimates for two intercepts (the first for severity = 0 and and the second for severity = 1), but one estimate for the slopes (the coefficients for each predictor). In contrast, the nominal logistic model has estimates for the two intercepts, and also has separate estimates for the slopes for each level of the response.

▼ **Parameter Estimates**

Term	Estimate	Std Error	ChiSquare	Prob>ChiSq
Intercept[0]	−0.1163012	0.0203592	32.63	<.0001*
Intercept[1]	4.42547899	0.0496809	7934.9	<.0001*
Alcohol[No]	−0.2090611	0.0174487	143.56	<.0001*
Weather[No]	−0.1293157	0.0139465	85.97	<.0001*

FIGURE 10.24   Parameter Estimates for *Ordinal* Logistic Regression Model for Accident Severity with two predictors.

▼ **Parameter Estimates**

Term	Estimate	Std Error	ChiSquare	Prob>ChiSq
Intercept	3.65328871	0.0953827	1467.0	<.0001*
Alcohol[No]	−0.591057	0.0591182	99.96	<.0001*
Weather[No]	−0.3868429	0.0873699	19.60	<.0001*
Intercept	3.73181765	0.0952068	1536.4	<.0001*
Alcohol[No]	−0.4075349	0.0586724	48.25	<.0001*
Weather[No]	−0.264972	0.0874458	9.18	0.0024*
For log odds of 0/2, 1/2				

FIGURE 10.25   Parameter Estimates for *Nominal* Logistic Regression Model for Accident Severity with two predictors.

Logistic regression for ordinal and *multinomial* (more than two unordered classes) data is performed in *Fit Y by X* (for one predictor) and *Fit Model* (for more than one predictor). The type of model generated is determined by the modeling type of the response variable in JMP.

PROBLEMS

10.1 Financial Condition of Banks. The file Banks.jmp includes data on a sample of 20 banks. The "Financial Condition" column records the judgment of an expert on the financial condition of each bank. This dependent variable takes one of two possible values—*weak* or *strong*—according to the financial condition of the bank. The predictors are two ratios used in the financial analysis of banks: TotLns&Lses/Assets is the ratio of total loans and leases to total assets and TotExp/Assets is the ratio of total expenses to total assets. The goal is to use the two ratios for classifying the financial condition of a new bank.

Run a logistic regression model (on the entire dataset) that models the status of a bank as a function of the two financial measures provided. Use the *Value Ordering* column property to specify the *success* class as *weak* (i.e., move it to the top of the value list). *Note:* In JMP 13, specify the target level when fitting the model.

a. Write the estimated equation that associates the financial condition of a bank with its two predictors in two forms:

  i. The logit as a function of the predictors.

  ii. The probability as a function of the predictors.

b. Consider a new bank whose total loans and leases/assets ratio = 0.6 and total expenses/assets ratio = 0.11. From your logistic regression model, estimate the following quantities for this bank (save the probability formula to the data table, and enter these values in a new row): the logit, the probability of being financially weak, and the classification of the bank. Confirm the probability using the *Profiler*.

c. The cutoff probability value of 0.5 is used to classify banks as being financially weak or strong. What is the misclassification rate for weak banks that are incorrectly classified as strong? Use the *Alternate Cutoff Confusion Matrix* add-in to find the cutoff value that minimizes the misclassification rate for banks that are financially weak. What is this cutoff value and the misclassification rate?

d. When a bank that is in poor financial condition is misclassified as financially strong, the misclassification cost is much higher than when a financially strong bank is misclassified as weak. To minimize the expected cost of misclassification, should the cutoff value for classification (which is currently at 0.5) be increased or decreased?

e. Interpret the estimated coefficient for the total loans and leases to total assets ratio (TotLns&Lses/Assets) in terms of the odds of being financially weak.

10.2 Identifying Good System Administrators. A management consultant is studying the roles played by experience and training in a system administrator's ability to complete a set of tasks in a specified amount of time. In particular, she is interested in discriminating between administrators who are able to complete given tasks within a specified time and those who are not. Data are collected on the performance of 75 randomly selected administrators. They are stored in the file `SystemAdminis-trators.jmp`.

The variable Experience measures months of full-time system administrator experience, while Training measures the number of relevant training credits. The dependent variable Completed task is either Yes or No, according to whether or not the administrator completed the tasks.

a. Create a scatterplot of Experience versus Training using color or symbol to differentiate programmers who complete the task from those who did not complete it. Which predictor(s) appear(s) potentially useful for classifying task completion?

b. Run a logistic regression model with both predictors using the entire dataset as training data. Model the probability of Completed Task = Yes. Among those who complete the task, what is the percentage of programmers who are incorrectly classified as failing to complete the task?

c. To decrease the percentage in part (b), should the cutoff probability be increased or decreased?

d. How much experience must be accumulated by a programmer with four years of training before his or her estimated probability of completing the task exceeds 50%?

10.3 Sales of Riding Mowers. A company that manufactures riding mowers wants to identify the best sales prospects for an intensive sales campaign. In particular, the manufacturer is interested in classifying households as prospective owners or nonowners on the basis of Income (in \$1000s) and Lot Size (in 1000 ft^2). The marketing expert looked at a random sample of 24 households, given in the file `RidingMowers.jmp`. Use all the data to fit a logistic regression of ownership on the two predictors.

a. What percentage of households in the study were owners of a riding mower?

b. Create a scatterplot of Income versus Lot Size using color or symbol to differentiate owners from nonowners. From the scatterplot, which class seems to have the higher average income, owners or nonowners?

c. Among nonowners, what is the percentage of households classified correctly?

d. To increase the percentage of correctly classified nonowners, should the cutoff probability be increased or decreased?

e. What are the odds that a household with a \$60K income and a lot size of 20,000 ft^2 is an owner?

f. What is the classification of a household with a \$60K income and a lot size of 20,000 ft^2?

g. What is the minimum income that a household with 16,000 ft^2 lot size should have before it is classified as an owner?

10.4 Competitive Auctions on eBay.com. The file `eBayAuctions.jmp` contains information on 1972 auctions transacted on eBay.com during May–June 2004. The goal is to use these data to build a model that will distinguish competitive auctions from noncompetitive ones. A competitive auction is defined as an auction with at least two bids placed on the item being auctioned. The data include variables that describe the item (auction category), the seller (his or her eBay rating), and the auction terms that the seller selected (auction duration, opening price, currency, day of week of auction close). In addition we have the price at which the auction closed. The goal is to predict whether or not the auction will be competitive.

Data preprocessing. Split the data into training and validation datasets using a 60 : 40% ratio.

a. Create a tabular summary of the average of the binary dependent variable (Competitive?) as a function of the various categorical variables. Use the information in the tables to reduce the number of for the categorical variables. For example, categories that appear most similar with respect to the distribution of competitive auctions could be combined using *Recode*.

b. Run a logistic model with all predictors (or recoded predictors). Model the probability that Competitive? = 1.

c. If we want to predict at the start of an auction whether it will be competitive, we cannot use the information on the closing price. Run a logistic model with all predictors as above, excluding price. How does this model compare to the full model with respect to accurate prediction?

d. Interpret the meaning of the coefficient for closing price. Does closing price have a practical significance? Is it statistically significant for predicting competitiveness of auctions? (Use a 10% significance level.)

e. Use mixed stepwise selection to find the model with the best fit to the training data. Which predictors are used?

f. Use stepwise selection and an exhaustive search to find the model with the lowest predictive error rate (use the validation data). Which predictors are used?

g. What is the danger in the best predictive model that you found?

h. Explain why the best-fitting model and the best predictive models are the same or different.

i. If the major objective is accurate classification, what cutoff value should be used?

j. Based on these data, what auction settings by the seller (duration, opening price, ending day, currency) would you recommend as being most likely to lead to a competitive auction?

11

NEURAL NETS

In this chapter we describe neural networks, a flexible data-driven method that can be used for classification or prediction. Although considered a "black box" in terms of interpretability, neural nets have been highly successful in terms of predictive accuracy. We discuss the concepts of "nodes" and "layers" (input layers, output layers, and hidden layers) and how they connect to form the structure of a network. We then explain how a neural network is fitted to data using a numerical example. Because overfitting is a major danger with neural nets, we present a strategy for avoiding it. We describe the different parameters that a user must specify and explain the effect of each on the process. Finally, we discuss the usefulness of neural nets and their limitations.

11.1 INTRODUCTION

Neural networks, also called *artificial neural networks*, are models for classification and prediction. The neural network is based on a model of biological activity in the brain, where neurons are interconnected and learn from experience. Neural networks mimic the way that human experts learn. The learning and memory properties of neural networks resemble the properties of human learning and memory, and neural nets also have a capacity to generalize from particulars.

A number of successful applications have been reported in financial settings (see Trippi and Turban, 1996) such as bankruptcy predictions, currency market trading, picking stocks and commodity trading, detecting fraud in credit card and monetary transactions, and customer relationship management (CRM). There have also been a number of successful applications of neural nets in engineering applications. One of the best known is ALVINN, an autonomous vehicle driving application for normal speeds on highways. Using as input a 30×32 grid of pixel intensities from a fixed camera on the vehicle, the classifier provides the direction of steering. The response variable is a categorical one with 30 classes, such as *sharp left*, *straight ahead*, and *bear right*.

Data Mining for Business Analytics: Concepts, Techniques, and Applications with JMP Pro®, First Edition.
Galit Shmueli, Peter C. Bruce, Mia L. Stephens, and Nitin R. Patel.

The main strength of neural networks is their high predictive performance. Their structure supports capturing very complex relationships between predictors and a response, which is often not possible with other predictive models.

11.2   CONCEPT AND STRUCTURE OF A NEURAL NETWORK

The idea behind neural networks is to combine the input information in a very flexible way that captures complicated relationships between the input variables and the response. For instance, recall that in linear regression models the form of the relationship between the response and the predictors is specified directly by the user. In many cases the exact form of the relationship is very complicated, or is generally unknown. In linear regression modeling we might try different transformations of the predictors, interactions between predictors, and so on, but the specified form of the relationship remains linear. In comparison, in neural networks the user is not required to specify the correct form. Instead, the network tries to learn about such relationships from the data. In fact, linear regression and logistic regression can be thought of as special cases of very simple neural networks that have only input and output layers and no hidden layers.

Although researchers have studied numerous different neural network architectures, the most successful applications in data mining of neural networks have been *multilayer feedforward networks*. These are networks in which there is an *input layer* consisting of nodes (sometimes called *neurons*) that simply accept the input values, and successive layers of nodes that receive input from the previous layers. The outputs of nodes in each layer are inputs to nodes in the next layer. The last layer is called the *output layer*. Layers between the input and output layers are known as *hidden layers*. A feedforward network is a fully connected network with a one-way flow and no cycles. Figure 11.1 shows a diagram for this architecture, with two hidden layers and one node in the output layer representing the target value to be predicted. In a classification problem with m classes, there would be m output nodes.

11.3   FITTING A NETWORK TO DATA

To illustrate how a neural network is fitted to data, we start with a very small illustrative example. Although the method is by no means useful in such a small example, it is useful for explaining the main steps and operations, for showing how computations are done, and for integrating all the different aspects of neural network data fitting. Later, we discuss a more realistic example.

Example 1: Tiny Dataset

Consider the following very small dataset. Table 11.1 includes information on a tasting score for a certain processed cheese. The two predictors are scores for fat and salt, indicating the relative presence of fat and salt in the particular cheese sample (where 0 is the minimum

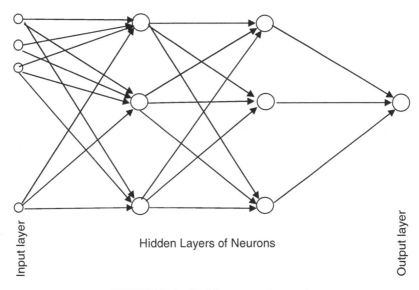

Input layer

Hidden Layers of Neurons

Output layer

FIGURE 11.1   Multilayer neural network.

amount possible in the manufacturing process, and 1 the maximum). The output variable is the cheese sample's consumer taste preference, where "like" or "dislike" indicate whether the consumer likes the cheese or not. Data are in `Tiny Dataset.jmp`.

Figure 11.2 describes a typical neural net that could be used for predicting cheese preference (like/dislike) by new consumers, based on these data. We numbered the nodes in the example from 1 to 7. Nodes 1 and 2 belong to the input layer, nodes 3 to 5 belong to the hidden layer, and nodes 6 and 7 belong to the output layer. The values on the connecting arrows are called *weights*, and the weight on the arrow from node i to node j is denoted by $w_{i,j}$. The *bias* values, denoted by θ_j, serve as an intercept for the output from node j. These are all explained in further detail below.

TABLE 11.1   Tiny Example on Tasting Scores for Six Consumers and Two Predictors

Obs.	Fat Score	Salt Score	Acceptance
1	0.2	0.9	like
2	0.1	0.1	dislike
3	0.2	0.4	dislike
4	0.2	0.5	like
5	0.4	0.5	like
6	0.3	0.8	like

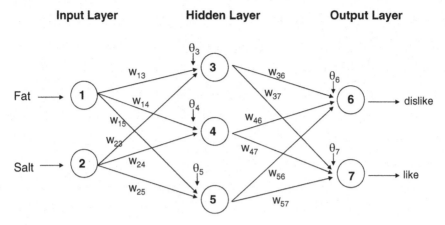

FIGURE 11.2 Neural network for the tiny example. Circles represent nodes ("neurons"), W on arrows are weights, and are bias values.

Computing Output of Nodes

We discuss the input and output of the nodes separately for each of the three types of layers (input, hidden, and output). The main difference is the function used to map from the input to the output of the node.

Input nodes take as input the values of the predictors. Their output is the same as the input. If we have p predictors, the input layer will include p nodes. In our example there are two predictors, and therefore the input layer (shown in Figure 11.2) has two nodes, each feeding into each node of the hidden layer. Consider the first observation: The input into the input layer is $fat = 0.2$ and $salt = 0.9$, and the output of this layer is also $x_1 = 0.2$ and $x_2 = 0.9$.

Hidden layer nodes take as input the output values from the input layer. The hidden layer in this example consists of three nodes, each receiving input from all the input nodes. To compute the output of a hidden layer node, we compute a weighted sum of the inputs and apply a certain function to it. More formally, for a set of input values $x_1, x_2, \ldots, x_p$, we compute the output of node j by taking the weighted sum[1] $\theta_j + \sum_{i=1}^{p} w_{ij} x_i$, where $\theta_j, w_{1,j}, \ldots, w_{p,j}$ are weights that are initially set randomly, then adjusted as the network "learns." Note that θ_j, also called the *bias* of node j, is a constant that controls the level of contribution of node j.

In the next step we take a function g of this sum. The function g, also called a *transfer function* or *activation function*, is some monotone function, such as a linear function $[g(s) = bs]$, an exponential function $[g(s) = \exp(bs)]$, a logistic function $\left[g(s) = \frac{1}{1+e^{-s}}\right]$, and a hyperbolic tangent (TanH) function $\left[g(s) = -1 + \frac{2}{1+e^{-s}}\right]$. The last two functions are by far the most popular in neural networks. Both are sigmoidal functions, namely they have an S shape. The logistic function maps values to the interval [0,1]. The practical value of the logistic function arises from the fact that it has a squashing effect on very small or very large values but is almost linear in the range where the value of the function is between 0.1

---

[1]Other options exist for combining inputs, such as taking the maximum or minimum of the weighted inputs rather than their sum, but they are much less popular.

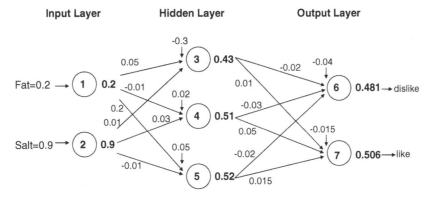

FIGURE 11.3   Computing node outputs (in boldface type) using the first observation in the tiny example and a logistic function.

and 0.9. The TanH maps values to the interval $[-1, 1]$, and is a centered and scaled version of the logistic function.

If we use a logistic function, we can write the output of node j in the hidden layer as

$$\text{Output}_j = g\left(\theta_j + \sum_{i=1}^{p} w_{ij}x_i\right) = \frac{1}{1 + e^{-(\theta_j + \sum_{i=1}^{p} w_{ij}x_i)}} \qquad (11.1)$$

Initializing the Weights   The values of θ_j and w_{ij} are initialized to small, usually random, numbers (typically, but not always, in the range 0.00 ± 0.05). In JMP, normally distributed random values are used for initial weights. Such values represent a state of no knowledge by the network, similar to a model with no predictors. The initial weights are used in the first round of training.

Returning to our example, suppose that the initial weights for node 3 are $\theta_3 = -0.3$, $w_{1,3} = 0.05$, and $w_{2,3} = 0.01$ (as shown in Figure 11.3). Using the logistic function, we can compute the output of node 3 in the hidden layer (using the first observation) as

$$\text{Output}_3 = \frac{1}{1 + e^{-[-0.3 + (0.05)(0.2) + (0.01)(0.9)]}} = 0.43$$

Figure 11.3 shows the initial weights, inputs, and outputs for observation 1 in our tiny example. If there is more than one hidden layer, the same calculation applies, except that the input values for the second, third, and so on, hidden layers would be the output of the preceding hidden layer. This means that the number of input values into a certain node is equal to the number of nodes in the preceding layer. (If there were an additional hidden layer in our example, its nodes would receive input from the three nodes in the first hidden layer.) Finally, the output layer obtains input values from the (last) hidden layer. It applies the same function as above to create the output. In other words, it takes a weighted average of its input values and then applies the function g. In our example, output nodes 6 and 7 receive input from the three hidden layer nodes. We can compute the output of these nodes by

$$\text{Output}_6 = \frac{1}{1 + e^{-[-0.04+(-0.02)(0.43)+(-0.03)(0.51)+(0.015)(0.52)]}} = 0.481$$

$$\text{Output}_7 = \frac{1}{1 + e^{-[-0.015+(0.01)(0.430)+(0.05)(0.507)+(0.015)(0.511)]}} = 0.506$$

These two numbers are *almost* the propensities P(Y = Dislike | Fat = 0.2, Salt = 0.9) and P(Y = Like | Fat = 0.2. Salt = 0.9). The last step involves normalizing these two values so that they add up to 1. In other words,

$$P(Y = \text{Dislike}) = \text{Output}_6/(\text{Output}_6 + \text{Output}_7)$$
$$= 0.481/(0.481 + 0.506) = 0.49$$
$$P(Y = \text{Like}) = 1 - P(Y = \text{Dislike})$$
$$= 0.506/(0.481 + 0.506) = 0.51$$

For classification we use a cutoff value (for a binary response) on the propensity (or estimated probability). Using a cutoff of 0.5, we would classify the record above as "like." When the neural net is use for classification with more than two classes, the class with the highest propensity is selected.

Relation to Linear and Logistic Regression   Consider a neural network with a single output node and no hidden layers. For a dataset with p predictors, the output node receives $x_1, x_2, \ldots, x_p$, takes a weighted sum of these and applies the g function. The output of the neural network is therefore $g\left(\theta + \sum_{i=1}^{p} w_i x_i\right)$.

First, consider a numerical output variable y. If g is the identity function [$g(s) = s$], the output is simply

$$\hat{y} = \theta + \sum_{i=1}^{p} w_i x_i$$

This is exactly equivalent to the formulation of a multiple linear regression! This means that a neural network with no hidden layers, a single output node, and an identity function g searches only for linear relationships between the response and the predictors.

Now consider a binary output variable Y. If g is the logistic function, the output is simply

$$\hat{P}(Y = 1) = \frac{1}{1 + e^{-\left(\theta + \sum_{i=1}^{p} w_i x_i\right)}}$$

which is equivalent to the logistic regression formulation!

In both cases, although the formulation is equivalent to the linear and logistic regression models, the resulting estimates for the weights (*coefficients* in linear and logistic regression) will differ. The neural net estimation method is different from the method used to calculate coefficients in linear regression (least squares), or the method used in logistic regression (maximum likelihood). We return to the topic of model estimation in a few moments.

Preprocessing the Data

When using a logistic activation function, neural networks perform best when the predictors and the response variables are on a scale of [0,1]. For a numerical variable X that takes values in the range $[a, b]$ where $a < b$, we normalize the measurements by subtracting a and dividing by $b - a$. The normalized measurement is then

$$X_{\text{norm}} = \frac{X - a}{b - a}$$

Another operation that can improve the performance of the neural network is transforming highly skewed predictors. In business applications there tend to be many highly right-skewed variables (e.g., income). Taking a log transform of a right-skewed variable will usually spread out the values more symmetrically.

ACTIVATION FUNCTIONS AND DATA PROCESSING FEATURES IN JMP PRO

The three activation functions used in the hidden layer in JMP Pro are *TanH* (the default), *Linear*, and *Gaussian*. In the model launch dialog, *Transform Covariates* transforms all continuous variables to near normality, so transforming the measurements in advance is not required. The *Robust Fit* option can also be used to minimize the impact of response outliers (available for continuous responses only). For categorical variables, the neural platform automatically creates indicator variables (again, this is done behind the scenes).

Training the Model

Training the model involves estimating the weights θ_j and w_{ij} that lead to the best predictive results. The process that we described earlier (Section 11.3) for computing the neural network output for an observation is repeated for all the observations in the training set. For each observation the model produces a prediction which is then compared with the actual response value. Their difference is the error for the output node.

Popular approaches to estimating weights in neural networks involve using the errors iteratively to update the estimated weights. In particular, the error for the output node is distributed across all the hidden nodes that led to it, so that each node is assigned "responsibility" for part of the error. Each of these node-specific errors is then used for updating the weights. The most popular method for using model errors to update weights is *Back Propagation of Errors*.

In the next section, we describe *Back Propagation of Errors*. There we discuss neural model fitting in JMP Pro and illustrate with the tiny data example. Rather than using back propagation of error, JMP Pro uses a global function of the errors (e.g., sum of squared errors) for estimating the weights.

Back Propagation of Error   *Back propagation* is an algorithm for using model errors to update weights ("learning"). As the name implies, errors are computed from the last layer (the output layer) back to the hidden layers.

Let us denote by $\hat{y}_k$ the output from output node k. The error associated with output node k is computed by

$$\text{err}_k = \hat{y}_k(1 - \hat{y}_k)(y_k - \hat{y}_k)$$

Notice that this is similar to the ordinary definition of an error $(y_k - \hat{y}_k)$ multiplied by a correction factor. The weights are then updated as follows:

$$\theta_j^{\text{new}} = \theta_j^{\text{old}} + l \times \text{err}_j \qquad (11.2)$$
$$w_{i,j}^{\text{new}} = w_{i,j}^{\text{old}} + l \times \text{err}_j$$

where l is a *learning rate* or *weight decay* parameter, a constant ranging typically between 0 and 1, which controls the amount of change in weights from one iteration to the other.

In our example the error associated with output node 7 for the first observation is $(0.506)(1 - 0.506)(1 - 0.506) = 0.123$. For output node 6 the error is $0.481(1 - 0.481)(1 - 0.481) = 0.129$. These errors are then used to compute the errors associated with the hidden layer nodes, and those weights are updated accordingly using a formula similar to (11.2).

Two methods for updating the weights are case updating and batch updating. In *case updating*, the weights are updated after each observation is run through the network (called a *trial*). For example, if we used case updating in the tiny example, the weights would first be updated after running observation 1 as follows: Using a learning rate of 0.5, the weights $\theta_7, w_{3,7}, w_{4,7}$, and $w_{5,7}$ are updated to

$$\theta_7 = -0.015 + (0.5)(0.123) = 0.047$$
$$w_{3,7} = 0.01 + (0.5)(0.123) = 0.072$$
$$w_{4,7} = 0.05 + (0.5)(0.123) = 0.112$$
$$w_{5,7} = 0.015 + (0.5)(0.123) = 0.077$$

Similarly, we obtain updated weights $\theta_6 = 0.025, w_{3,6} = 0.045, w_{4,6} = 0.035$, and $w_{5,6} = 0.045$. These new weights are next updated after the second observation is run through the network, the third, and so on, until all observations are used. This is called one *epoch*, *sweep*, or *iteration* through the data. Typically, there are many epochs.

In *batch updating*, the entire training set is run through the network before each updating of weights takes place. As a result the errors err_k in the updating equation is the sum of the errors from all observations. In practice, case updating tends to yield more accurate results than batch updating, but it requires a longer runtime. This is a serious consideration, since even in batch updating, hundreds or even thousands of sweeps through the training data are executed. When does the updating stop? The most common conditions are one of the following:

1. The new weights are only incrementally different from those of the preceding iteration.
2. The misclassification rate reaches a required threshold.
3. The limit on the number of runs is reached.

Neural Network Model Fitting in JMP   The neural platform in JMP uses a different approach to optimizing the weights. JMP treats this as a nonlinear optimization problem, and finds optimal values for the weights and bias values that minimize a function of the combined errors.[2] This approach is used, in part, because it produces similar results to back propagation of error but is generally much faster. It can also be used for both continuous and categorical responses.

Neural networks are highly flexible, and generally have excellent predictive capabilities. However, a drawback is that neural networks have a tendency to overfit the data. JMP minimizes overfitting in two ways: utilizing a penalty parameter and requiring model crossvalidation. Both the penalty parameter and crossvalidation are integrated into the fitting process. The neural fitting algorithm searches iteratively for the optimal value of the penalty parameter AND the optimal weights that minimize error.[3] The algorithm stops searching when the function of the errors on the crossvalidation data is no longer improving.

While we omit the technical details, the iterative process works as follows:

1. Set the penalty parameter to 0.
2. Set the starting values for the weights (typically random normal values).
3. Vary the penalty parameter within an increasingly narrowing range of values.
4. For each value of the penalty parameter, the algorithm searches for weights that minimize the model errors.
5. The model error for the crossvalidation data is monitored, and the algorithm stops when the crossvalidation error is no longer improving. The model with the lowest error on the crossvalidation data is selected.

Additional information is available in the JMP documentation under the *Help* menu (search for *neural* in the book *Specialized Models*) or at www.jmp.com/support/help/Neural_Networks.shtmlx.

---

[2] JMP Pro uses the *maximum likelihood function* to help find the optimal weight and bias values. In general, the likelihood function represents the agreement between the parameter estimates and the observed data. Maximizing the likelihood function provides parameter estimates that are most consistent with the data (Sall, 2012). The algorithm used in the neural platform searches for weights that minimize the *negative loglikelihood* of the data.

[3] We use the term error in this section to refer to the negative loglikelihood, which is the function of the errors that is minimized.

FITTING A NEURAL NETWORK IN JMP PRO

Choose the *Neural* option from the *Analyze > Modeling* menu to fit a neural network. There are several options that the user can choose, which are described in more detail in the JMP documentation:

- Some form of model crossvalidation is required in the neural platform. The default is a one-third random holdback. For small data sets, KFold crossvalidation is recommended. Here the data set is divided into K subsets, or folds. One of the folds is held out as a validation subset while the other K-1 folds are used to train the model. This process rotates through each of the folds, so that each fold acts as the validation set one time.
- The default neural network is one layer with three nodes and the TanH activation function.
- Options to transform covariates, to minimize the influence of response outliers (robust fit, with continuous responses), to apply different penalty methods, and to start the fitting algorithm multiple times (each start is a tour) are found under *Fitting Options*.
- Options for *Boosting*, which fits an additive sequence of models, are also provided (see the boosting discussion in Section 11.4).
- The red triangle for the model provides additional options, such as *Diagram* (to display the neural model), *Estimates* (to show the estimated weights), and *Save Formula* (to save the prediction formula, probabilities, and classifications or predicted values to the data table).

Fitting a Neural Network for the Tiny Data Example in JMP   Let us examine the output from running a neural network in JMP on the tiny data. Following Figures 11.2 and 11.3, we used a single hidden layer with three nodes and the default TanH activation function. Since validation is required in the JMP Pro neural platform, we fit this model with a validation column, with four observations in the training set and two observations in the validation set.

Classification matrices for the training and validation sets are provided in Figure 11.4. We can see that the network correctly classifies both observations in the validation set, but misclassifies one observation in the training set as a "dislike." In Figure 11.5 we see the saved formulas and probabilities, and the resulting model classifications (under *Most Likely Acceptance*). Observation 4, in the training set, was the only misclassification. But, notice how close the probabilities for observation 1 are to the 0.5 cutoff, indicating that a small change in the cutoff value would result in a different classification. This unstable result is not surprising, since the number of observations is far too small to estimate the 13 weights. However, for purposes of illustration, we discuss the remainder of the output.

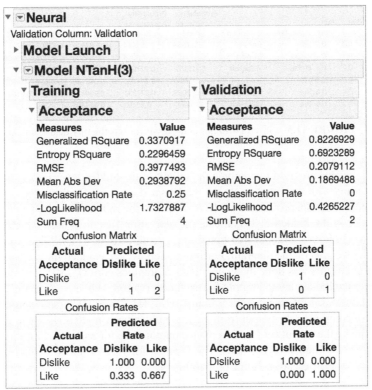

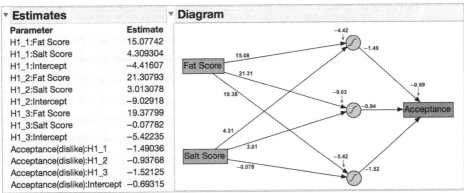

FIGURE 11.4   Output for neural network for the tiny data example. Estimates and Diagram were displayed using red triangle options. The final weights in the neural network diagram were added manually.

The final network weights are shown on the left in Figure 11.4, under *Estimates*. The first group of estimates, starting with H1, show the weights and bias values that connect the input layer and the hidden layer. The first three estimates, starting with H1_1, are the weights and bias value for the first node, the second three values (starting with H1_2) are for the second node, and the last three values (starting with H1_3) are the estimates for the third node. The *bias values* (intercepts) are the weights θ_3, θ_4, and θ_5 in Figure 11.2. These

	Obs	Fat Score	Salt Score	Acceptance	Validation	Probability (Acceptance = dislike)	Probability (Acceptance = like)	H1_1	H1_2	H1_3	Most Likely Acceptance
1	1	0.2	0.9	like	Training	0.447556	0.552444	0.84514	−0.77307	−0.66871	like
2	2	0.1	0.1	dislike	Training	0.949447	0.050553	−0.84508	−0.99728	−0.94094	dislike
3	3	0.2	0.4	dislike	Validation	0.722072	0.277928	0.160176	−0.94482	−0.65781	dislike
4	4	0.2	0.5	like	Training	0.655339	0.344661	0.360128	−0.92614	−0.66001	dislike
5	5	0.4	0.5	like	Training	0.022069	0.977931	0.954915	0.462325	0.816079	like
6	6	0.3	0.8	like	Validation	0.09597	0.90403	0.944404	−0.11269	0.162926	like

FIGURE 11.5   Saved Formulas and Classifications the tiny data example.

estimated weights are used to compute the output of the hidden layer nodes. The last four values, starting with *Acceptance*, are the weights and intercept for the output node.

The neural diagram provided by JMP is shown on the right in Figure 11.4. The symbols in the nodes in the hidden layer indicate the activation functions used for each node. In this case we used the default TanH function for all three nodes.[4] JMP doesn't display the final model weights on the diagram. For illustration, we have added the estimated weights to the diagram (this was done manually, using a graphics editor). JMP uses these weights to compute the hidden layer's output for each observation. When we save the formula for the model, these values are saved in columns H1_1, H1_2, and H1_3 (see Figure 11.5). The formula for computing the output from the first hidden layer node is shown in Figure 11.6.

To compute the output for the Acceptance output node for each observation, we use the outputs from the hidden layer nodes. The formula for *P(Acceptance = like)* is shown in Figure 11.7. For observation 1, this probability is

$$\text{Output} = \frac{1}{1 + e^{-[-0.693 + (-1.49)(0.845) + (-0.938)(-0.773) + (-1.521)(-0.668)]}} = 0.552$$

The probabilities for the other five observations are calculated in the same way, computing the hidden layer outputs and then plugging these outputs into the computation for the output layer.

Using the Output for Prediction and Classification

In the case of a binary response ($m = 2$), JMP uses just one output node with a cutoff value to map a numerical output value to one of the two classes. Although we typically use a cutoff of 0.5 with other classifiers, in neural networks there is a tendency for values to cluster around 0.5 (from above and below). An alternative is to use the validation set to determine a cutoff that produces reasonable predictive performance. Recall that a formula column can be used to change the cutoff for classification, and the *Alternate Cutoff Confusion Matrix Add-in* can be used to compare error rates for different cutoff values (available from the JMP user community at community.jmp.com).

When the neural net is used for classification and we have m classes, the diagram displays only one node. JMP estimates the weights for $m − 1$ classes, but provides output for all m

---

[4] JMP displays a single output node rather than showing output nodes for each of m classes, as seen earlier. When the response is categorical, the logistic function is used to calculate the estimated probability for each class from the output of the hidden layer(s). When the response is continuous, a linear function of the output from the hidden layer(s) is used to calculate the predicted response.

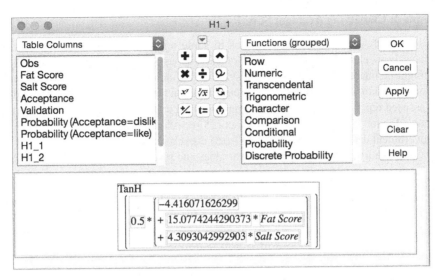

FIGURE 11.6   Formula to compute output from the first hidden layer node.

classes. How do we translate these *m* outputs into a classification rule? Usually the class with the largest probability value determines the neural network's classification.

When the neural network is used for predicting a numerical rather than a categorical response, the same fitting process is used and the model launch options are identical. But, rather than calculating the probability of an outcome the model calculates a predicted value.

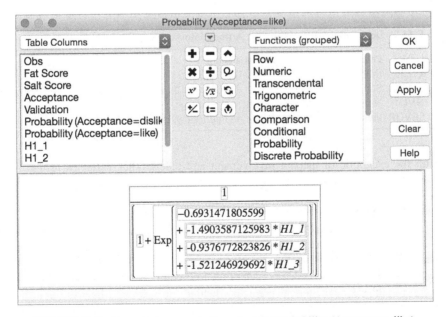

FIGURE 11.7   Formula to compute the output for Probability (Acceptance=like).

Example 2: Classifying Accident Severity

Let us apply the network training process to some real data: US automobile accidents that have been classified by their level of severity as *no injury*, *injury*, or *fatality*. A firm might be interested in developing a system for quickly classifying the severity of an accident, based on initial reports and associated data in the system (some of which rely on GPS-assisted reporting). Such a system could be used to assign emergency response team priorities. Figure 11.8 shows a small extract (15 records, four predictor variables) from a US government database. The data are in `Accidents NN.jmp`.

The explanation of the four predictor variables and response is given in Table 11.2.

	Row Id.	ALCHL_I	PROFIL_I_R	SUR_COND	VEH_INVL	MAX_SEV_IR	Validation
1	1	2	0	1	1	0	Training
2	2	2	1	1	1	2	Validation
3	3	1	0	1	1	0	Validation
4	4	2	0	2	2	1	Training
5	5	2	1	1	2	1	Training
6	6	2	0	1	1	0	Training
7	7	2	0	2	1	2	Validation
8	8	2	1	2	1	1	Validation
9	9	2	1	1	1	1	Training
10	10	2	0	1	1	0	Training
11	11	2	1	1	1	1	Validation
12	12	1	1	1	2	1	Training
13	13	1	1	1	2	2	Validation
14	14	2	0	1	2	2	Validation
15	15	2	0	1	1	0	Validation

FIGURE 11.8   Subset from the accident data, for a high-fatality region.

TABLE 11.2   Description of Variables for Automobile Accident Example

ALCHL_I	Presence (1) or absence (2) of alcohol
PROFIL_I_R	Profile of the roadway: level (1), other (0)
SUR_COND	Surface condition of the road: dry (1), wet (2), snow/slush (3), ice (4), unknown (9)
VEH_INVL	Number of vehicles involved
MAX_SEV_IR	Presence of injuries/fatalities: no injuries (0), injury (1), fatality (2)

With the exception of alcohol involvement and a few other variables in the larger database, most of the variables are ones that we might reasonably expect to be available at the time of the initial accident report, before accident details and severity have been determined by first responders. A data mining model that could predict accident severity on the basis of these initial reports would have value in allocating first responder resources.

We fit the neural network, using a training set with 599 observations and a validation set with 400 observations. The neural net architecture for this classification problem has four nodes in the input layer, one for each of the four predictors (note that JMP will create indicator variables for categorical predictors behind the scenes). Many different neural models are possible. We use a single hidden layer and experiment with the number of nodes using the default TanH activation function. We run several models, starting with three nodes and slowly increasing the number of nodes in the model. Our largest models has eight nodes. We examine the resulting confusion matrices, and find that five nodes gives a good balance between improving the predictive performance on the training set without deteriorating the performance on the validation set. In other words, networks with more than five nodes in the hidden layer performed as well as the five-node network, but add undesirable complexity.

The *Model Launch* dialog in the neural platform in JMP Pro allows you to easily compare models of different complexity. For example, we can run models with different numbers of hidden layers, different numbers of nodes in the hidden layers, and different activation functions and compare the misclassification rates or other validation statistics. For illustration, in Figure 11.9 we show the results of the single layer five-node model (top) described above and the results for a much more complex model (bottom). The complex model has two hidden layers, nine nodes in each layer, and uses the three activation functions available in JMP Pro (TanH, Linear, and Gaussian). The validation misclassification rate for the second model is slightly better (0.12 vs. 0.125), but at the cost of a great deal of added complexity.

To put model complexity into some perspective, let us consider the model weights that must be estimated for each of these models (in Figure 11.9). For our five-node model there are five connections from each of the four nodes in the input layer to each node in the hidden layer, for a total of $4 \times 5 = 20$ connections between the input layer and the hidden layer. There are also 5 connections between the hidden layer and the output layer. However, two of our predictors are binary categorical variables, and the other two predictors and the response have more than two levels. This results in a total of 67 parameters that must be estimated for this model. (To display these parameter estimates, select *Show Estimates* from the red triangle for the model.) To estimate all of the weights in the more complicated model (bottom, in Figure 11.9), we must estimate 209 parameters! (This can be seen by right-clicking on the estimates table and selecting *Make Into Data Table*.)

Note: If you are following along in JMP Pro your model results may be different. You can obtain reproducible results by setting the random seed prior to creating neural models. Recall that the *Random Seed Reset* add-in is available from the JMP User Community (community.jmp.com). In JMP 13, an option to set the random seed is available in the neural Model Launch dialog.

For all neural models, our results depend largely on how we build the model, and there are a few pitfalls to avoid. We discuss these next.

Avoiding Overfitting

As previously discussed, a weakness of the neural network is that it can easily overfit the data, causing the error rate on validation data (and most important, on new data) to be too large. JMP Pro guards against overfitting by using a penalty parameter in the optimization process and by requiring model crossvalidation. Recall that the neural fitting algorithm searches for both optimal weights and the optimal value of the penalty parameter. At each iteration the algorithm keeps track of the errors on the crossvalidation data. JMP uses an early stopping rule to terminate the fitting algorithm when the crossvalidation error is no longer improving.

To illustrate the effect that the penalty parameter has on preventing overfitting, compare the confusion matrices of the five-node network (top, in Figure 11.9) with those from another five-node neural network fit with no penalty parameter (Figure 11.10). The initial model performs slightly better on the training data (a misclassification rate of 0.132 vs.

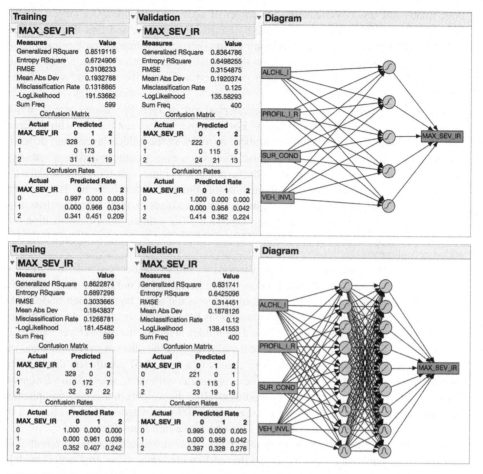

FIGURE 11.9 Neural network models for accident data, with five nodes in one hidden layer (top) and two hidden layer with several nodes and different activation functions (bottom).

0.14), but it performs much better than the no penalty neural network on the validation set (0.125 vs. 0.1475).

11.4   USER INPUT IN JMP PRO

One of the time-consuming and complex aspects of training a model is that we first need to decide on a network architecture. This means specifying the number of hidden layers, the number of nodes in each layer, and the activation function. The usual procedure is to make intelligent guesses using past experience and to conduct several trial-and-error runs on different architectures. Algorithms exist that grow the number of nodes selectively during training or trim them in a manner analogous to what is done in classification and regression trees (see Chapter 11). Research continues on such methods. As of now, no automatic method seems clearly superior to the trial-and-error approach. The neural model launch dialog, with available options, is shown in Figure 11.11. A few general guidelines for choosing an architecture follow.

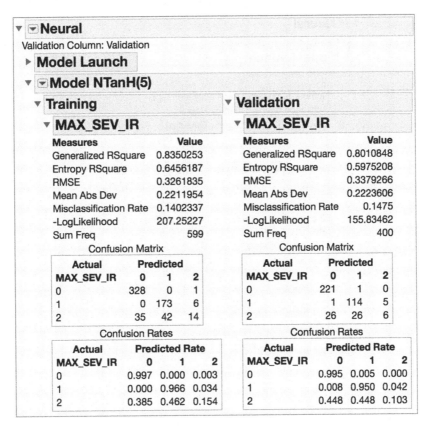

FIGURE 11.10   JMP Pro neural network for accident data with 5 nodes and no penalty.

Number of hidden layers. The most popular choice for the number of hidden layers is one. A single hidden layer is usually sufficient to capture even very complex relationships between the predictors.

Number of nodes. The number of nodes in the hidden layers also determines the level of complexity of the relationship between the predictors that the network captures. The trade-off is between under- and overfitting. On the one hand, using too few nodes might not be sufficient to capture complex relationships (e.g., recall the special cases of a linear relationship in linear and logistic regression, in the extreme case of zero nodes or no hidden layer). On the other hand, too many nodes might lead to overfitting. A rule of thumb is to start with p (number of predictors) nodes and gradually decrease or increase a bit while checking for overfitting. Another approach is to start with the default neural model, with one layer and three nodes, and then run a much more complex model with two layers and several nodes and different activation functions. If the fit statistics don't improve substantially with the more complex model, then a simpler model may suffice.

Number of tours. This field, in the *Fitting Options* section of the Model Launch dialog, tells JMP how many times to restart the model-fitting algorithm. Since the initial weights are based on random starting values, the algorithm can lead to different final

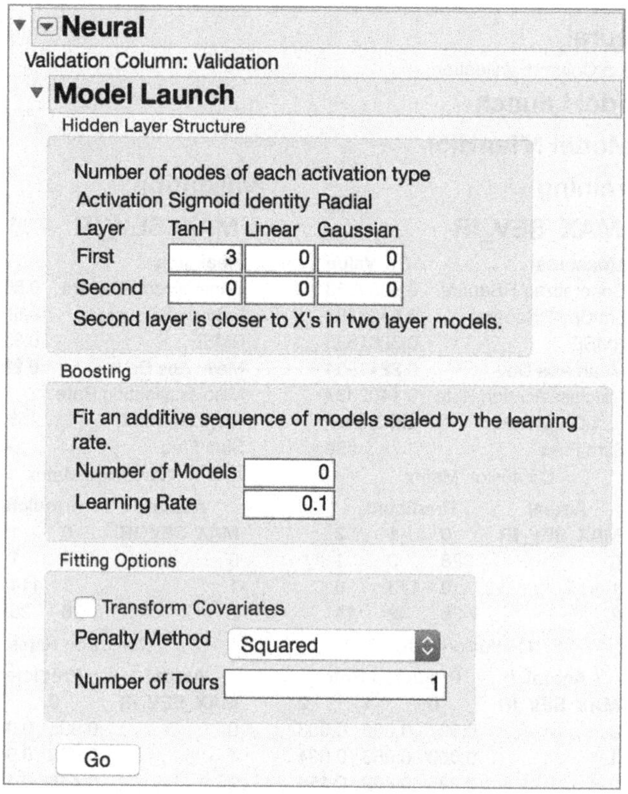

FIGURE 11.11   JMP Pro Neural Model Launch Dialog.

models. When a value (other than 1) is entered into this field, JMP keeps track of the best model after each *tour*, and the model with the lowest -loglikelihood on the validation data is chosen.

Missing value coding (or Informative Missing). This option, which is a check box in the initial Neural dialog window, tells JMP how to handle missing values. If this option is selected, missing values for categorical variables are treated as a separate category. For continuous variables the mean is imputed for missing values, and a new column, indicating whether or not the value is missing, is included in the model.

In addition to the choice of architecture, the user should pay attention to the *choice of predictors.* Since neural networks are highly dependent on the quality of their input, predictors should be chosen based domain knowledge, and variable selection and dimension reduction techniques should first be applied. We return to this point in the discussion of advantages and weaknesses below.

Another option that can be selected in the neural platform JMP Pro is *Boosting*, which is also available for classification and regression trees (see Chapter 9) and is discussed further in Chapter 13. Boosting is an automated way to quickly construct a large additive neural network model by fitting a sequence of smaller models. First, a very simple model is fit, with one layer and one or two nodes. Then, the simple model is fit on the scaled residuals from this model, and so on. All of the models are combined to form the larger final model. The number of models can be specified, and the validation data is used to stop the fitting process when an optimal model is found (or the specified number of models is reached). A *learning rate* for boosted models can also be specified. This parameter takes a value in the range [0, 1], and is used to avoid overfitting. Learning rates close to 1 converge more quickly on a final model, but also have a higher tendency to overfit data.

UNSUPERVISED FEATURE EXTRACTION AND DEEP LEARNING

Most of the data we have dealt with in this book has been numeric or categorical, and the available predictor variables or features have been inherently informative (e.g., the size of a car's engine, or a car's gas mileage). It has been relatively straightforward to discover their impact on the target variable, and focus on the most meaningful ones. With some data, however, we begin with a huge mass of values that are not "predictors" in the same sense. With image data, one "variable" would be the color and intensity of the pixel in a particular position, and one 2-inch square image might have 40,000 pixels. We are now in the realm of "big data."

How can we derive meaningful higher level features, such as edges, curves, shapes? And even higher level features such as faces? *Deep learning* has made big strides in this area. *Deep learning networks* (DLN) refer to neural nets with many hidden layers used to self-learn features from complex data. DLNs are especially effective at capturing local structure and dependencies in complex data, such as in images or audio. This is done by using neural nets in an unsupervised way, where the data are used as both the input (with some added noise) and the output. Using the multiple hidden layers, the DLN's learning is hierarchical—typically from a single pixel to edges, from edges to higher level features, and so on. The network then "pools" similar high-level features across images to build up clusters of similar features. If they appear often enough in large numbers of images, distinctive and regular structures like faces, vehicles, and houses will emerge as highest level features. See Le at al. (2012) for a technical description of how this method yielded labels not just of faces but of different types of faces (e.g., cat faces, of which there are many on the Internet).

Other applications include speech and handwriting recognition. Facebook and Google are effectively using deep learning to identify faces from images, and high- end retail stores use facial recognition technology involving deep learning to identify VIP customers and give them special sales attention (www.npr.org/sections/alltechconsidered/2013/07/21/203273764/high-end-stores-use-facial-recognition-tools-to-spot-vipsJuly 21, 2013).

While DLNs are not new, they have only gained momentum recently, thanks to dramatic improvements in computing power, allowing us to deal both with the huge amount of data and the extreme complexity of the networks, and also thanks to the abundance of unlabeled data such as images and audio. As with ANN, DLN suffers from overfitting and slow runtime. Solutions include using regularization (removing "useless" nodes by zeroing out their coefficients) and tricks to improve over the back-propagation algorithm.

11.5   EXPLORING THE RELATIONSHIP BETWEEN PREDICTORS AND RESPONSE

Neural networks are known to be "black boxes" in the sense that their output does not shed light on the patterns in the data that it models (like our brains). In fact, that is one of the biggest criticism of the method. However, in some cases it is possible to learn more about the relationships that the network captures, by conducting a sensitivity analysis on validation set. This is done by setting all predictor values to their mean (using the *categorical profiler*, available from the red triangle for the model), and obtaining the network's prediction. Then the process is repeated by setting each predictor sequentially to its minimum (and then maximum) value. By comparing the predictions from different levels of the predictors, we can get a sense of which predictors affect predictions more and in what way.

UNDERSTANDING NEURAL MODELS IN JMP PRO

As we have seen with other types of models, the JMP *Profiler*, or *Categorical Profiler* for categorical responses, provides a graphical way to explore the relationships between the predictors and the response(s). This option is available under the red triangle for the model, and from within the *Model Comparison* platform if prediction formulas for the model have been saved to the data table.

The *Assess Variable Importance* option in the *Profiler* can be also used to explore the variables in the neural model. This feature uses simulation to provide measures of the sensitivity of variables to inputs in the model.

When formulas are saved for different neural (or other) models, the *Model Comparison* platform (from *Analyze > Modeling*) provides an efficient way to compare validation statistics for the different models.

11.6   ADVANTAGES AND WEAKNESSES OF NEURAL NETWORKS

As mentioned in Section 11.1, the main advantage of neural networks is their good predictive performance. They are known to have high tolerance to noisy data and the ability to capture highly complicated relationships between the predictors and a response. Their weakest point is in providing insight into the structure of the relationship, hence their black-box reputation.

Several considerations and dangers should be kept in mind when using neural networks. First, although they are capable of generalizing from a set of examples, extrapolation is still a serious danger. If the network sees only cases in a certain range, its predictions outside this range can be completely invalid.

Second, neural networks do not have a built-in variable selection mechanism. This means that there is need for careful consideration of predictors. Classification and regression trees (see Chapter 9) and other dimension reduction techniques (e.g., principal components analysis in Chapter 13) are often used to first identify key predictors. The *Assess Variable Importance* option in JMP Pro can also be used to identify a subset of variables to use in a new (and smaller) neural model.

Third, the extreme flexibility of the neural network relies heavily on having sufficient data for training purposes. As our tiny example shows, a neural network performs poorly when the training set size is insufficient, even when the relationship between the response and predictors is very simple. A related issue is that in classification problems, the network requires sufficient records of the minority class. This is achieved by oversampling, as explained in Chapter 5.

Fourth, a technical problem is the risk of obtaining weights that lead to a local optimum rather than the global optimum, in the sense that the weights converge to values that do not provide the best fit to the training data. We described several parameters that are used to try to avoid this situation (e.g., controlling the learning rate). However, there is no guarantee that the resulting weights are indeed the optimal ones.

Finally, a practical consideration that can determine the usefulness of a neural network is the timeliness of computation. Neural networks are relatively heavy on computation time, requiring a longer runtime than other classifiers. This runtime grows greatly when the number of predictors is increased (as there will be many more weights to compute). The fitting algorithm in JMP Pro (a quasi Newton nonlinear optimization algorithm) was implemented to reduce overall computation time. In applications where real-time or near-real-time prediction is required, runtime should be measured to make sure that it does not cause unacceptable delay in the decision-making process.

PROBLEMS

11.1 Credit Card Use. Consider the following hypothetical bank data on consumers' use of credit card credit facilities in Table 11.3. Create JMP data, and create a neural network like that used for the tiny data example. Use the default validation method (Holdback Portion).

   a. How does the model perform on the validation set?

   b. How many parameters are estimated to build this model?

   c. Use the *Categorical Profiler* to explore how the predicted response changes as you change values of the predictors. Describe what you observe.

   d. Fit another neural network to this same data.

       i. Describe the differences between this model and the first model you generated.

       ii. Why do the results differ?

TABLE 11.3 Data for Credit Card Example and Variable Descriptions

Years	Salary	Used Credit
4	43	0
18	65	1
1	53	0
3	95	0
15	88	1
6	112	1

Note : Years = Number of years that a customer has been with the bank; Salary = customer's salary (in thousands of dollars); Used Credit 1 = customer has left an unpaid credit card balance at the end of at least one month in the prior year; and 0 = balance was paid off at the end of each month.

11.2 Neural Net Evolution. A neural net typically starts out with random weights; hence it produces essentially random predictions in the first iteration. Describe how the neural net evolves (in JMP) to produce a more accurate prediction?

11.3 Car Sales. Consider again the data on used cars (`ToyotaCorolla.jmp`) with 1436 records and details on 38 attributes, including Price, Age, KM, HP, and other specifications. The goal is to predict the price of a used Toyota Corolla based on its specifications.

a. Determine which variables to include, and use the neural platform in JMP Pro to fit a model. Use the validation column for validation, and use the default values in the Neural model launch dialog. Record the RMSE for the training data and the validation data, and save the formula for the model to the data table (use the *Save Fast Formulas* option, which will save the formula as one column in the data table). Repeat the process, changing the number of nodes (and only this) to 5, 10, and 25.

  i. Using your recorded values, what happens to the RMSE for the training data as the number of nodes increases?

  ii. What happens to the RMSE for the validation data?

  iii. Comment on the appropriate number of nodes for the model.

  iv. Use the *Model Comparison* platform to compare these four models (use the Validation column as either a *By* variable or as a *Group* variable, and focus only on the validation data). Here, *RASE* is reported rather than RMSE. Compare RASE and AAE (average absolute error) values for these four models. Which model has the lowest "error"?

b. Conduct a similar experiment to assess the effect of changing the number of layers in the network as well as the activation functions.

11.4 Direct Mailing to Airline Customers. East-West Airlines has entered into a partnership with the wireless phone company Telcon to sell the latter's service via direct mail. The file `EastWestAirlinesNN.jmp` contains a subset of a data sample of who has already received a test offer. About 13% accepted.

You are asked to develop a model to classify East-West customers as to whether they purchased a wireless phone service contract (target variable Phone_Sale), a model that can be used to predict classifications for additional customers.

a. Create a validation column (stratified on Phone_Sale). Then run a neural net model on these data. Request lift curves, and interpret the meaning (in business terms) of the lift curve for the validation set.

b. Comment on the difference between the training and validation lift curves.

c. Run a second neural net model on the data, this time setting the number of tours to 20. Comment now on the difference between this model and the model you ran earlier.

d. Run a third neural net model on the data, with one tour. This time add a second hidden layer, different activation functions, and several nodes. Comment on the difference between this model and the first model you ran, and how overfitting might have affected results.

e. What sort of information, if any, is provided about the effects and importance of the various variables?

f. For this assignment, we did not ask you to set the random seed when creating the validation column or when building the models.

    i. Comment on why we might set the random seed before creating the validation column.

    ii. Comment on why we might set the random seed before building each neural network model.

12

DISCRIMINANT ANALYSIS

In this chapter we describe the method of discriminant analysis, which is a model-based approach to classification. We discuss the main principle where classification is based on the distance of an observation from each of the class averages. We explain the underlying measure of "statistical distance," which takes into account the correlation between predictors. The output of a discriminant analysis procedure generates estimated "discriminant functions," which are then used to produce discriminant scores that can be translated into classifications or propensities (probabilities of class membership). Finally, we discuss the underlying model assumptions, the practical robustness to some, and the advantages of discriminant analysis when the assumptions are reasonably met (e.g., the sufficiency of a small training sample).

12.1 INTRODUCTION

Discriminant analysis is another classification method. Like logistic regression, it is a classical statistical technique that can be used for classification and profiling. It uses sets of measurements on different classes of items to classify new items into one of those classes (*classification*). Common uses of the method have been in classifying organisms into species and subspecies; classifying applications for loans, credit cards, and insurance into low- and high-risk categories; classifying customers of new products into early adopters, early majority, late majority, and laggards; classifying bonds into bond rating categories; classifying skulls of human fossils; as well as in research studies involving disputed authorship, decision on college admission, medical studies involving alcoholics and nonalcoholics, and methods to identify human fingerprints. Discriminant analysis can also be used to highlight aspects that distinguish the classes (*profiling*).

We return to two examples that were described in earlier chapters, riding mowers and personal loan acceptance (Universal Bank). In each of these, the response has two classes. We close with a third example involving more than two classes.

Data Mining for Business Analytics: Concepts, Techniques, and Applications with JMP Pro®, First Edition.
Galit Shmueli, Peter C. Bruce, Mia L. Stephens, and Nitin R. Patel.

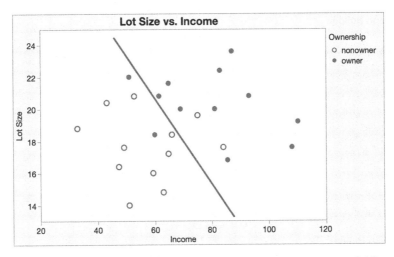

FIGURE 12.1   Scatterplot of Lot Size vs. Income for 24 owners and nonowners of riding mowers. The (ad hoc) line tries to separate owners from nonowners.

Example 1: Riding Mowers

We return to the example from Chapter 7, where a riding mower manufacturer would like to find a way to classify families in a city into those likely to purchase a riding mower and those not likely to buy one. A pilot random sample of 12 owners and 12 nonowners in the city is undertaken. The data are given in Chapter 7 (Table 7.1), and a scatterplot is shown in Figure 12.1.

We can think of a linear classification rule as a line that separates the two-dimensional region into two parts, with most of the owners in one half-plane and most nonowners in the complementary half-plane. A good classification rule would separate the data so that the fewest points are misclassified: the ad hoc line shown in Figure 12.1 seems to do a good job in discriminating between the two classes as it makes four misclassifications out of 24 points. Can we do better?

Example 2: Personal Loan Acceptance (Universal Bank)

Riding mowers is a classic example and is useful in describing the concept and goal of discriminant analysis. However, in today's business applications, the number of observations is generally much larger, and the separation into classes is much less distinct. To illustrate this, we return to the Universal Bank example described in Chapter 9, where the bank's goal is to identify new customers most likely to accept a personal loan. For simplicity, we consider only two predictor variables: the customer's annual income (Income, in $000s), and the average monthly credit card spending (CCAvg, in $000s). The top graph in Figure 12.2 shows the acceptance of a personal loan by a random subset of 200 customers from the bank's database as a function of Income and CCAvg (the random sample was selected in *Graph Builder* using the *Sampling* option from the red triangle). We use a logarithmic scale on both axes to enhance visibility because there are many points condensed in the low-income, low-CCAvg spending area (right-click on an axis and use *Axis Settings* to

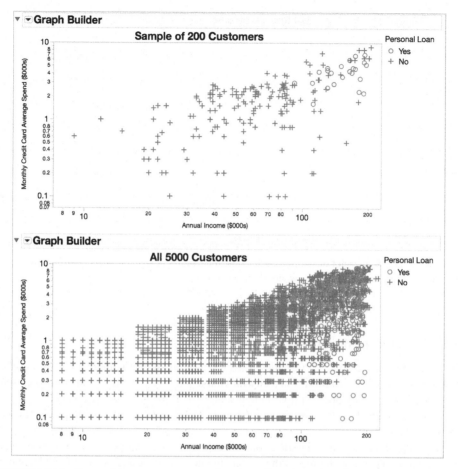

FIGURE 12.2 Personal loan acceptance as a function of income and credit card spending for 200 (top) and 5000 (bottom) customers of the Universal Bank. Axes are in logarithmic scale.

change the scale). Even for this small subset, the separation is not clear. The second figure shows all 5000 customers and the added complexity of dealing with large numbers of observations.

12.2 DISTANCE OF AN OBSERVATION FROM A CLASS

Finding the best separation between observations involves measuring their distance from their class. The general idea is to classify an observation to the class to which it is closest. Suppose that we are required to classify a new customer of Universal Bank as being an acceptor or a nonacceptor of their personal loan offer, based on an income of x. From the bank's database we find that the average income for loan acceptors was \$144.75K and for nonacceptors \$66.24K. We can use Income as a predictor of loan acceptance via a simple *Euclidean distance rule*: if x is closer to the average income of the acceptor class than to the average income of the nonacceptor class, classify the customer as an acceptor; otherwise, classify the customer as a nonacceptor. In other words, if $|x - 144.75| < |x - 66.24|$, classification = acceptor; otherwise, nonacceptor. Moving from a single predictor variable (Income) to two or more predictor variables, the equivalent of the mean of a class is the *centroid* of a class. This is simply the vector of means $\bar{\mathbf{x}} = \left[\bar{x}_1, \ldots, \bar{x}_p\right]$. The Euclidean distance between an observation with p measurements $\mathbf{x} = \left[x_1, \ldots, x_p\right]$ and the centroid $\bar{\mathbf{x}}$ is defined as the root of the sum of the squared differences between the individual values and the means:

$$D_{\text{Euclidean}}(\mathbf{x}, \bar{\mathbf{x}}) = \sqrt{(x_1 - \bar{x}_1)^2 + \cdots + (x_p - \bar{x}_p)^2} \tag{12.1}$$

Using the Euclidean distance has three drawbacks. First, the distance depends on the units we choose to measure the predictor variables. We will get different answers if we decide to measure income in dollars, for instance, rather than in thousands of dollars.

Second, Euclidean distance does not take into account the variability of the predictor variables. For example, when we compare the variability in income in the two classes, we find that for acceptors the standard deviation is lower than for nonacceptors ($31.6K vs. $40.6K). Therefore the income of a new customer might be closer to the acceptors' average income in dollars, but because of the large variability in income for nonacceptors, this customer is just as likely to be a nonacceptor. For this reason we want the distance measure to take into account the variance of the different variables and measure a distance in standard deviations rather than in the original units. This is equivalent to z-scores.

Third, Euclidean distance ignores the correlation between the predictor variables. This is often a very important consideration, especially when we are using many predictor variables to separate classes. In fact, there will often be variables that alone are useful discriminators between classes, but in the presence of other predictor variables are practically redundant, as they capture the same effects as the other variables.

A solution to these drawbacks is to use a measure called *statistical distance* (or *Mahalanobis distance*). Let us denote by S the covariance matrix between the p-variables. The definition of a statistical distance is

$$D_{\text{Statistical}}(\mathbf{x}, \bar{\mathbf{x}}) = [\mathbf{x} - \bar{\mathbf{x}}]' S^{-1} [\mathbf{x} - \bar{\mathbf{x}}]$$

$$= [(x_1 - \bar{x}_1), (x_2 - \bar{x}_2), \ldots, (x_p - \bar{x}_p)] S^{-1} \begin{bmatrix} x_1 - \bar{x}_1 \\ x_2 - \bar{x}_2 \\ \vdots \\ x_p - \bar{x}_p \end{bmatrix}$$

$$\tag{12.2}$$

(The notation $\prime$, which represents the *transpose operation*, simply turns the column vector into a row vector. S^{-1} is the inverse matrix of S, which is the p-dimension extension to division.)

When there is a single predictor ($p = 1$), this reduces to a z-score, since we subtract the mean and divide by the standard deviation. The statistical distance takes into account not only the predictor averages but also the spread of the predictor values and the correlations between the different predictors. To compute a statistical distance between an observation and a class, we must compute the predictor averages (the centroid) and the covariances between each pair of predictors. These are used to construct the distances. The method of discriminant analysis uses *squared* statistical distance as the basis for finding a separating line (or, if there are more than two variables, a separating hyperplane) that is equally distant

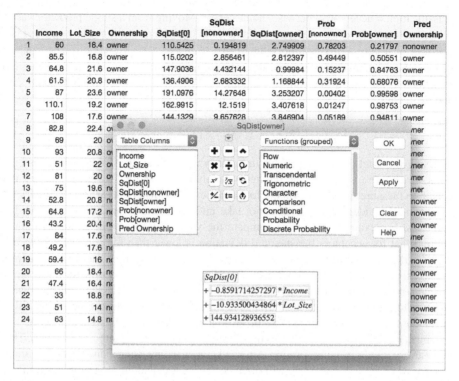

	Income	Lot_Size	Ownership	SqDist[0]	SqDist [nonowner]	SqDist[owner]	Prob [nonowner]	Prob[owner]	Pred Ownership
1	60	18.4	owner	110.5425	0.194819	2.749909	0.78203	0.21797	nonowner
2	85.5	16.8	owner	115.0202	2.856461	2.812397	0.49449	0.50551	owner
3	64.8	21.6	owner	147.9036	4.432144	0.99984	0.15237	0.84763	owner
4	61.5	20.8	owner	136.4906	2.683332	1.168844	0.31924	0.68076	owner
5	87	23.6	owner	191.0976	14.27648	3.253207	0.00402	0.99598	owner
6	110.1	19.2	owner	162.9915	12.1519	3.407618	0.01247	0.98753	owner
7	108	17.6	owner	144.1329	9.657628	3.846904	0.05189	0.94811	owner
8	82.8	22.4	ow						mer
9	69	20	ow						mer
10	93	20.8	ow						mer
11	51	22	ow						mer
12	81	20	ow						mer
13	75	19.6	n						mer
14	52.8	20.8	n						nowner
15	64.8	17.2	n						nowner
16	43.2	20.4	n						nowner
17	84	17.6	n						mer
18	49.2	17.6	n						nowner
19	59.4	16	n						nowner
20	66	18.4	n						nowner
21	47.4	16.4	n						nowner
22	33	18.8	n						nowner
23	51	14	n						nowner
24	63	14.8	n						nowner

FIGURE 12.3 Discriminant analysis output for riding mower data, displaying the formula for the squared distance for Owner (saved as a formula to the data table).

from the different class means.[1] It is based on measuring the squared statistical distances of an observation to each of the classes and allocating it to the closest class.

12.3 FROM DISTANCES TO PROPENSITIES AND CLASSIFICATIONS

In JMP, discriminant functions are estimated to compute the squared distance of each record from the centroids of each class. Returning to the riding mowers example, in Figure 12.3 we see the formula for the squared distances for owner (*SqDist[owner]*). To view these functions in JMP, the formulas must be saved to the data table (from the red triangle select *Score Options > Save Formulas*). Note that the values in column *SqDist[0]* are the squared distances of each observation to the overall centroid of the data (this formula is stored in a format that allows JMP to calculate the squared distances more quickly).

A household is classified into the class of *owners* if the owners squared distance (*or distance score*) is lower than the nonowner squared distance, and into *nonowners* if the reverse is the case. These discriminant functions are specified in a way that can be

---

[1] An alternative approach finds a separating line or hyperplane that is "best" at separating the different clouds of points. In the case of two classes, the two methods coincide.

▼ Squared Distances to each group

Row	Actual	nonowner	owner
1	owner	0.1948191	2.7499086
2	owner	2.8564614	2.8123966
3	owner	4.4321438	0.9998403
4	owner	2.6833323	1.1688444
5	owner	14.276475	3.2532075
6	owner	12.1519	3.4076181
7	owner	9.657628	3.846904
8	owner	9.5429788	1.2460889
9	owner	2.2294449	0.4677979
10	owner	8.6820571	0.85306
11	owner	4.4655294	3.1714345
12	owner	4.1881874	0.0210131
13	nonowner	2.558766	0.2225031
14	nonowner	2.3486534	2.5781727
15	nonowner	0.2120977	3.6894197
16	nonowner	2.1349846	4.9170735
17	nonowner	2.6637394	1.6640699
18	nonowner	0.2579409	6.2343006
19	nonowner	0.6264304	7.0706829
20	nonowner	0.50748	1.8598059
21	nonowner	0.9075089	9.129141
22	nonowner	2.2209497	9.5603279
23	nonowner	3.6821947	14.951055
24	nonowner	1.8811406	9.4881782

FIGURE 12.4 Squared distances for riding mower data. To view this table in JMP, select *Score Options > Show Distances to each group* from the red triangle.

generalized easily to more than two classes. The values given for the functions are simply the weights to be associated with each variable in the linear function in a manner analogous to multiple linear regression. For instance, the first household (highlighted in Figure 12.3) has an income of \$60K and a lot size of 18.4K ft^2. The *owner* distance score is therefore $110.54 + (-0.86)(60) + (-10.93)(18.4) + 144.93 = 2.75$. The nonowner score for this record is 0.19. Since the distance score for nonowner is lower than the distance score for owner, the household is (mis)classified by the model as a nonowner. The squared distances for all 24 households are given in Figure 12.4.

An alternative way for classifying an observation into one of the classes is to compute the probability of belonging to each of the classes and assigning the observation to the most likely class. If we have two classes, we need only compute a single probability for each observation (e.g., of belonging to *owners*). Using a cutoff of 0.5 is equivalent to assigning the observation to the class with the lowest distance score. The advantage of this approach is that we obtain propensities, which can be used for goals such as ranking. For example, we can generate ROC curves, and can save the formulas and use the *Model Comparison* platform to compare lift curves and other statistics (the options for ROC curves and saving formulas are available from *Score Options*, under the top red triangle).

Let us assume that there are *m* classes. To compute the probability of belonging to a certain class *k* for an observation, we need to compute the squared distances for the different

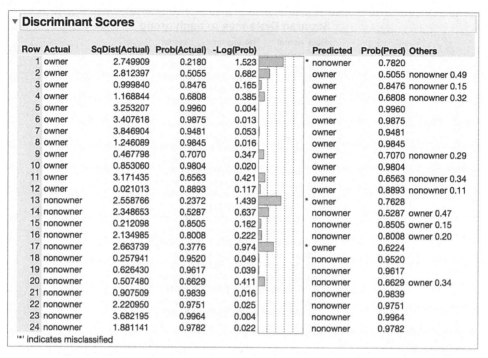

FIGURE 12.5 Output for riding mower data, displaying the propensities (estimated probability) of ownership (under *Prob(Actual)*).

classes, $d_1^2, d_2^2, \ldots, d_m^2$ for that observation and combine them using the following formula:

$$P(\text{Observation belongs to class } k) = \frac{1}{1 + \sum_{k \neq m} e^{(-[d_k^2 - d_m^2]/2)}} \tag{12.3}$$

In JMP, these probabilities are computed automatically, as can be seen in the *Discriminant Scores* table in Figure 12.5. Probabilities for all groups can be displayed by selecting *Show Probabilities to each Group* from *Score Options* under the red triangle, or by saving the formulas to the data table.

We now have three misclassified observations, compared to four in our original (ad hoc) classification (see Figure 12.1). This can be seen in the scatterplot matrix in Figure 12.6, which was produced using the *Scatterplot Matrix* option from the top red triangle in the Discriminant analysis window.[2]

---

[2]The density ellipses graphically display the correlation between the two variables for each group. The covariance of the data in each ellipse is a weighted average of the covariances for the two groups. The discriminant function is the line that best separates the two density ellipses. The scatterplot matrix in Figure 12.6 was modified for illustration. It was rescaled, a legend was added to better visualize the points (right-click in the middle of the graph and select *Row Legend*), marker colors were changed, and the line was drawn (using the *Line* tool from the toolbar) to better visualize the class separation.

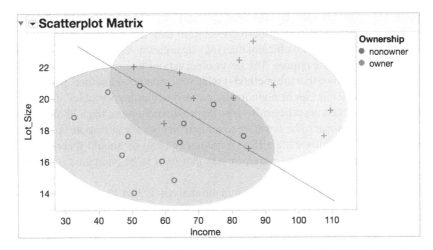

FIGURE 12.6 Class separation obtained from the discriminant model (compared to ad hoc line).

LINEAR DISCRIMINANT ANALYSIS IN JMP

The *Discriminant* platform in JMP is available from the *Analyze > Multivariate Methods* platform. The predictor variables and the validation column must be coded as numeric. The default output includes:

- The *Canonical Plot*, which shows the squared distances from each observation to the centroids for each group.
- The table of *Discriminant Scores*, which includes squared distances, probabilities, and predicted classes.
- The *Score Summaries* table, which provides overall statistics and confusion matrices.

Additional options, such as distances to each group and probabilities for each group, are available under the red triangle. To save the probability formulas and the predicted classifications to the data table, use *Score Options > Save Formulas* from the top red triangle.

12.4   CLASSIFICATION PERFORMANCE OF DISCRIMINANT ANALYSIS

The discriminant analysis method relies on two main assumptions: First, it assumes that the measurements in all classes come from a multivariate normal distribution. When this assumption is reasonably met, discriminant analysis is a more powerful tool than other classification methods, such as logistic regression. In fact, Efron (1975) showed that discriminant analysis is 30% more efficient than logistic regression if the data are multivariate normal, in the sense that we require 30% fewer observations to arrive at the same results. In practice, it has been shown that this method is relatively robust to departures from normality in the sense that predictors can be nonnormal or coded as dummy variables. This is true as long as the smallest class is sufficiently large (approximately more than 20 observations). This method is also known to be sensitive to outliers in both the univariate space of single predictors and in the multivariate space. Exploratory analysis should therefore be used to identify whether there are any extreme cases that need to be addressed (i.e. corrected, re-coded, or eliminated).

The second assumption behind discriminant analysis is that the correlation structure between the different predictors within a class is the same across classes. This can be roughly checked by computing the correlation matrix between the predictors for each class and comparing matrices. If the correlations differ substantially across classes, the classifier will tend to classify cases into the class with the largest variability. When the correlation structure differs significantly and the dataset is very large, an alternative is to use quadratic discriminant analysis (under *Discriminant Method* select *Quadratic, Different Covariances*).[3]

Notwithstanding the caveats embodied in these statistical assumptions, recall that in a predictive modeling environment, the ultimate test is whether the model works effectively. So a reasonable approach is to conduct some exploratory analysis with respect to normality and correlation, train and evaluate a model, and then, depending on classification accuracy and what you learned from the initial exploration, circle back and explore further to learn whether outliers should be examined or choice of variables revisited.

With respect to the evaluation of classification accuracy, we once again use the general measures of performance that were described in Chapter 5 (judging the performance of a classifier), with the principal measure based on the misclassification rate and the confusion matrix. The same argument for using the validation set for evaluating performance still holds. For example, in the riding mower example, families 1, 13, and 17 are misclassified, as can be seen in Figure 12.5 (misclassified observations are marked with an asterisk). This means that the model yields an error rate of 12.5% for these data. However, this rate is a biased estimate—it is overly optimistic because we have used the same data for fitting the classification functions and for estimating the error. Therefore, as with all other models, we test performance on a validation set that includes data that were not involved in estimating the classification functions.

The confusion matrix for the mower data, along with other summary statistics, is shown in Figure 12.7. Since our dataset is small, we have not used a validation set in this example.

---

[3] In practice, quadratic discriminant analysis has not been found useful except when the difference in the correlation matrices is large and the number of observations available for training and testing is large. The reason is that the quadratic model requires estimating many more parameters that are all subject to error [for c classes and p variables, the total number of parameters to be estimated for all the different correlation matrices is $cp(p + 1)/2$].

▼ Score Summaries

Source	Count	Number Misclassified	Percent Misclassified	Entropy RSquare	-2LogLikelihood
Training	24	3	12.5000	0.53433	15.4934

Training

Actual Ownership	Predicted nonowner	owner
nonowner	10	2
owner	1	11

FIGURE 12.7   Score summaries and Confusion Matrix for mower data.

12.5 PRIOR PROBABILITIES

So far we have assumed that our objective is to minimize the classification error. The method presented above assumes that the chances of encountering an observation from either class is the same. If the probability of encountering an observation for classification in the future is not equal for the different classes, we should modify our functions to reduce our expected (long-run average) error rate. The modification is done as follows: Let us denote by p_k the prior or future probability of membership in class k (in the two-class case we have p_1 and $p_2 = 1 - p_1$). We modify the formula for the estimated probabilities for each class by adding $-2\log(p_k)$.[4] The formula for computing probabilities, where p_k is the prior probability of membership for group k, becomes

$$P(\text{Observation belongs to class } k) = \frac{1}{1 + \sum_{k \neq m} e^{(-[(d_k^2 - 2log(p_k)) - (d_m^2 - 2log(p_m))]/2)}}$$

(12.4)

To illustrate this, suppose that the percentage of riding mower owners in the population is 15%, compared to 50% in the sample. This means that the model should classify fewer households as owners. To account for this distortion, we adjust the constants in the discriminant functions (see Figure 12.3) by adding $-2\log(0.15)$ = 3.794 to the squared distance formula for owner and $-2\log(0.85) = 0.325$ to the formula for nonowner. For example, for family 1, the adjusted owner distance score is now $110.54 + (-0.86)(60) + (-10.93)(18.4) + 144.93 + 3.794 = 6.54$ (see Figure 12.8), and the adjusted nonowner score is now 0.519. It is therefore classified as nonowner.

To see how the adjustment can affect classifications, consider family 13, which was misclassified as an owner in the case involving equal probability of class membership.

---

[4] JMP sets the prior class probabilities as the proportions that are encountered in the dataset. This is based on the assumption that a random sample will yield a reasonable estimate of membership probabilities. However, the priors can be specified within the discriminant analysis platform using *Specify Priors*.

Discriminant Scores

Row	Actual	SqDist(Actual)	Prob(Actual)	-Log(Prob)		Predicted	Prob(Pred)	Others
1	owner	6.544149	0.0469	3.060		* nonowner	0.9531	
2	owner	6.606637	0.1528	1.878		* nonowner	0.8472	
3	owner	4.794080	0.4954	0.702		* nonowner	0.5046	
4	owner	4.963084	0.2734	1.297		* nonowner	0.7266	
5	owner	7.047447	0.9776	0.023		owner	0.9776	
6	owner	7.201858	0.9332	0.069		owner	0.9332	
7	owner	7.641144	0.7633	0.270		owner	0.7633	nonowner 0.24
8	owner	5.040329	0.9179	0.086		owner	0.9179	
9	owner	4.262038	0.2986	1.209		* nonowner	0.7014	
10	owner	4.647300	0.8984	0.107		owner	0.8984	nonowner 0.10
11	owner	6.965675	0.2521	1.378		* nonowner	0.7479	
12	owner	3.815253	0.5864	0.534		owner	0.5864	nonowner 0.41
13	nonowner	2.883804	0.6379	0.449		nonowner	0.6379	owner 0.36
14	nonowner	2.673691	0.8641	0.146		nonowner	0.8641	owner 0.14
15	nonowner	0.537136	0.9699	0.031		nonowner	0.9699	
16	nonowner	2.460022	0.9579	0.043		nonowner	0.9579	
17	nonowner	2.988777	0.7747	0.255		nonowner	0.7747	owner 0.23
18	nonowner	0.582979	0.9912	0.009		nonowner	0.9912	
19	nonowner	0.951468	0.9930	0.007		nonowner	0.9930	
20	nonowner	0.832518	0.9176	0.086		nonowner	0.9176	
21	nonowner	1.232547	0.9971	0.003		nonowner	0.9971	
22	nonowner	2.545988	0.9955	0.004		nonowner	0.9955	
23	nonowner	4.007233	0.9994	0.001		nonowner	0.9994	
24	nonowner	2.206178	0.9961	0.004		nonowner	0.9961	

'*' indicates misclassified

FIGURE 12.8   Discriminant scores for mower data, with the prior probability for owners set at 0.15.

When we account for the lower probability of owning a mower in the population, family 13 is classified properly as a nonowner (its owner distance score is above the nonowner score). Figure 12.8 shows the distance scores for all households, adjusting for priors. Note that six of the 24 observations are now misclassified.

12.6   CLASSIFYING MORE THAN TWO CLASSES

Example 3: Medical Dispatch to Accident Scenes

Ideally, every automobile accident call to 911 results in the immediate dispatch of an ambulance to the accident scene. However, in some instances the dispatch might be delayed (e.g., at peak accident hours or in some resource-strapped towns or shifts). In such an event, the 911 dispatchers must make decisions about which units to send based on sketchy information. It is useful to augment the limited information provided in the initial call with additional information in order to classify the accident as minor injury, serious injury, or death. For this purpose we can use data that were collected on automobile accidents in the United States in 2001 that involved some type of injury. For each accident, additional information is recorded, such as day of week, weather conditions, and road type. Figure 12.9 shows a small sample of observations with 10 measurements of interest. These measurements were extracted from a larger set of possible measurements, some of which were dummy coded. Data are in `Accidents1000 DA.jmp`.

	RushHour	WRK_ZONE	WKDY	INT_HWY	LGTCON_day	LEVEL	SPD_LIM	SUR_COND _dry	TRAF_WAY _two_way	WEATHER _adverse	MAX_SEV
1	1	0	1	1	0	1	70	0	0	1	no-injury
2	1	0	1	0	0	0	70	0	0	1	no-injury
3	1	0	1	0	0	0	65	0	0	1	nonfatal
4	1	0	1	0	0	0	55	0	1	0	nonfatal
5	1	0	0	0	0	0	35	0	0	1	no-injury
6	1	0	1	0	0	1	35	0	0	1	no-injury
7	0	0	1	1	0	1	70	0	0	1	nonfatal
8	0	0	1	0	0	1	35	0	1	1	no-injury
9	1	0	1	0	0	0	25	0	0	1	nonfatal
10	1	0	1	0	0	0	35	0	0	1	nonfatal
11	1	0	1	0	0	0	30	0	0	1	nonfatal
12	1	0	1	0	0	0	60	0	0	0	no-injury
13	1	0	1	0	0	0	40	0	1	0	no-injury
14	0	0	1	0	0	1	45	1	0	0	nonfatal
15	1	0	1	0	0	1	35	1	1	0	nonfatal

FIGURE 12.9 Sample of 15 automobile accidents from the 2001 Department of Transportation database. Each accident is classified as one of three injury types (no-injury, nonfatal, or fatal).

USING CATEGORICAL PREDICTORS IN DISCRIMINANT ANALYSIS IN JMP

All predictors in discriminant analysis must be continuous—categorical predictors must first be converted to continuous indicator (dummy) variables. To create indicator columns in JMP use the *Make Indicator Columns* utility under *Cols > Utilities*, and make sure the modeling type is continuous.

The goal is to see how well the predictors can be used to classify injury type correctly. To evaluate this, a sample of 1000 observations was drawn and partitioned into training and validation sets, and a discriminant analysis was performed. Distance scores for the first 15 observations are shown in Figure 12.10.

The output structure is very similar to that for the two-class case. The only difference is that each observation now has three discriminant functions (one for each injury type), and the confusion matrices for the training and validation sets are 3×3 to account for all the combinations of correct and incorrect classifications (see *Score Summaries* at the top of Figure 12.11).

The rule for classification is still to classify an observation to the class that has the lowest corresponding distance score (or highest probability). The distance scores are computed, as before, using the discriminant function coefficients. The saved formula for the squared distances for *fatal* accidents is shown in Figure 12.12. For each accident, the propensity (probability) of each class can also be calculated. The saved formula for the probability for *fatal* accidents is shown in Figure 12.13. The probabilities of an accident involving no injuries or nonfatal injuries are computed in a similar manner.

Squared distances and membership probabilities can be obtained directly in the discriminant platform in JMP for all observations and classes (see Figure 12.11). For the first

Discriminant Analysis

Discriminant Method: Linear
Classification: MAX_SEV
Validation: Validation

Canonical Plot

Discriminant Scores

Row	Actual	SqDist(Actual)	Prob(Actual)	-Log(Prob)		Predicted	Prob(Pred)	Others
1	no-injury	15.3283	0.5454	0.606		no-injury	0.5454	fatal 0.13 nonfatal 0.32
2	no-injury	14.8251	0.6751	0.393	v	no-injury	0.6751	nonfatal 0.30
3	nonfatal	14.5141	0.3115	1.166	v *	no-injury	0.6687	
4	nonfatal	21.5454	0.0159	4.142	*	fatal	0.9548	
5	no-injury	15.6533	0.5332	0.629		no-injury	0.5332	nonfatal 0.44
6	no-injury	13.5957	0.5754	0.553		no-injury	0.5754	nonfatal 0.41
7	nonfatal	15.4285	0.2691	1.313	v *	no-injury	0.4343	fatal 0.30
8	no-injury	11.6947	0.5265	0.642	v	no-injury	0.5265	nonfatal 0.40
9	nonfatal	15.9223	0.4029	0.909	*	no-injury	0.5940	
10	nonfatal	12.7587	0.3799	0.968	*	no-injury	0.6152	
11	nonfatal	14.1062	0.3914	0.938	v *	no-injury	0.6047	
12	no-injury	21.0122	0.0491	3.015	*	fatal	0.9258	
13	no-injury	18.5104	0.0538	2.922	v *	fatal	0.9125	
14	nonfatal	10.0670	0.4702	0.755	v	nonfatal	0.4702	no-injury 0.43
15	nonfatal	9.8636	0.5094	0.674	v	nonfatal	0.5094	no-injury 0.44

FIGURE 12.10 Discriminant scores for the first 15 observations of the accidents data (validation data are marked with a v).

accident in the training set, the highest probability (and the lowest squared distance) is for no injuries, and therefore it is classified as a *no-injury* accident. The actual and predicted classifications are shown in Figure 12.10 (and are provided when the formulas are saved to the data table). The first two observations are correctly classified as no-injury, while the third accident, which was a nonfatal accident, is incorrectly classified as no-injury.

12.7 ADVANTAGES AND WEAKNESSES

Discriminant analysis is considered more of a statistical classification method than a data mining method. This is reflected in its absence or short mention in many data mining resources. However, it is very popular in social sciences and has shown good performance. The use and performance of discriminant analysis are similar to those of multiple linear regression. The two methods therefore share several advantages and weaknesses.

Like linear regression, discriminant analysis searches for the optimal weighting of predictors. In linear regression the weighting is with relation to the response, whereas in discriminant analysis it is with relation to separating the classes. Both use the same estimation method of least squares, and the resulting estimates are robust to local optima.

In both methods an underlying assumption is normality. In discriminant analysis we assume that the predictors are approximately from a multivariate normal distribution. Although this assumption is violated in many practical situations (e.g., with commonly used binary predictors), the method is surprisingly robust. According to Hastie et al. (2001), the reason might be that data can usually support only simple separation boundaries, such as linear boundaries. However, for continuous variables that are found to be very skewed (as can be seen through a histogram), transformations such as the log transform can improve

Score Summaries

Source	Count	Number Misclassified	Percent Misclassified	Entropy RSquare	-2LogLikelihood
Training	600	303	50.5000	−0.3226	1167.13
Validation	400	228	57.0000	−0.4788	

Training

Actual	Predicted		
MAX_SEV	fatal	no-injury	nonfatal
fatal	3	0	2
no-injury	30	117	145
nonfatal	27	99	177

Validation

Actual	Predicted		
MAX_SEV	fatal	no-injury	nonfatal
fatal	1	0	2
no-injury	23	68	109
nonfatal	24	70	103

▸ Groups

Probabilities to each group

Row	Actual	fatal	no-injury	nonfatal
1	no-injury	0.13299	0.54538	0.32162
2	no-injury	0.02486	0.67508	0.30006
3	nonfatal	0.01979	0.66869	0.31152
4	nonfatal	0.95484	0.02926	0.01589
5	no-injury	0.02191	0.53316	0.44493
6	no-injury	0.01326	0.57541	0.41133
7	nonfatal	0.29662	0.43433	0.26905
8	no-injury	0.07309	0.52649	0.40042
9	nonfatal	0.00307	0.59399	0.40294
10	nonfatal	0.00492	0.61519	0.37990
11	nonfatal	0.00388	0.60472	0.39140
12	no-injury	0.92584	0.04906	0.02510
13	no-injury	0.91252	0.05383	0.03366
14	nonfatal	0.09962	0.43019	0.47020
15	nonfatal	0.05057	0.44001	0.50942

Squared Distances to each group

Row	Actual	fatal	no-injury	nonfatal
1	no-injury	18.150681	15.328269	16.384485
2	no-injury	21.428485	14.825068	16.446744
3	nonfatal	20.026318	12.98638	14.514108
4	nonfatal	13.353984	20.324376	21.54541
5	no-injury	22.037369	15.653277	16.015088
6	no-injury	21.136033	13.595659	14.267062
7	nonfatal	15.233423	14.470694	15.428501
8	no-injury	15.643756	11.694723	12.242125
9	nonfatal	25.678266	15.146165	15.9223
10	nonfatal	21.453731	11.794671	12.758704
11	nonfatal	23.331703	13.236122	14.106207
12	no-injury	15.136994	21.012248	22.352824
13	no-injury	12.849558	18.510388	19.449576
14	nonfatal	13.170614	10.244805	10.066957
15	nonfatal	14.48343	10.156557	9.8636269

FIGURE 12.11 Score Summaries (with confusion matrices), Probabilities and Squared Distances for the first 15 observations.

```
SqDist[0]
+ −1.8451247017583 * RushHour
+ −1.035721888051 * WRK_ZONE
+ −9.5602988939122 * WKDY
+ 3.68375653127554 * INT_HWY
+ −7.4140248430762 * LGTCON_day
+ −5.2537875463818 * LEVEL
+ −1.0102634442539 * SPD_LIM
+ −19.997720150439 * SUR_COND_dry
+ −14.215953228658 * TRAF_WAY_two_way
+ −19.376042203325 * WEATHER_adverse
+ 41.6166356364959
```

FIGURE 12.12 The formula for the squared distance for fatal accidents, saved to the data table.

performance. In addition, the method's sensitivity to outliers commands exploring the data for extreme values and removing those observations from the analysis.

$$\frac{1}{\left[1 + \left[\text{Exp} \left[-0.5 * \left(\textit{SqDist[fatal]} - \textit{SqDist[nonfatal]} \right) \right] \right] + \text{Exp} \left[-0.5 * \left(\textit{SqDist[no-injury]} - \textit{SqDist[nonfatal]} \right) \right] \right]}$$

FIGURE 12.13 The formula for the probability of fatal accidents, saved to the data table

An advantage of discriminant analysis as a classifier (it is like logistic regression in this respect) is that it provides estimates of single-predictor contributions.[5] This is useful for obtaining a ranking of the importance of predictors, and for variable selection.

Finally, the method is computationally simple, parsimonious, and especially useful for small datasets. With its parametric form, discriminant analysis makes the most out of the data and is therefore especially useful where the data are few (as explained in Section 12.4).

PROBLEMS

12.1 Personal Loan Acceptance. Universal Bank is a relatively young bank growing rapidly in terms of overall customer acquisition. The majority of these customers are liability customers with varying sizes of relationship with the bank. The customer base of asset customers is quite small, and the bank is interested in expanding this base rapidly to bring in more loan business. In particular, it wants to explore ways of converting its liability customers to personal loan customers.

A campaign the bank ran for liability customers last year showed a healthy conversion rate of over 9% successes. This has encouraged the retail marketing department to devise smarter campaigns with better target marketing. The goal of our analysis is to model the previous campaign's customer behavior to analyze what combination of factors make a customer more likely to accept a personal loan. This will serve as the basis for the design of a new campaign.

The file `UniversalBank.jmp` contains data on 5000 customers. The data include customer demographic information (e.g., age, income), the customer's relationship with the bank (e.g., mortgage, securities account), and the customer response to the last personal loan campaign (Personal Loan). Among these 5000 customers, only 480 (= 9.6%) accepted the personal loan that was offered to them in the previous campaign.

Data preparation: Partition the data (60% training and 40% validation), and turn categorical predictors into continuous dummy variables.

a. Compute summary statistics for the predictors separately for loan acceptors and nonacceptors. For continuous predictors, compute the mean and standard deviation. For the original categorical predictors, compute the percentages or proportions. Are there predictors where the two classes differ substantially?

---

[5]Comparing predictor contribution requires normalizing all the predictors before running discriminant analysis. Then compare each coefficient across the discriminant functions: coefficients with large differences indicate a predictor with high separation power.

b. Perform a discriminant analysis that models Personal Loan as a function of the remaining predictors (excluding ZIP Code). The *success* class as 1 (loan acceptance). Examine the model performance on the validation set.

    i. What is the misclassification rate on the validation set?

    ii. Is one type of misclassification more likely than the other?

    iii. Select three customers who were misclassified as acceptors and three who were misclassified as nonacceptors. The goal is to determine why they are misclassified. First, examine their probability of being classified as acceptors: is it close to the threshold of 0.5? If not, compare their predictor values to the summary statistics of the two classes to determine why they were misclassified.

c. As in many marketing campaigns, it is more important to identify customers who will accept the offer rather than customers who will not accept it. Therefore, a good model should be especially accurate at detecting acceptors. Examine the ROC curve and lift curve for the validation set and interpret them in light of this ranking goal. (Hints: ROC curve is a red triangle option under *Score Options*. To view the lift curve, save the formulas to the data table, and use *Analyze > Modeling > Model Comparison*, with the validation column in the *By* field.)

d. Compare the results from the discriminant analysis with those from a logistic regression. Examine the confusion matrices, the ROC and lift curves (again, using *Model Comparison*). Which method performs better on your validation set in detecting the acceptors?

e. The bank is planning to continue its campaign by sending its offer to 1000 additional customers. Suppose that the cost of sending the offer is $1 and the profit from an accepted offer is $50. What is the expected profitability of this campaign? (Compute this manually–refer to the discussion in Chapter 5.)

f. The cost of misclassifying a loan acceptor customer as a nonacceptor is much higher than the opposite misclassification cost. To minimize the expected cost of misclassification, should the cutoff value for classification (which is currently at 0.5) be increased or decreased?

12.2 Identifying Good System Administrators. A management consultant is studying the roles played by experience and training in a system administrator's ability to complete a set of tasks in a specified amount of time. In particular, she is interested in discriminating between administrators who are able to complete given tasks within a specified time and those who are not. Data are collected on the performance of 75 randomly selected administrators. They are stored in the file SystemAdministrators.jmp.

Using these data, the consultant performs a discriminant analysis. The variable Experience measures months of full time system administrator experience, while Training measures number of relevant training credits. The dependent variable Completed is either *Yes* or *No*, according to whether or not the administrator completed the tasks.

a. Create a scatterplot of Experience vs. Training using color or symbol to differentiate administrators who completed the tasks from those who did not complete them. See if you can identify a line that separates the two classes with minimum misclassification.

b. Run a discriminant analysis with both predictors using the entire dataset as training data. Among those who completed the tasks, what is the percentage of administrators who are classified incorrectly as failing to complete the tasks?

c. How would you classify an administrator with four months of experience and 6 credits of training.

d. How much experience must be accumulated by a administrator with 4 training credits before his or her estimated probability of completing the tasks exceeds 50%?

e. Compare the classification accuracy of this model to that resulting from a logistic regression model.

12.3 Detecting Spam Email (from the UCI Machine Learning Repository). A team at Hewlett-Packard collected data on a large number of email messages from their postmaster and personal email for the purpose of finding a classifier that can separate email messages that are spam vs. nonspam (aka "ham"). The spam concept is diverse: it includes advertisements for products or websites, "make money fast" schemes, chain letters, pornography, and so on. The definition used here is "unsolicited commercial email."

The file Spambase.jmp contains information on 4601 email messages, among which 1813 are tagged "spam." The predictors include 57 attributes, most of them are the average number of times a certain word (e.g., mail, George) or symbol (e.g., #, !) appears in the email. A few predictors are related to the number and length of capitalized words.

a. To reduce the number of predictors to a manageable size, examine how each predictor differs between the spam and nonspam emails by comparing the spam-class average and nonspam-class average. (Hint: Use the *Graph Builder* with the *Column Switcher*.) Which are the 11 predictors that appear to vary the most between spam and nonspam emails? From these 11, which words or signs occur more often in spam?

b. Partition the data into training and validation sets, then perform a discriminant analysis on the training data using only the 11 predictors.

c. If we are interested mainly in detecting spam messages, is this model useful? Use the confusion matrix, ROC curve and lift curve for the validation set for the evaluation.

d. In the sample, almost 40% of the email messages were tagged as spam. However, suppose that the actual proportion of spam messages in these email accounts is 10%. How does this information change the distance scores, the predicted probabilities, and the misclassifications?

e. A spam filter that is based on your model is used, so that only messages that are classified as nonspam are delivered while messages that are classified as spam are quarantined. Consequently, misclassifying a nonspam email (as spam) has much heftier results. Suppose that the cost of quarantining a nonspam email is 20 times that of not detecting a spam message. Assume that the proportion of spam is reflected correctly by the sample proportion. To explore costs of misclassification, save the discriminant formula to the data table, then create a formula in the data table to calculate the costs of misclassification. Summarize these costs—how costly is it to quarantine nonspam email vs. not detecting a spam message?

13

COMBINING METHODS: ENSEMBLES
AND UPLIFT MODELING

In this chapter we look at two useful approaches that combine methods for improving predictive power: *ensembles* and *uplift modeling*. An ensemble combines multiple supervised models into a "super-model." The previous chapters in this part of the book introduced different supervised methods for prediction and classification. Earlier, in Chapter 5, we learned about evaluating predictive performance, which can be used to compare several models and choose the best one. This is facilitated by the *Model Comparison* platform in JMP Pro. An ensemble is based on the powerful notion of *combining models*. Instead of choosing a single predictive model, we can combine several models to achieve improved predictive accuracy. We explain the underlying logic and why ensembles can improve predictive accuracy, and then introduce popular approaches for combining models, including simple averaging, bagging, and boosting.

In *uplift modeling* we combine supervised modeling with A-B testing, which is a simple type of randomized experiment. We describe the basics of A-B testing and how it is used along with predictive models in persuasion messaging not to predict outcomes but to predict who should receive which message treatment.

13.1 ENSEMBLES

Ensembles played a major role in the million-dollar Netflix Prize contest[1] that started in 2006. At the time, Netflix, the largest DVD rental service in the United States, wanted to improve their movie recommendation system (from www.netflixprize.com):

---

[1] Adapted with permission from *Practical Time Series Forecasting: A Hands-On Guide*, Second Edition, by Galit Shmueli. Copyright © 2011 Axelrod-Schnall Publishers.

Data Mining for Business Analytics: Concepts, Techniques, and Applications with JMP Pro®, First Edition.
Galit Shmueli, Peter C. Bruce, Mia L. Stephens, and Nitin R. Patel.
© 2017 John Wiley & Sons, Inc. Published 2017 by John Wiley & Sons, Inc.

> Netflix is all about connecting people to the movies they love. To help customers
> find those movies, weve developed our world-class movie recommendation system:
> Cinematch^{SM} ... And while Cinematch is doing pretty well, it can always be made
> better.

In a bold move, the company decided to share a large amount of data on movie ratings by
their users, and set up a contest, open to the public, aimed at improving their recommendation system:

> We provide you with a lot of anonymous rating data, and a prediction accuracy bar that
> is 10% better than what Cinematch can do on the same training data set.

During the contest, an active leader-board showed the results of the competing teams. An
interesting behavior started appearing: different teams joined forces to create combined, or
ensemble predictions, which proved more accurate than the individual predictions. The winning team, called "BellKor/s Pragmatic Chaos" combined results from the "BellKor" and
"Big Chaos" teams alongside additional members. In a 2010 article in *Chance* magazine,
the Netflix Prize winners described the power of their ensemble approach:

> An early lesson of the competition was the value of combining sets of predictions from
> multiple models or algorithms. If two prediction sets achieved similar RMSEs, it was
> quicker and more effective to simply average the two sets than to try to develop a
> new model that incorporated the best of each method. Even if the RMSE for one set
> was much worse than the other, there was almost certainly a linear combination that
> improved on the better set.

Why Ensembles Can Improve Predictive Power

The principle of combining methods is popular for reducing risk. For example, in finance,
portfolios are created for reducing investment risk. The return from a portfolio is typically
less risky because the variation is smaller than each of the individual components.

In predictive modeling, "risk" is equivalent to variation in prediction error. The more
variable our prediction errors, the more volatile our predictive model. Consider predictions
from two different models for a set of n observations. $e_{1,i}$ is the prediction error for the ith
observation by method 1 and $e_{2,i}$ is the prediction for the same observation by method 2.

Suppose that each model produces prediction errors and that are, on average, zero (for
some observations the model overpredicts and for some it underpredicts, but on average,
the error is zero):

$$E(e_{1,i}) = E(e_{2,i}) = 0$$

If for each observation we take an average of the two predictions, $\overline{y}_i = \hat{y}_{1,i} + \hat{y}_{2,i}/2$, then
the expected average error will also be zero:

$$E\left(y_i - \overline{y_i}\right) = E\left(y_i - \frac{\hat{y}_{1,i} + \hat{y}_{2,i}}{2}\right) \tag{13.1}$$

$$= E\left(\frac{y_i - \hat{y}_{1,i}}{2} + \frac{y_i - \hat{y}_{2,i}}{2}\right) = E\left(\frac{e_{1,i} + e_{2,i}}{2}\right) = 0$$

This means that the ensemble has the same average error as the individual models. Now let us examine the variance of the ensemble's prediction errors:

$$Var\left(\frac{e_{1,i} + e_{2,i}}{2}\right) = \frac{1}{4}\left(Var(e_{1,i}) + Var(e_{2,i})\right) + 2 \times \frac{1}{2}Cov\left(e_{1,i}, e_{2,i}\right) \qquad (13.2)$$

This variance can be lower than each of the individual variances $Var(e_{1,i})$ and $Var(e_{2,i})$ under some circumstances. A key component is the covariance (or equivalently, correlation) between the two prediction errors. The case of no correlation leaves us with a quantity that can be smaller than each of the individual variances. The variance of the average prediction error will be even smaller when the two prediction errors are negatively correlated.

In summary, using an average of two predictions can potentially lead to smaller error variance, and therefore better predictive power. These results generalize to more than two methods; you can combine results from multiple prediction methods or classifiers.

THE WISDOM OF CROWDS

In his book *The Wisdom of Crowds*, James Surowiecki recounts how Francis Galton, a prominent statistician from the 19th century, watched a contest at a county fair in England. The contest's objective was to guess the weight of an ox. Individual contest entries were highly variable, but the mean of all the estimates was surprisingly accurate—within 1% of the true weight of the ox. On balance, the errors from multiple guesses tended to cancel one another out. You can think of the output of a predictive model as a more informed version of these guesses. Averaging together multiple guesses will yield a more accurate answer than the vast majority of the individual guesses. Note that in Galton's story there were a few (lucky) individuals who scored better than the average. An ensemble estimate will not always be more accurate than all the individual estimates in all cases, but it will be more accurate most of the time.

Simple Averaging

The simplest approach for creating an ensemble is to combine the predictions, classifications, or propensities from multiple models. For example, we might have a linear regression model, a regression tree, and a k-NN algorithm. We use all of three methods to score, say, a test set. We then combine the three sets of results.

The three models can also be variations that use the same algorithm. For example, we might have three linear regression models, each using a different set of predictors.

Combining Predictions   In prediction tasks where the outcome variable is numerical, we can combine the predictions from the different methods by simply taking an average. In the example above, for each observation in the test set we have three predictions (one from each model). The ensemble prediction is then the average of the three values.

One alternative to a simple average is taking the median prediction, which would be less affected by extreme predictions. Another possibility is computing a weighted average, where weights are proportional to a quantity of interest. For instance, weights can be proportional to the accuracy of the model, or if different data sources are used, the weights can be proportional to the quality of the data.

Ensembles for prediction are useful not only in cross-sectional prediction, but also in time series forecasting (see Chapters 15–17). In forecasting, the same approach of combining future forecasts from multiple methods can lead to more precise predictions. One example is the weather forecasting application Forecast.io (www.forecast.io), which describes their algorithm as follows:

> Forecast.io is backed by a wide range of data sources, which are aggregated together statistically to provide the most accurate forecast possible for a given location.

Combining Classifications   Combining the results from multiple classifiers can be done using "voting." if for each record we have multiple classifications, a simple rule would be to choose the most popular class among these classifications. For example, we could use a classification tree, a Naive Bayes classifier, and discriminant analysis for classifying a binary outcome. For each observation we then generate three predicted classes. Simple voting would choose the most common class among the three.

As in prediction, we can assign heavier weights to scores from some models, based on considerations such as model accuracy or data quality. This would be done by setting a "majority rule" that is different from 50%.

Combining Propensities   Similar to predictions, propensities can be combined by taking a simple (or weighted) average. Recall that some algorithms, such as Naive Bayes, produce biased propensities and should therefore not be simply averaged with propensities from other methods.

Bagging

Another form of ensembles is based on averaging across multiple random data samples. Bagging, short for "Bootstrap aggregating," comprises of two steps:

1. Generate multiple random samples (by sampling with replacement from the original data)—this method is called "bootstrap sampling."
2. Running an algorithm on each sample and producing scores.

Bagging improves the performance stability of a model and helps avoid overfitting by separately modeling different data samples and then combining the results. It is therefore especially useful for algorithms such as trees and neural networks. In Chapter 9 we described random forests, an ensemble based on bagged trees.

Boosting

Boosting is a slightly different approach to creating ensembles. Here the goal is to directly improve areas in the data where our model makes errors. The steps in Boosting are:

1. Fit a model to the data.
2. Draw a sample from the data so that misclassified observations (or observations with large prediction errors) have higher probabilities of selection.

3. Fit the model to the new sample.

4. Repeat steps 2–3 multiple times.

CREATING ENSEMBLE MODELS IN JMP PRO

There are a number of ways of creating ensemble models in JMP Pro:

- Simple model averaging is available in the *Model Comparison* platform (under *Analyze > Modeling*).
- To combine predictions, classifications, or propensities, formulas for models can be saved to the data table. These saved models (or their results) can then be averaged or combined using a formula column in the data table.
- JMP provides a variant of model averaging in the *Fit Model > Stepwise* platform. Here several models are fit, and the coefficients in the final model are weighted averages of the coefficients from each of the fitted models. The weights are used to minimize overfitting.
- Bagging, or bootstrap aggregation, is available in JMP Pro as a method (*Bootstrap Forest*) in the *Partition* platform.
- Boosting is available as a method (*Boosted Tree*) in the *Partition* platform, and is an option in the *Neural* platform.

For a discussion of Bootstrap Forest and Boosted Tree, see Chapter 9.

Advantages and Weaknesses of Ensembles

Combining scores from multiple models is aimed at generating more precise predictions (lowering the prediction error variance). The approach is most useful when the combined models generate prediction errors that are negatively associated, but it can also be useful when the correlation is low. Ensembles can use simple averaging, weighted averaging, voting, medians, and so forth. Models can be based on the same algorithm or on different algorithms, using the same sample or different samples. Ensembles have become a major strategy for participants in data mining contests, where the goal is to optimize some predictive measure. In that sense, ensembles also provide an operational way to obtain solutions with high predictive power in a fast way, by engaging multiple teams of "data crunchers" working in parallel and combining their results.

Ensembles that are based on different data samples help avoid overfitting. However, remember that you can also overfit the data with an ensemble if you tweak it (e.g., choosing the "best" weights when using a weighted average).

The major disadvantage of an ensemble is the resources that it requires: computationally, as well as in terms of software availability and the analyst's skill and time investment. Implementing ensembles that combine results from different algorithms requires developing each of the models and evaluating them. Boosting-type ensembles and bagging-type ensembles do not require such effort, but they do have a computational cost (although boosting has been parallelized in JMP Pro and computation time is generally not an issue). Ensembles that rely on multiple data sources require collecting and maintaining multiple data sources. And finally, ensembles are "black-box" methods, in that the relationship between the predictors and the output variable usually becomes nontransparent.

13.2   UPLIFT (PERSUASION) MODELING

Long before the advent of the Internet, sending messages directly to individuals (i.e., direct mail) held a big share of the advertising market. Direct marketing affords the marketer the ability to invite and monitor direct responses from consumers. This, in turn, allows the marketer to learn whether the messaging is paying off. A message can be tested with a small section of a large list and, if it pays off, the message can be rolled out to the entire list. With predictive modeling, we have seen that the rollout can be targeted to that portion of the list that is most likely to respond or behave in a certain way. None of this was possible with media advertising (television, radio, newspaper, magazine).

Direct response also made it possible to test one message against another and find out which does better.

A-B Testing

A-B testing is the marketing industry's term for a standard scientific experiment in which results can be tracked by an individual. The idea is to test one treatment against another, or a treatment against a control. "Treatment" is simply the term for the intervention you are testing: in a medical trial it is typically a drug, device, or other therapy; in marketing it is typically an offering to a consumer—for example, an email or a web page shown to a consumer. A general display ad in a magazine would not generally qualify, unless it had a specific call to action that allows the marketer to trace the action (e.g., purchase) to a given ad, plus the ability to split the magazine distribution randomly and provide a different offer to each segment.

An important element of A-B testing is randomization—the treatments are assigned or delivered to individuals randomly. This way any difference between treatment A and treatment B can be attributed to the treatment (unless it is due to chance).

Uplift

An A-B test tells you which treatment does better on average but says nothing about which treatment does better for which individual. A classic example is in political campaigns. Consider the following scenario: The campaign director for Smith, a Democratic congressional candidate, would like to know which voters should be called to encourage to support Smith. Voters who tend to vote Democratic but are not activists might be more inclined to vote for Smith if they got a call. Active Democrats are probably already supportive of him, and therefore a call to them would be wasted. Calls to Republicans are not only wasteful, but they could be harmful.

Campaigns now maintain extensive data on voters to help guide decisions about outreach to individual voters. Prior to the 2008 Obama campaign, the practice was to make rule-based decisions based on expert political judgment. Since 2008, it has increasingly been recognized that rather than relying on judgment or supposition to determine whether an individual should be called, it is best to use the data to develop a model that can predict whether a voter will respond positively to outreach.

Gathering the Data

US states maintain publicly available files of voters, as part of the transparent oversight process for elections. The voter file contains data such as name, address, and date of birth. Political parties have "poll-watchers" at elections to record who votes, so they have additional data on which elections voters voted in. Census data for neighborhoods can be appended, based on voter address. Finally, commercial demographic data can be purchased and matched to the voter data. Table 13.1 shows a small extract of data derived from the voter file for the US state of Delaware. The actual data used in this problem is in the file `VoterPersuasion.jmp` and contains 10,000 records and many additional variables beyond those shown in Table 13.1. The *Notes* section, in the top left corner of the data table, contains additional information regarding the data and the variables (double-click on *Notes* to view).

TABLE 13.1   Data on voters (small subset of variables and records) and Data Dictionary

Voter	Age	NH_White	Comm_PT	H_F1	Reg_Days	PR_Pelig	E_Elig	Political_C
1	28	61	0	0	3997	0	20	1
2	23	67	3	0	300	0	0	1
3	57	64	4	0	2967	0	0	0
4	70	53	2	1	16620	100	90	1
5	37	76	2	0	3786	0	20	0

Data Dictionary

Age	voter age in years
NH_White	neighborhood average of % non hispanic white in household
Comm_PT	neighborhood % of workers who take public transit
H_F1	single female household (1 = yes)
Reg_Days	days since voter registered at current address
PR_Pelig	voted in what % of nonpresidential primaries
E_Pelig	voted in what % of any primaries
Political_C	is there a political contributor in the home (1 = yes)

First, the campaign director conducts a survey of 10,000 voters to determine their inclination to vote Democratic. Then she conducts an experiment, randomly splitting the sample of 10,000 voters in half and mailing a flyer promoting Smith to half the list (treatment A), and nothing to the other half (treatment B). The control group that gets no flyer is essential, since other campaigns or news events might cause a shift in opinion. The goal is to measure the change in opinion after the flyer goes out, relative to the no-flyer control group.

The next step is conducting a post-flyer survey of the same sample of 10,000 voters, to measure whether each voter's opinion of Smith has shifted in a positive direction. A binary variable, Moved_AD, will be added to the collected data, indicating whether opinion has moved in a Democratic direction (1) or not (0).

Table 13.2 summarizes the results of the survey, by comparing the movement in a Democratic direction for each of the treatments. Overall, the flyer (*Flyer = 1*) is modestly effective.

TABLE 13.2 Results of sending a pro-Democratic flyer to voters

Treatment	# Voters	# Moved Dem.	% Moved Dem.
Flyer = 1 (flyer sent)	5000	2012	40.2%
Flyer = 0 (no flyer sent)	5000	1722	34.4%

Movement in a Democratic direction among those who got no message is 34.4%. This probably reflects the approach of the election, the heightening campaign activity, and the reduction in the "no opinion" category. It also illustrates the need for a control group. Among those who did get the flyer, the movement in a Democratic direction is 40.2%. So, overall, the lift from the flyer is 5.8%.

We can now append two variables to the voter data shown earlier in Table 13.1: *Flyer* (whether they received the flyer (1) or not (0)) and *Moved_AD* (whether they moved in a Democratic direction (1) or not (0)). The augmented data are shown in Table 13.3.

Any classification method can be used to create an uplift model. In the next section we discuss how to develop an uplift model based on a logistic regression model. Then we show how to use the JMP Pro *Uplift* platform, which builds uplift models based on classification trees.

A Simple Model

We start by developing a predictive model with *Moved_AD* as the target (output variable) and various predictor variables, including *Flyer*. Table 13.4 shows the first few lines from the output of a logistic regression model that was used to predict Moved_AD.

However, our interest is not just how the flyer did overall, nor is it whether we can predict the probability that a voter's opinion will move in a favorable direction. Rather, our goal is to predict how much (positive) impact the flyer will have on a specific voter. That way the campaign can direct its limited resources toward the voters who are the most persuadable—those for whom mailing the flyer will have the greatest positive effect.

TABLE 13.3 Target variable (Moved_AD) and treatment variable (Flyer) added to voter data

Voter	Age	NH_White	Comm_PT	H_F1	Reg_Days	PR_Pelig	E_Elig	Political_C	Flyer	Moved_AD
1	28	61	0	0	3997	0	20	1	1	0
2	23	67	3	0	300	0	0	1	1	1
3	57	64	4	0	2967	0	0	0	0	0
4	70	53	2	1	16620	100	90	1	0	0
5	37	76	2	0	3786	0	20	0	1	0

TABLE 13.4 Classifications and propensities from predictive model (small extract)

Voter	Flyer	Actual Moved_AD	Predicted Moved_AD	Predicted Prob.
1	0	1	1	0.5975
2	1	1	1	0.5005
3	0	0	0	0.2235
4	0	0	0	0.3052
5	1	0	0	0.4140

Modeling Individual Uplift

To answer the question about the flyer's impact on each voter, we need to model the effect of the flyer at the individual voter level. For each voter, uplift is defined as follows:

Uplift = increase in propensity of favorable opinion after getting flyer.

An uplift model can be built (manually) using a standard predictive model. To build an uplift model, we follow the steps below to estimate the change in probability of "success" (propensity) that comes from receiving the treatment (the flyer, in this example). This approach to uplift modeling is known as the *two model* approach:

1. Randomly split a data sample into treatment and control groups, conduct an A-B test, and record the outcome (in our example: Moved_AD).
2. Partition the combined data into training and validation sets, build a predictive model with this outcome as the target variable, and include a predictor variable that denotes treatment status (in our example: Flyer).
3. Score this predictive model on the validation set. This will yield, for each validation record, its propensity of success given its treatment.
4. Reverse the value of the treatment variable and re-score the same model to that partition. This will yield for each validation record its propensity of success had it received the other treatment.
5. Uplift is estimated for each individual by $P(\text{Success}|\text{Treatment} = 1)$ - $P(\text{Success}|\text{Treatment} = 0)$
6. For new data where no experiment has been performed, simply add a synthetic predictor variable for treatment and assign first a "1," score the model, then a "0," and score the model again. Estimate uplift for the new record(s) as above.

Continuing with the small voter example, in the results from step 3 shown in Table 13.4, the right column gives the propensities from the model. Next we re-estimate the logistic model, but with the values of the treatment variable Flyer reversed for each row. Table 13.5 shows the propensities with variable Flyer reversed (you can see the reversed values in column Flyer). Finally, in step 5, we calculate the uplift for each voter.

Table 13.6 shows the uplift for each voter–the success (Moved_AD = 1) propensity given Flyer = 1 minus the success propensity given Flyer = 0.

TABLE 13.5   Classifications and propensities from predictive model (small extract) with Flyer values reversed

Voter	Flyer	Actual Moved_AD	Predicted Moved_AD	Predicted Prob.
1	1	1	1	0.6908
2	0	1	1	0.3996
3	1	0	0	0.3022
4	1	0	0	0.3980
5	0	0	0	0.3194

TABLE 13.6   Uplift: change in propensities from sending message vs. not sending message

Voter	Prob. if Flyer = 1	Prob. if Flyer = 0	Uplift
1	0.6908	0.5975	0.0933
2	0.5005	0.3996	0.1009
3	0.3022	0.2235	0.0787
4	0.3980	0.3052	0.0928
5	0.4140	0.3194	0.0946

Using the Results of an Uplift Model

Once we have estimated the uplift for each individual, the results can be ordered by uplift. The flyer could then be sent to all those voters with a positive uplift or, if resources are limited, only to a subset—those with the greatest uplift.

Uplift modeling is used mainly in marketing and, more recently, in political campaigns. It has two main purposes:

- To determine whether to send someone a persuasion message, or just leave them alone.
- When a message is definitely going to be sent, to determine which message, among several possibilities, to send.

Technically this amounts to the same thing— "send no message" is simply another category of treatment, and an experiment can be constructed with multiple treatments, such as, no message, message A, and message B. However, practitioners tend to think of the two purposes as distinct, and tend to focus on the first. Marketers want to avoid sending discount offers to customers who would make a purchase anyway, or renew a subscription anyway. Political campaigns, likewise, want to avoid calling voters who would vote for their candidate in any case. And both parties especially want to avoid sending messages or offers where the effect might be antagonistic—where the uplift is negative.

Creating Uplift Models in JMP Pro

The built-in utility for uplift modeling in JMP Pro is based on the *Partition* platform, which is used to fit classification and regression trees (see Chapter 9). While the models built using the *Partition* platform find splits to optimize a prediction, models built using the *Uplift* platform find splits to maximize a treatment difference, or uplift. Note that this approach is different from (and somewhat more complex than) the *two-model* approach described in the previous section.

Essentially, here is what JMP Pro does: for a given node, it considers all possible splits. It models the data in each node as a linear model with two effects, the split and the treatment, and an interaction term that captures the relationship between the split and the treatment. The LogWorth is used to measure the significance of the interaction terms and to select the best split. As with classification trees, when a validation column is used, the model is built on the training data and the final model is selected based on the maximum validation RSquare.

USING THE *UPLIFT* PLATFORM IN JMP PRO

- Select *Uplift* from the *Analyze > Consumer Research* menu in JMP Pro. The completed dialog for our voter persuasion example and the variables listed in Table 13.3 is shown in Figure 13.1.
- The initial uplift analysis window is shown in Figure 13.1. The overall rates for Moved_AD (for the training data) are shown in the bottom left corner. The uplift from the treatment is reported under *Trt Diff*.
- To create the uplift model, use the *Split* button (repeatedly) or use the *Go* button to automatically build the model. When the *Go* button is used, the final number of splits is determined by the maximum value of RSquare for the validation data. (For details, refer to the discussion in Chapter 9.)
- A variety of options for interpreting the model are available from the top red triangle. The *Leaf Report* displays the rates and the uplift for each of the splits in the tree. The *Uplift Graph* sorts the uplift (on the training data) from the highest uplift to the lowest.
- Use the top red triangle to save the difference formula (the uplift) and the prediction formula with classifications to the data table. Saving the difference formula calculates the uplift for both the training and validation data, and can be used to score new data.

The completion of this analysis is left to the reader as one of the cases in Chapter 18.

13.3 SUMMARY

In practice, the methods discussed in this book are often used not in isolation, but as building blocks in an analytic process whose goal is always to inform and provide insight.

In this chapter we looked at two ways that multiple models are deployed. In ensembles, multiple models are weighted and combined to produce improved predictions. In uplift modeling, the results of A-B testing are folded into the predictive modeling process as a predictor variable to guide choices not just about whether to send an offer or persuasion message but also as to who to send it to.

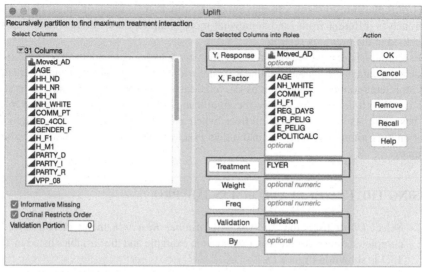

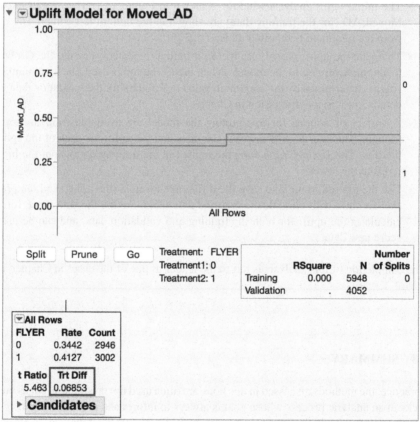

FIGURE 13.1   Uplift Model Specification Dialog in JMP Pro (top) and Initial Uplift Analysis (bottom) using the voter persuasion data. The *Show Points* Display Option has been turned off in the graph.

PROBLEMS

13.1 Acceptance of Consumer Loan. Universal Bank has begun a program to encourage its existing customers to borrow via a consumer loan program. The bank has promoted the loan to 5000 customers, of whom 480 accepted the offer. The data are available in file `UniversalBank.jmp`. The bank now wants to develop a model to predict which customers have the greatest probability of accepting the loan, in order to reduce promotion costs and send the offer only to a subset of its customers.

We will develop several models, then combine them in an ensemble.

a. Partition the data: 60% training, 40% validation.

b. Fit models the following models to the data: (1) logistic regression, (2) classification tree, and (3) neural network. Use Personal Loan as the target variable, and do not include Zip Code in the models.

   i. Report the confusion matrix and misclassification (error) rates for each of the three models.

   ii. Save the formula for each model to the data table and report the first 10 rows of these saved columns.

c. Add a new column to this data table, and use the *formula editor* to create a formula for the average of the predicted probabilities for the target class (Personal Loan = Yes) from the three models.

   i. Add a second new column, and create a formula to classify the outcome (Personal Loan) based on the average of the predicted probabilities. Use a cutoff of 0.5.

   ii. Use *Analyze > Tabulate* to create a confusion matrix for the classifications generated above, and note the overall error rate.

d. Compare the error rates for the three individual methods and the ensemble method.

13.2 eBay Auctions—Boosting and Bagging. Using the eBay auction data (file `eBayAuctions.jmp`) with variable *Competitive* as the target, partition the data into training (50%), validation (30%), and test sets (20%).

a. Create a classification tree. Use the Go button to create the model. Looking at the test set, what is the overall accuracy? What is the lift at portion = 0.10? Save the prediction formula to the data table.

b. Run the same tree, but first select the *Boosted Tree* method. Use the default settings. For the test set, what is the overall accuracy? What is the lift at portion = 0.10? Save the prediction formula to the data table.

c. Now try the same tree with the *Bootstrap Forest* method selected, and accept the default settings. What is the lift at portion = 0.10? Again, save the prediction formula to the data table.

d. Compare the three models using the *Model Comparison* platform under *Analyze > Modeling*. In the *Model Comparison* dialog, use the validation column as the *By* variable, but leave everything else blank. Compare the misclassification rates for the three models. Which model has the best accuracy? Compare the lift curves for the three models. Which model does the best job of sorting the response?

    e. In the *Model Comparison* analysis window, select *Model Averaging* from the top red triangle. Return to the data table, and view the formulas for the new columns that have been saved. Describe how the models are being averaged.

13.3 **Hair Care Product–Uplift Modeling.** This problem uses the data set in `Hair Care Product.jmp`, located in the *Sample Data Library* in JMP (under the *Help* menu). In this hypothetical case, a promotion for a hair care product was sent to some members of a buyers club. Purchases were then recorded for both the members who got the promotion and those who did not.

    a. What is the purchase propensity (probability)

        i. among those who received the promotion?

        ii. among those who did not receive the promotion?

    b. Fit an uplift model using JMP Pro Uplift platform with *Purchase* as the target (*Y*, Response), Promotion as the treatment, Validation as the validation column, and the other variables as *X*, Factors. Compare the rates in the initial uplift analysis to those reported above.

    c. Click the *Split* button one time. Which variable does the model split on? What is the uplift in the two branches? Interpret this value.

    d. Click the *Go* button, and request a *Leaf Report* and the *Uplift Graph*.

        i. What is the highest uplift value? Interpret this value.

        ii. What are the set of conditions with the highest uplift? With the lowest uplift?

        iii. From the context of future promotions and increasing purchases of the product, how would you use this information?

    e. Save the difference formula and the prediction formula to the data table. Report the predicted class and propensities for the first 10 records in the validation set.

PART V

MINING RELATIONSHIPS AMONG RECORDS

14

CLUSTER ANALYSIS

This chapter is about the popular unsupervised learning task of clustering, where the goal is to segment the data into a set of homogeneous clusters of observations for the purpose of generating insight. Separating a dataset into clusters of homogeneous records is also useful for improving performance of supervised methods, by modeling each cluster separately rather than the entire, heterogeneous dataset. Clustering is used in a vast variety of business applications, from customized marketing to industry analysis. We describe two popular clustering approaches: hierarchical clustering and k-means clustering. In hierarchical clustering, observations are sequentially grouped to create clusters, based on distances between observations and distances between clusters. We describe how the algorithm works in terms of the clustering process and mention several common distance metrics used. Hierarchical clustering also produces a useful graphical display of the clustering process and results, called a dendrogram. We present dendrograms and illustrate their usefulness. k-means clustering is widely used in large dataset applications. In k-means clustering, observations are allocated to one of a prespecified set of clusters, according to their distance from each cluster. We describe the k-means clustering algorithm and its computational advantages. Finally, we present techniques that assist in generating insight from clustering results.

14.1 INTRODUCTION

Cluster analysis is used to form groups or clusters of similar observations based on several measurements made on these observations. The key idea is to characterize the clusters in ways that would be useful for the aims of the analysis. This idea has been applied in many areas, including astronomy, archaeology, medicine, chemistry, education, psychology, linguistics, and sociology. Biologists, for example, have made extensive use of classes and subclasses to organize species. A spectacular success of the clustering idea in chemistry was Mendeleev's periodic table of the elements.

Data Mining for Business Analytics: Concepts, Techniques, and Applications with JMP Pro®, First Edition.
Galit Shmueli, Peter C. Bruce, Mia L. Stephens, and Nitin R. Patel.
© 2017 John Wiley & Sons, Inc. Published 2017 by John Wiley & Sons, Inc.

One popular use of cluster analysis in marketing is for *market segmentation:* customers are segmented based on demographic and transaction history information, and a marketing strategy is tailored for each segment. In countries such as India, where customer diversity is extremely location-sensitive, chain stores often perform market segmentation at the store level, rather than at a chain-wide (called "micro segmentation"). Another use is for *market structure analysis:* identifying groups of similar products according to competitive measures of similarity. In marketing and political forecasting, clustering of neighborhoods using US postal zip codes has been used successfully to group neighborhoods by lifestyles. Claritas, a company that pioneered this approach, grouped neighborhoods into 40 clusters using various measures of consumer expenditure and demographics. Examining the clusters enabled Claritas to come up with evocative names, such as "Bohemian Mix," "Furs and Station Wagons," and "Money and Brains," for the groups that captured the dominant lifestyles. Knowledge of lifestyles can be used to estimate the potential demand for products (e.g., sports utility vehicles) and services (e.g., pleasure cruises). Similarly, sales organizations will derive customer segments and give them names— "personas"—to focus sales efforts.

In finance, cluster analysis can be used for creating *balanced portfolios*: Given data on a variety of investment opportunities (e.g., stocks), one may find clusters based on financial performance variables such as return (daily, weekly, or monthly), volatility, beta, and other characteristics, such as industry and market capitalization. Selecting securities from different clusters can help create a balanced portfolio. Another application of cluster analysis in finance is for *industry analysis*: For a given industry, we are interested in finding groups of similar firms based on measures such as growth rate, profitability, market size, product range, and presence in various international markets. These groups can then be analyzed in order to understand industry structure and to determine, for instance, who is a competitor.

An interesting and unusual application of cluster analysis, described in Berry and Linoff (1997), is the design of a new set of sizes for army uniforms for women in the US Army. The study came up with a new clothing size system with only 20 sizes, where different sizes fit different body types. The 20 sizes are combinations of five measurements: chest, neck, and shoulder circumference, sleeve outseam, and neck-to-buttock length (for further details, see McCullugh et al., 1998). This example is important because it shows how a completely new insightful view can be gained by examining clusters of records.

Cluster analysis can be applied to huge amounts of data. For instance, Internet search engines use clustering techniques to cluster queries that users submit. These can then be used for improving search algorithms. The objective of this chapter is to describe the key ideas underlying the most commonly used techniques for cluster analysis and to lay out their strengths and weaknesses.

Typically, the basic data used to form clusters are a table of measurements on several variables, where each column represents a variable and a row represents an observation. Our goal is to form groups of records so that similar observations are in the same group. The number of clusters may be prespecified or determined from the data.

Example: Public Utilities

Table 14.1 gives corporate data on 22 public utilities in the United States (the definition of each variable is given in the table footnote, and the data are in Utilities.jmp). We are interested in forming groups of similar utilities. The observations to be clustered are

TABLE 14.1   Data on 22 Public Utilities

Company	Fixed	RoR	Cost	Load	Demand	Sales	Nuclear	Fuel Cost
Arizona Public Service	1.06	9.2	151	54.4	1.6	9,077	0.0	0.628
Boston Edison Co.	0.89	10.3	202	57.9	2.2	5,088	25.3	1.555
Central Louisiana Co.	1.43	15.4	113	53.0	3.4	9,212	0.0	1.058
Commonwealth Edison Co.	1.02	11.2	168	56.0	0.3	6,423	34.3	0.700
Consolidated Edison Co. (NY)	1.49	8.8	192	51.2	1.0	3,300	15.6	2.044
Florida Power & Light Co.	1.32	13.5	111	60.0	-2.2	11,127	22.5	1.241
Hawaiian Electric Co.	1.22	12.2	175	67.6	2.2	7,642	0.0	1.652
daho Power Co.	1.10	9.2	245	57.0	3.3	13,082	0.0	0.309
Kentucky Utilities Co.	1.34	13.0	168	60.4	7.2	8,406	0.0	0.862
Madison Gas & Electric Co.	1.12	12.4	197	53.0	2.7	6,455	39.2	0.623
Nevada Power Co.	0.75	7.5	173	51.5	6.5	17,441	0.0	0.768
New England Electric Co.	1.13	10.9	178	62.0	3.7	6,154	0.0	1.897
Northern States Power Co.	1.15	12.7	199	53.7	6.4	7,179	50.2	0.527
Oklahoma Gas & Electric Co.	1.09	12.0	96	49.8	1.4	9,673	0.0	0.588
Pacific Gas & Electric Co.	0.96	7.6	164	62.2	-0.1	6,468	0.9	1.400
Puget Sound Power & Light Co.	1.16	9.9	252	56.0	9.2	15,991	0.0	0.620
San Diego Gas & Electric Co.	0.76	6.4	136	61.9	9.0	5,714	8.3	1.920
The Southern Co.	1.05	12.6	150	56.7	2.7	10,140	0.0	1.108
Texas Utilities Co.	1.16	11.7	104	54.0	-2.1	13,507	0.0	0.636
Wisconsin Electric Power Co.	1.20	11.8	148	59.9	3.5	287	41.1	0.702
United Illuminating Co.	1.04	8.6	204	61.0	3.5	6,650	0.0	2.116
Virginia Electric & Power Co.	1.07	9.3	174	54.3	5.9	10,093	26.6	1.306

Noted: Fixed = fixed-charge covering ratio (income/debt), RoR = rate of return on capital, Cost = cost per kilowatt capacity in place, Load = annual load factor, Demand = peak kilowatthour demand growth from 1974 to 1975, Sales = sales (kilowatthour use per year), Nuclear = percent nuclear, Fuel Cost = total fuel costs (cents per kilowatthour).

the utilities, and the clustering will be based on the eight measurements on each utility. An example where clustering would be useful is a study to predict the cost impact of deregulation. To do the requisite analysis, economists would need to build a detailed cost model of the various utilities. It would save a considerable amount of time and effort if we could cluster similar types of utilities and build detailed cost models for just one "typical" utility in each cluster and then scale up from these models to estimate results for all utilities.

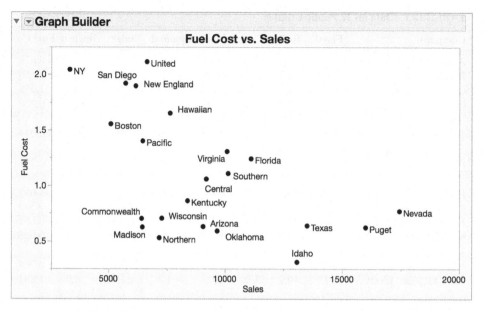

FIGURE 14.1   Scatterplot of Fuel Cost vs. Sales for the 22 utilities

For simplicity, let us consider only two of the measurements: *Sales* and *Fuel Cost*. Figure 14.1 shows a scatterplot of these two variables, with labels marking each company (labels were added by selecting all rows in the data table and using *Rows > Label* and by selecting the Company column and using *Cols > Label*). At first glance there appear to be two or three clusters of utilities: one with utilities that have high fuel costs, a second with utilities that have lower fuel costs and relatively low sales, and a third with utilities with low fuel costs but high sales.

We can therefore think of cluster analysis as a more formal algorithm that measures the distance between records, and according to these distances (here, nonhierarchical distances), forms clusters.

Two general types of clustering algorithms for a dataset of *n* observations are hierarchical and nonhierarchical clustering:

Hierarchical methods can be either agglomerative or divisive. Agglomerative methods begin with *n* clusters and sequentially merge similar clusters until a single cluster is obtained. Divisive methods work in the opposite direction, starting with one cluster that includes all observations. Hierarchical methods are especially useful when the goal is to arrange the clusters into a natural hierarchy.

Nonhierarchical methods, such as *k*-means, are used for a prespecified number of clusters with observations assigned to each cluster. These methods are generally less computationally intensive and are therefore preferred with very large datasets.

We concentrate here on the two most popular methods: hierarchical agglomerative clustering and *k*-means clustering. In both cases we need to define two types of distances: distance between two observations and distance between two clusters. In both cases there are a variety of metrics that can be used.

14.2   MEASURING DISTANCE BETWEEN TWO OBSERVATIONS

We denote by d_{ij} a *distance metric*, or *dissimilarity measure*, between observations i and j. For observation i, we have the vector of p measurements $(x_{i1}, x_{i2}, \ldots, x_{ip})$, and for observation j, we have the vector of measurements $(x_{j1}, x_{j2}, \ldots, x_{jp})$. For example, we can write the measurement vector for Arizona Public Service as $[1.06, 9.2, 151, 54.4, 1.6, 9077, 0, 0.628]$.

Distances can be defined in multiple ways, but in general, the following properties are required:

Nonnegative: $d_{ij} \geq 0$

Self-proximity: $d_{ii} = 0$ (the distance from an observation to itself is zero)

Symmetry: $d_{ij} = d_{ji}$

Triangle inequality: $d_{ij} \leq d_{ik} + d_{kj}$ (the distance between any pair cannot exceed the sum of distances between the other two pairs)

Euclidean Distance

The most popular distance measure is the *Euclidean distance*, d_{ij}, which between two observations, i and j, is defined by

$$d_{ij} = \sqrt{(x_{i1} - x_{j1})^2 + (x_{i2} - x_{j2})^2 + \cdots + (x_{ip} - x_{jp})^2}$$

For instance, the Euclidean distance between Arizona Public Service and Boston Edison Co. can be computed from the raw data by

$$d_{12} = \sqrt{(1.06 - 0.89)^2 + (9.2 - 10.3)^2 + (151 - 202)^2 + \cdots + (0.628 - 1.555)^2}$$
$$= 3989.408$$

Normalizing Numerical Measurements

The measure computed above is highly influenced by the scale of each variable, so that variables with larger scales (e.g., *Sales*) have a much greater influence over the total distance. It is therefore customary to *normalize* (or *standardize*) continuou measurements before computing the Euclidean distance. This converts all measurements to the same scale. Normalizing a measurement means subtracting the average and dividing by the standard deviation (normalized values are also called *z-scores*).[1] For instance, the average sales amount across the 22 utilities is 8914.045 and the standard deviation is 3549.984. The normalized sales for Arizona Public Service is therefore $(9077 - 8914.045)/3549.984 = 0.046$.

Returning to the simplified utilities data with only two measurements (Sales and Fuel Cost), we first normalize the measurements (see Table 14.2), and then compute the

---

[1]When there are outliers, *robust* standardization can be applied. This reduces the influence of outliers on the estimates of the means and standard deviations used to standardize the data. In JMP, this is called *Standardize Robustly*.

TABLE 14.2   Original and Normalized Measurements for Sales and Fuel Cost

Company	Sales	Fuel Cost	NormSales	NormFuel
Arizona Public Service	9,077	0.628	0.0459	−0.8537
Boston Edison Co.	5,088	1.555	−1.0778	0.8133
Central Louisiana Co.	9,212	1.058	0.0839	−0.0804
Commonwealth Edison Co.	6,423	0.7	−0.7017	−0.7242
Consolidated Edison Co. (NY)	3,300	2.044	−1.5814	1.6926
Florida Power & Light Co.	11,127	1.241	0.6234	0.2486
Hawaiian Electric Co.	7,642	1.652	−0.3583	0.9877
Idaho Power Co.	13,082	0.309	1.1741	−1.4273
Kentucky Utilities Co.	8,406	0.862	−0.1431	−0.4329
Madison Gas & Electric Co.	6,455	0.623	−0.6927	−0.8627
Nevada Power Co.	17,441	0.768	2.4020	−0.6019
New England Electric Co.	6,154	1.897	−0.7775	1.4283
Northern States Power Co.	7,179	0.527	−0.4887	−1.0353
Oklahoma Gas & Electric Co.	9,673	0.588	0.2138	−0.9256
Pacific Gas & Electric Co.	6,468	1.4	−0.6890	0.5346
Puget Sound Power & Light Co.	15,991	0.62	1.9935	−0.8681
San Diego Gas & Electric Co.	5,714	1.92	−0.9014	1.4697
The Southern Co.	10,140	1.108	0.3453	0.0095
Texas Utilities Co.	13,507	0.636	1.2938	−0.8393
Wisconsin Electric Power Co.	7,287	0.702	−0.4583	−0.7206
United Illuminating Co.	6,650	2.116	−0.6378	1.8221
Virginia Electric & Power Co.	10,093	1.306	0.3321	0.3655
Mean	8,914.05	1.10	0.00	0.00
Standard deviation	3,549.98	0.56	1.00	1.00

Euclidean distance between each pair. Table 14.3 gives these pairwise distances for the first five utilities. A similar table can be constructed for all 22 utilities

TABLE 14.3   Distance Matrix Between Pairs of the First Five Utilities, Using Euclidean Distance and Normalized Measurements

	Arizona	Boston	Central	Commonwealth	Consolidated
Arizona	0				
Boston	2.01	0			
Central	0.77	1.47	0		
Commonwealth	0.76	1.58	1.02	0	
Consolidated	3.02	1.01	2.43	2.57	0

Other Distance Measures for Numerical Data

It is important to note that the choice of the distance measure plays a major role in cluster analysis. The main guideline is domain dependent: What exactly is being measured? How are the different measurements related? What scale should it be treated as (numerical, ordinal, or nominal)? Are there outliers? Finally, depending on the goal of the analysis, should the clusters be distinguished mostly by a small set of measurements, or should they be separated by multiple measurements that weight moderately?

Although Euclidean distance is the most widely used distance, it has three main features that need to be kept in mind. First, as mentioned above, it is highly scale dependent. Changing the units of one variable (e.g., from cents to dollars) can have a huge influence on the results. Standardizing is therefore a common solution. But unequal weighting should be considered if we want the clusters to depend more on certain measurements and less on others. The second feature of Euclidean distance is that it completely ignores the relationship between the measurements. Thus, if the measurements are in fact strongly correlated, a different distance (e.g., the statistical distance, described below) is likely to be a better choice. Third, Euclidean distance is sensitive to outliers. If the data are believed to contain outliers and careful removal is not a choice, the use of more robust distances (e.g., the Manhattan distance, described below) is preferred.

For these reasons additional popular distance metrics are often used.

Correlation-Based Similarity Sometimes it is more natural or convenient to work with a similarity measure between observations rather than distance, which measures dissimilarity. A popular similarity measure is the square of the correlation coefficient, r_{ij}^2, where the correlation coefficient is defined by

$$r_{ij} \equiv \frac{\sum_{m=1}^{p} (x_{im} - \overline{x}_m)(x_{jm} - \overline{x}_m)}{\sqrt{\sum_{m=1}^{p} (x_{im} - \overline{x}_m)^2 \sum_{m=1}^{p} (x_{jm} - \overline{x}_m)^2}}$$

Such measures can always be converted to distance measures. In the example above we could define a distance measure $d_{ij} = 1 - r_{ij}^2$.

Statistical Distance (also called *Mahalanobis Distance*) This metric has an advantage over the other metrics mentioned in that it takes into account the correlation between measurements. With this metric, measurements that are highly correlated with other measurements do not contribute as much as those that are uncorrelated or mildly correlated. The statistical distance between observations i and j is defined as

$$d_{i,j} = \sqrt{(\mathbf{x}_i - \mathbf{x}_j)' S^{-1} (\mathbf{x}_i - \mathbf{x}_j)},$$

where $\mathbf{x}_i$ and $\mathbf{x}_j$ are p-dimensional vectors of the measurements values for observations i and j, respectively and S is the covariance matrix for these vectors. (', a transpose operation, simply turns a column vector into a row vector). S^{-1} is the inverse matrix of S, which is the p-dimension extension to division. (For further information on statistical distance, see Chapter 12).

Manhattan Distance ("City Block") This distance looks at the absolute differences rather than squared differences, and is defined by

$$d_{ij} = \sum_{m=1}^{p} | x_{im} - x_{jm} |$$

Maximum Coordinate Distance This distance looks only at the measurement on which observations i and j deviate most. It is defined by

$$d_{ij} = \max_{m=1,2,\ldots,p} |x_{im} - x_{jm}|$$

Distance Measures for Categorical Data

In the case of measurements with binary values, it is more intuitively appealing to use similarity measures than distance measures. Suppose that we have binary values for all the x_{ij}'s, and for observations i and j we have the following 2×2 table:

		Observation j		
		0	1	
Observation i	0	a	b	$a+b$
	1	c	d	$c+d$
		$a+c$	$b+d$	n

where a denotes the number of variables for which observations i and j do not have that attribute (they each have value 0 on that attribute), d is the number of variables for which the two observations have the attribute present, and so on. The most useful similarity measures in this situation are the following:

Matching coefficient: $(a+d)/n$.

Jaquard's coefficient: $d/(b+c+d)$. This coefficient ignores zero matches. This is desirable when we do not want to consider two people to be similar simply because a large number of characteristics are absent in both. For example, if *owns a Corvette* is one of the variables, a matching "yes" would be evidence of similarity, but a matching "no" tells us little about whether the two people are similar.

Distance Measures for Mixed Data

When the measurements are mixed (some continuous and some binary), a similarity coefficient suggested by Gower is very useful. *Gower's similarity measure* is a weighted average of the distances computed for each variable, after scaling each variable to a [0,1] scale. It is defined as

$$s_{ij} = \frac{\sum_{m=1}^{p} w_{ijm} s_{ijm}}{\sum_{m=1}^{p} w_{ijm}}$$

where s_{ijm} is the similarity between observations i and j on measurement m, and w_{ijm} is a binary weight given to the corresponding distance.

The similarity measures s_{ijm} and weights w_{ijm} are computed as follows:

1. For continuous measurements, $s_{ijm} = 1 - \frac{|x_{im} - x_{jm}|}{\max(x_m) - \min(x_m)}$ and $w_{ijm} = 1$ unless the value for measurement m is unknown for one or both of the observations, in which case $w_{ijm} = 0$.
2. For binary measurements, $s_{ijm} = 1$ if $x_{im} = x_{jm} = 1$ and 0 otherwise. $w_{ijm} = 1$ unless $x_{im} = x_{jm} = 0$.
3. For nonbinary categorical measurements, $s_{ijm} = 1$ if both observations are in the same category, and otherwise $s_{ijm} = 0$. As in continuous measurements, $w_{ijm} = 1$ unless the category for measurement m is unknown for one or both of the observations, in which case $w_{ijm} = 0$.

14.3   MEASURING DISTANCE BETWEEN TWO CLUSTERS

We define a cluster as a set of one or more observations. How do we measure distance between clusters? The idea is to extend measures of *distance between observations* into *distances between clusters*. Consider cluster *A*, which includes the *m* observations $A_1, A_2, \ldots, A_m$, and cluster *B*, which includes *n* observations $B_1, B_2, \ldots, B_n$. The most widely used measures of distance between clusters are minimum and maximum, average, and centrol distances

Minimum Distance

The distance between the pair of observations A_i and B_j that are closest:

$$\min(\text{distance}(A_i, B_j)), \quad i = 1, 2, \ldots, m; \quad j = 1, 2, \ldots, n.$$

Maximum Distance

The distance between the pair of observations A_i and B_j that are farthest:

$$\max(\text{distance}(A_i, B_j)), \quad i = 1, 2, \ldots, m; \quad j = 1, 2, \ldots, n.$$

Average Distance

The average distance of all possible distances between observations in one cluster and observations in the other cluster:

$$\text{Average}(\text{distance}(A_i, B_j)), \quad i = 1, 2, \ldots, m; \quad j = 1, 2, \ldots, n.$$

Centroid Distance

The distance between the two cluster centroids. A *cluster centroid* is the vector of measurement averages across all the observations in that cluster. For cluster A, this is the vector $\bar{x}_A = \left[(1/m \sum_{i=1}^{m} x_{1i}, \ldots, 1/m \sum_{i=1}^{m} x_{pi}) \right]$. The centroid distance between clusters A and B is

$$\text{distance}(\bar{x}_A, \bar{x}_B).$$

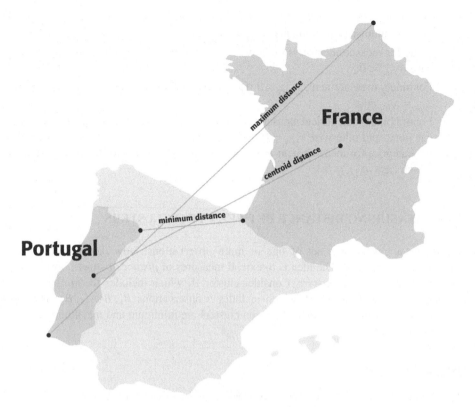

FIGURE 14.2 Two-dimensional representation of several different distance measures between Portugal and France

Minimum distance, maximum distance, and centroid distance are illustrated visually for two dimensions with a map of Portugal and France in Figure 14.2.

For instance, consider the first two utilities (Arizona, Boston) as cluster A, and the next three utilities (Central, Commonwealth, Consolidated) as cluster B. Using the normalized scores in Table 14.2 and the distance matrix in Table 14.3, we can compute each of the distances described above. Using Euclidean distance for each distance calculation, we get:

- The closest pair is Arizona and Commonwealth, and therefore the minimum distance between clusters A and B is 0.76.
- The farthest pair is Arizona and Consolidated, and therefore the maximum distance between clusters A and B is 3.02.
- The average distance is $(0.77 + 0.76 + 3.02 + 1.47 + 1.58 + 1.01)/6 = 1.44$.
- The centroid of cluster A is

$$\left[\frac{0.0459 - 1.0778}{2}, \frac{-0.8537 + 0.8133}{2} \right] = [-0.516, -0.020]$$

and the centroid of cluster B is

$$\left[\frac{0.0839 - 0.7017 - 1.5814}{3}, \frac{-0.0804 - 0.7242 + 1.6926}{3} \right]$$
$$= [-0.733, 0.296].$$

The distance between the two centroids is then

$$\sqrt{(-0.516 + 0.733)^2 + (-0.020 - 0.296)^2} = 0.38$$

In deciding among clustering methods, domain knowledge is key. If you have good reason to believe that the clusters might be chain- or sausage-like, minimum distance would be a good choice. This method does not require that cluster members all be close to one another, only that the new members being added be close to one of the existing members. An example of an application where this might be the case would be characteristics of crops planted in long rows, or disease outbreaks along navigable waterways that are the main areas of settlement in a region. Another example is laying and finding mines (land or marine). Minimum distance is also fairly robust to small deviations in the distances. However, adding or removing data can influence it greatly.

Maximum and average distance are better choices if you know that the clusters are more likely to be spherical (e.g., customers clustered on the basis of numerous attributes). If you do not know the probable nature of the cluster, these are good default choices, since most clusters tend to be spherical in nature.

We now move to a more detailed description of the two major types of clustering algorithms: hierarchical (agglomerative) and nonhierarchical.

14.4   HIERARCHICAL (AGGLOMERATIVE) CLUSTERING

The idea behind hierarchical agglomerative clustering is to start with each cluster comprising exactly one observation and then progressively agglomerating (combining) the two nearest clusters until there is just one cluster left at the end, which consists of all the observations.

Returning to the small example of five utilities and two measures (Sales and Fuel Cost) and using the distance matrix (Table 14.3), the first step in the hierarchical clustering would join Arizona and Commonwealth, which are the closest (using normalized measurements and Euclidean distance). Next we would recalculate a 4 × 4 distance matrix that would have the distances between these four clusters: {Arizona, Commonwealth}, {Boston}, {Central}, and {Consolidated}. At this point we use a measure of distance between clusters, such as the ones described in Section 14.3. Each of these distances (minimum, maximum, average, and centroid distance) can be implemented in the hierarchical scheme as described below.

HIERARCHICAL CLUSTERING IN JMP AND JMP PRO

Hierarchical clustering is available from the *Cluster* platform under *Analyze > Multivariate Methods*. The hierarchical clustering *Method* defines the distance measure used. The JMP Clustering dialog, with the available *Methods* and other options such as *Standardize Data* and *Standardize Robustly* (to reduce the influence of outliers), is shown in Figure 14.3. Note that in JMP 13 the Analyze menu structure has changed.

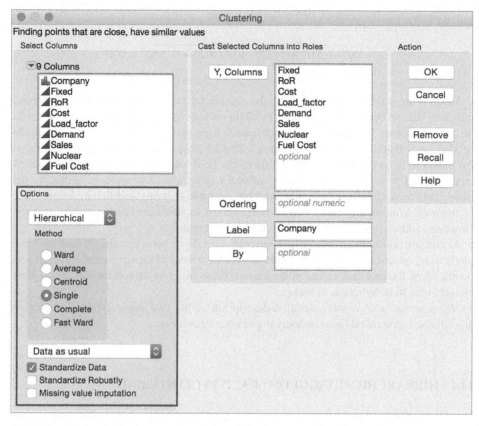

FIGURE 14.3 The Clustering dialog for the *Utilities* data, showing selections for *Hierarchical* clustering with the *Single* method and *Standardize Data*

HIERARCHICAL AGGLOMERATIVE CLUSTERING ALGORITHM:

1. Start with *n* clusters (each observation = cluster).
2. The two closest observations are merged into one cluster.
3. At every step, the two clusters with the smallest distance are merged. This means that either single observations are added to existing clusters or two existing clusters are combined.

Single Linkage

In *single-linkage clustering*, the distance measure that we use is the minimum distance (the distance between the nearest pair of observations in the two clusters, one observation in each cluster). In our utilities example we would compute the distances between each of

TABLE 14.4   Distance Matrix after Arizona and Commonwealth Consolidation Cluster Together, Using Single Linkage

	Arizona-Commonwealth	Boston	Central	Consolidated
Arizona-Commonwealth	0			
Boston	min(2.01,1.58)	0		
Central	min(0.77,1.02)	1.47	0	
Consolidated	min(3.02,2.57)	1.01	2.43	0

{Boston}, {Central}, and {Consolidated} with {Arizona, Commonwealth} to create the 4×4 distance matrix shown in Table 14.4.

The next step would consolidate {Central} with {Arizona,Commonwealth} because these two clusters are closest. The distance matrix will again be recomputed (this time it will be 3×3), and so on.

This method has a tendency to cluster together at an early stage observations that are distant from each other because of a chain of intermediate observations in the same cluster. Such clusters have elongated sausage-like shapes when visualized as objects in space.

Complete Linkage

In *complete linkage clustering*, the distance between two clusters is the maximum distance (between the farthest pair of observations). If we used complete linkage with the five-utilities example, the recomputed distance matrix would be equivalent to Table 14.4, except that the "min" function would be replaced with a "max."

This method tends to produce clusters at the early stages with observations that are within a narrow range of distances from each other. If we visualize them as objects in space, the observations in such clusters would have roughly spherical shapes.

Average Linkage

Average linkage clustering is based on the average distance between clusters (between all possible pairs of observations). If we used average linkage with the five-utilities example, the recomputed distance matrix would be equivalent to Table 14.4, except that the "min" function would be replaced with "average." This method is also called *Unweighted Pair-Group Method using Centroids* (UPGMC).

Note that unlike average linkage, the results of the single and complete linkage methods depend only on the ordering of the inter-observation distances. Linear transformations of the distances (and other transformations that do not change the ordering) do not affect the results.

Centroid Linkage

Centroid linkage clustering is based on centroid distance, where clusters are represented by their mean values for each variable, which forms a vector of means. The distance between two clusters is the distance between these two vectors. In average linkage, each pairwise distance is calculated, and the average of all such distances is calculated. In contrast, in centroid distance clustering, just one distance is calculated: the distance between group means. This method is also called *Unweighted Pair-Group Method using Centroids* (UPGMC).

Ward's Method

Ward's method is also agglomerative, in that it joins observations and clusters together progressively to produce larger and larger clusters, but it operates slightly differently from the general approach described above. Ward's method considers the "loss of information" that occurs when observations are clustered together. When each cluster has one observation, there is no loss of information and all individual values remain available. When observations are joined together and represented in clusters, information about an individual observation is replaced by the information for the cluster to which it belongs. To measure loss of information, Ward's method employs a measure "error sum of squares" (ESS) that measures the difference between individual observations and a group mean.

This is easiest to see in univariate data. For example, consider the values (2, 6, 5, 6, 2, 2, 2, 2, 0, 0, 0) with a mean of 2.5. Their ESS is equal to

$$(2 - 2.5)^2 + (6 - 2.5)^2 + (5 - 2.5)^2 + \ldots + (0 - 2.5)^2 = 50.5$$

The loss of information associated with grouping the values into a single group is therefore 50.5. Now group the observations into four groups: (0, 0, 0), (2, 2, 2, 2), (5), and (6, 6). The loss of information is the sum of the ESS's for each group, which is 0 (each observation in each group is equal to the mean for that group, so the ESS for each group is 0). Thus clustering the 10 observations into 4 clusters results in no loss of information, and this would be the first step in Ward's method. In moving to a smaller number of clusters Ward's method would choose the configuration that results in the smallest incremental loss of information.

Ward's method tends to result in convex clusters that are of roughly equal size, which can be an important consideration in some applications (e.g., in establishing meaningful customer segments).

JMP also offers a computational shortcut to Ward's method called *Fast Ward*, which is automatically used if there are more than 2000 records to be clustered.

Dendrograms: Displaying Clustering Process and Results

A *dendrogram* is a treelike diagram that summarizes the process of clustering. At the side are the observations. Similar observations are joined by lines whose horizontal length reflects the distance between the observations. Figure 14.4 shows the dendrogram that results from clustering all 22 utilities using the eight normalized measurements, Euclidian distance, and single linkage.

The number of clusters created can be changed by dragging the diamond-shaped icon within the dendrogram or by using a red triangle option. Visually, this means drawing a vertical line on a dendrogram. Observations with connections to the left the vertical line (i.e. their distance is smaller than the cutoff distance) belong to the same cluster. For example, setting the number of clusters to 6 in Figure 14.4 results in six clusters. Using the *Color Clusters* option from the red triangle makes it easier to see which observations make up the different clusters.

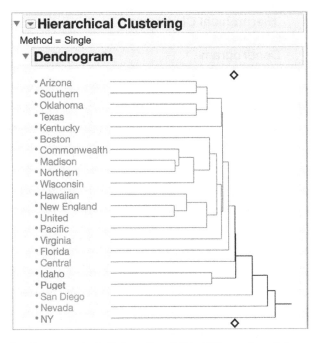

FIGURE 14.4   Dendrogram: *single linkage* for all 22 utilities, using all 8 measurements and 6 clusters

The six clusters are (from top to bottom on the dendrogram):

{All others}
{Central}
{Puget, Idaho}
{San Diego}
{Nevada}
{NY}

Note that if we wanted five clusters, they would be identical to the six above, with the exception that the first two clusters would be merged into one cluster (Central would be grouped into the first cluster). In general, all hierarchical methods have clusters that are nested within each other as the number of clusters decreases. This is a valuable property for interpreting clusters and is essential in certain applications, such as taxonomy of varieties of living organisms.

The average linkage dendrogram is shown in Figure 14.5. When we use average linkage with six clusters, the resulting six clusters are (from top to bottom in the dendrogram):

{Arizona, Southern, Oklahoma, Texas, Florida, Central, Kentucky}
{Commonwealth, Wisconsin, Madison, Northern, Virginia}
{Boston, New England, United, Pacific, Hawaiian}
{San Diego}
{NY}
{Idaho, Puget, Nevada}

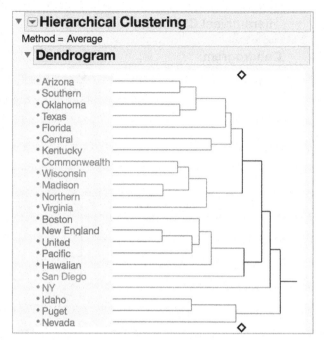

FIGURE 14.5   Dendrogram: *average linkage* for all 22 utilities, using all 8 measurements and 6 clusters

Validating Clusters

One important goal of cluster analysis is to come up with *meaningful clusters*. Since there are many variations that can be chosen, it is important to make sure that the resulting clusters are valid, in the sense that they really generate some insight. To see whether the cluster analysis is useful, consider each of the following aspects:

1. *Cluster interpretability.* Is the interpretation of the resulting clusters reasonable? To interpret the clusters, explore the characteristics of each cluster:

   a. Obtain summary statistics (e.g., average, min, max) from each cluster on each measurement that was used in the cluster analysis (use *Cluster Summary* from the red triangle to display means and standard deviations—see Figure 14.6).

   b. Examine the clusters for separation along some common feature (variable) that was not used in the cluster analysis.

   c. Label the clusters: based on the interpretation, try to assign a name or label to each cluster (use *Save Clusters* to save the cluster numbers to the data table, then use *Cols > Recode* to rename the clusters).

2. *Cluster stability.* Do cluster assignments change significantly if some of the inputs are altered slightly? Another way to check stability is to partition the data and see how well clusters formed based on one part apply to the other part:

   a. Partition the data, and hide and exclude partition B.

   b. Cluster partition A.

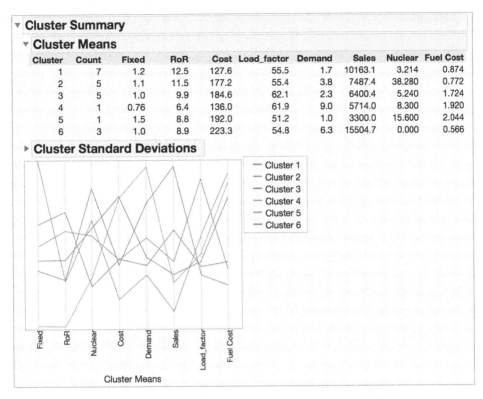

▼ Cluster Summary

▼ Cluster Means

Cluster	Count	Fixed	RoR	Cost	Load_factor	Demand	Sales	Nuclear	Fuel Cost
1	7	1.2	12.5	127.6	55.5	1.7	10163.1	3.214	0.874
2	5	1.1	11.5	177.2	55.4	3.8	7487.4	38.280	0.772
3	5	1.0	9.9	184.6	62.1	2.3	6400.4	5.240	1.724
4	1	0.76	6.4	136.0	61.9	9.0	5714.0	8.300	1.920
5	1	1.5	8.8	192.0	51.2	1.0	3300.0	15.600	2.044
6	3	1.0	8.9	223.3	54.8	6.3	15504.7	0.000	0.566

▶ Cluster Standard Deviations

FIGURE 14.6   Cluster Summary for hierarchical clustering with 6 clusters

    c. Use *Save Formula for Closest Cluster* to save the cluster numbers and cluster formulas to the data table. Cluster numbers for partition A will be saved to the data table, and clusters for partition B will be generated (each observation is assigned to the cluster with the closest centroid).

    d. Use *Graph Builder* or other tools to assess how consistent the cluster assignments are for the two partitions and for the dataset as a whole.

3. *Cluster separation.* Examine the ratio of between-cluster variation to within-cluster variation to see whether the separation is reasonable. There exist statistical tests for this task (an *F*-ratio), but their usefulness is somewhat controversial (and these tests are not available in JMP).

4. *Number of clusters.* The number of resulting clusters must be useful, given the purpose of the analysis. For example, suppose that the goal of the clustering is to identify categories of customers and assign labels to them for market segmentation purposes. If the marketing department can only manage to sustain three different marketing presentations, it would probably not make sense to identify more than three clusters. In some cases the *Distance* (or *Scree*) plot at the bottom of the dendrogram can be used to determine the number of clusters. An elbow or sharp upward bend in this plot may indicate the optimal number of clusters. The scree plot for the Utilities data, shown at the bottom of Figure 14.9, shows a relatively steady increase rather than a sharp bend. In this case, the scree plot isn't overly useful. Here the context of the

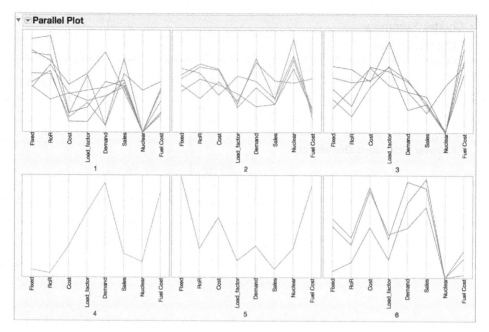

FIGURE 14.7   Parallel plot for the 22 utilities, from average linking with 6 clusters.

study and practical considerations are more important in determining the best number of clusters.

Returning to the utilities example, we notice that both methods (single and average linkage) identify New York and San Diego as singleton clusters. Also, both dendrograms imply that a reasonable number of clusters in this dataset is six. One insight that can be derived from the average linkage clustering is that clusters tend to group geographically. The four nonsingleton clusters form (approximately) a southern group, a northern group, an east/west seaboard group, and a west group.

We can further characterize each of the clusters by producing summary statistics (see Figure 14.6) and other graphical displays. A *parallel plot* (available from the red triangle) is useful for understanding the profiles of the clusters across the different variables (see Figure 14.7). For example, we can see that clusters 1, 3, and 6 use very little or no nuclear power, while cluster 2 is high in nuclear power usage.

Saving clusters to the data table allows us to use other graphical and exploratory tools to characterize the clusters and explore the clusters across the different variables, such as the *Graph Builder*. The *Column Switcher*, which is an option under the top red triangle in all JMP graphing and analysis platforms under *Script*, can be used in conjunction with the *Graph Builder* to explore clusters across the different variables. In Figure 14.8 we can see the cluster 6 has the highest sales (killowatthour use per year).

Two-Way Clustering

In addition to clustering the data (the observations), the variables can be clustered. Variable clustering groups measurements have similar characteristics or attributes. The resulting

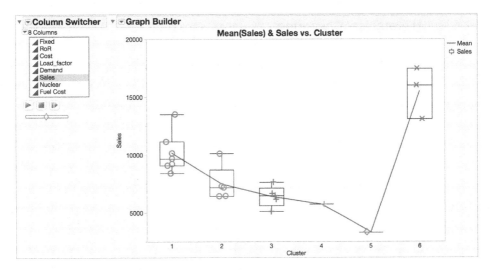

FIGURE 14.8   Exploring the saved clusters using the *Graph Builder* and the *Column Switcher*

dendrogram can be useful if the variables are on the same scale. The *Two Way Clustering* option (from the top red triangle) clusters both observations and variables. The resulting color map (or heatmap) shows the values of the data colored across the value range for each measurement (we've selected a white (low) to green (high) color theme in this example). Figure 14.9 shows a color map of the four clusters and two singletons (see Figure 14.5), highlighting the different profile that each cluster has in terms of the eight variables. We see, for instance, that cluster 2 (with Commonwealth and Wisconsin) is characterized by utilities with a high percentage of nuclear power, cluster 1 (the cluster at the top) is characterized by high fixed cost and RoR and low nuclear, and the last cluster has low nuclear but high cost per kilowatt, high demand, and high sales. Note that these variables aren't on the same scale, so we ignore the variable clustering.

Limitations of Hierarchical Clustering

Hierarchical clustering is very appealing in that it does not require specification of the number of clusters, and in this sense is purely data driven. The ability to represent the clustering process and results through dendrograms is also an advantage of this method, as it is easier to understand and interpret. There are, however, a few limitations to consider:

1. Hierarchical clustering requires the computation and storage of a $n \times n$ distance matrix. For very large datasets, this can be expensive and slow.

2. The hierarchical algorithm makes only one pass through the data. This means that observations that are allocated incorrectly early in the process cannot be reallocated subsequently.

3. Hierarchical clustering also tends to have low stability. Reordering data or dropping a few observations can lead to a different solution.

4. The choice of distance between clusters, single and complete linkage, are robust to changes in the distance metric (e.g., Euclidean, statistical distance) as long as the relative ordering is kept. In contrast, average linkage is more influenced by the choice

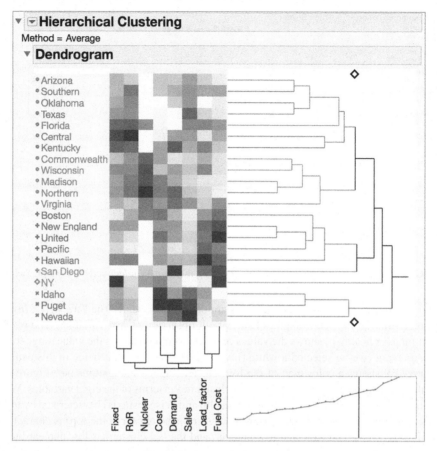

FIGURE 14.9   Color map for the 22 utilities. Rows are sorted by the 6 clusters from average linkage clustering. Darker denotes higher values for a measurement.

of distance metric, and might lead to completely different clusters when the metric is changed.

5. Hierarchical clustering is sensitive to outliers. (However, the *Standardize Robustly* option in the *Hierarchical Clustering* dialog can be used to reduce the influence of outliers on the estimated means and standard deviations.)

14.5   NONHIERARCHICAL CLUSTERING: THE *K*-MEANS ALGORITHM

A nonhierarchical approach to forming good clusters is to prespecify a desired number of clusters, k, and to assign each case to one of k clusters so as to minimize a measure of dispersion within the clusters. In other words, the goal is to divide the sample into a predetermined number k of nonoverlapping clusters so that clusters are as homogeneous as possible with respect to the measurements used.

A common measure of within-cluster dispersion is the sum of distances (or sum of squared Euclidean distances) of observations from their cluster centroid. The problem can be set up as an optimization problem involving integer programming, but because solving integer programs with a large number of variables is time-consuming, clusters are often computed using a fast, heuristic method that produces good (although not necessarily optimal) solutions. The *k*-means algorithm is one such method.

The *k*-means algorithm starts with an initial partition of the observations into *k* clusters. Subsequent steps modify the partition to reduce the sum of the distances of each observation from its cluster centroid. The modification consists of allocating each observation to the nearest of the *k* centroids of the previous partition. This leads to a new partition for which the sum of distances is smaller than before. The means of the new clusters are computed and the improvement step is repeated until the improvement is very small.

---

k-MEANS CLUSTERING ALGORITHM:

1. Start with *k* initial clusters (user chooses *k*).
2. At every step, each observation is reassigned to the cluster with the "closest" centroid.
3. Recompute the centroids of clusters that lost or gained an observation, and repeat step 2.
4. Stop when moving any more observations between clusters increases cluster dispersion.

Returning to the example with the five utilities and two measurements, let us assume that $k = 2$ and that the initial clusters are A = {Arizona, Boston} and B = {Central, Commonwealth, Consolidated}. The cluster centroids computed in Section 14.4 are

$$\bar{x}_A = [-0.516, -0.020] \text{ and } \bar{x}_B = [-0.733, 0.296].$$

The distance of each observation from each of these two centroids is shown in Table 14.5. We see that Boston is closer to cluster B, and that Central and Commonwealth are each closer to cluster A. We therefore move each of these observations to the other cluster and obtain A={Arizona,Central,Commonwealth} and B={Consolidated,Boston}. Recalculating the centroids gives

$$\bar{x}_A = [-0.191, -0.553] \text{ and } \bar{x}_B = [-1.33, 1.253].$$

TABLE 14.5   Distance of Each Observation from Each Centroid

	Distance from Centroid A	Distance from Centroid B
Arizona	1.0052	1.3887
Boston	1.0052	0.6216
Central	0.6029	0.8995
Commonwealth	0.7281	1.0207
Consolidated	2.0172	1.6341

TABLE 14.6   Distance of Each Observation from Each Newly Calculated Centroid

	Distance from Centroid A	Distance from Centroid B
Arizona	0.3827	2.5159
Boston	1.6289	0.5067
Central	0.5463	1.9432
Commonwealth	0.5391	2.0745
Consolidated	2.6412	0.5067

The distance of each observation from each of the newly calculated centroids is given in Table 14.6. At this point we stop because each observation is allocated to its closest cluster.

Initial Partition into k Clusters

The choice of the number of clusters can either be driven by external considerations (previous knowledge, practical constraints, etc.), or we can try a few different values for k and compare the resulting clusters. After choosing k, the n observations are partitioned into these initial clusters. The number of clusters in the data is generally not known, so it is a good idea to run the algorithm with different values for k that are near the number of clusters that one expects from the data. Note that the clusters obtained using different values of k will not be nested (unlike those obtained by hierarchical methods).

K-MEANS CLUSTERING IN JMP

k-Means clustering is an option in the *Clustering* platform. The dialog for k-means clustering in JMP is shown in Figure 14.10 (top). The *Columns Scaled Individually* option is selected to normalize the measurements. This dialog produces a *Control Panel* for indicating the upper and lower bound for the number of clusters to form (bottom, in Figure 14.10). A separate analysis is provided for each number. In this example, we form partitions starting with three clusters and ending with eight clusters. The *Single Step* option allows the user to form one partition at a time, starting with the lowest number of clusters specified.

The results of running the k-means algorithm for all 22 utilities and eight measurements with $k = 6$ are shown in Figure 14.11. The *Parallel Coordinate Plot* was selected using a red triangle option. As in the results from the hierarchical clustering, we see once again that San Diego and New York are singleton clusters. Other clusters are very similar to those that emerged in the hierarchical clustering with average linkage, and cluster 6 is identical. In JMP, the company can be seen by clicking on a line in any plot (since Company has been selected as a column label using *Cols > Label*).

To examine the clusters more closely, we save the clusters to the data table (using the red triangle). This saves both the cluster number and the standardized distance within the cluster. Now we can graph the distances using the *Graph Builder* (see Figure 14.12). We see that for clusters with more than one observation, cluster 4 has the largest within-cluster

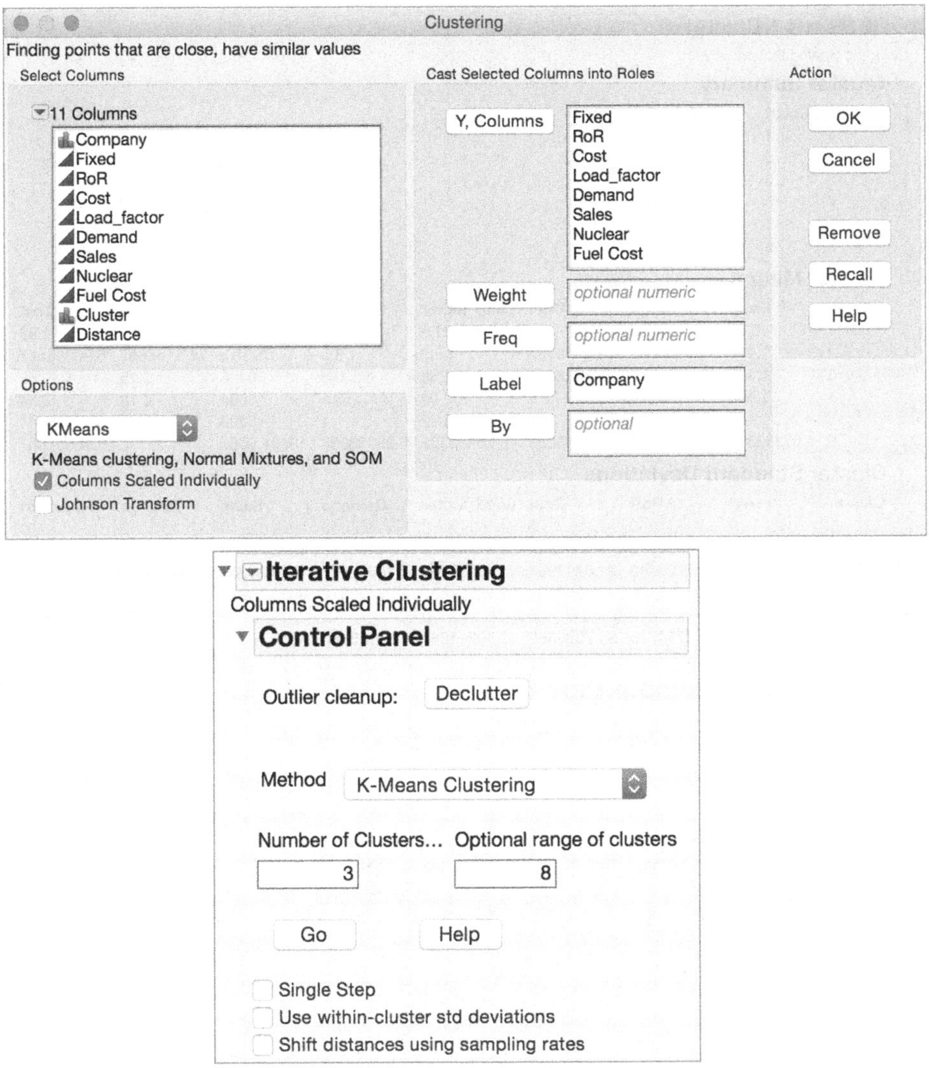

FIGURE 14.10   K-Means Dialog (top) and Control Panel (bottom) for specifying the number of clusters

variability. The labels in the graph were added by selecting all rows in the data table and using *Rows > Label* (after setting Company as a column label).

To characterize the resulting clusters, we examine the cluster centroids in the *Parallel Coordinate Plots* (or "profile plots") in Figure 14.13. This is provided at the bottom of the Parallel Coordinates Plot for the individual clusters. We can see, for instance, that cluster 5 has high average Fixed and Fuel Cost, and low average Sales and Load Factor. In contrast, cluster 1 has the low Nuclear and Sales, but the highest Demand.

We can also use *Tabulate* to inspect the information on the within-cluster dispersion for the saved clusters. From Figure 14.14 we see that cluster 3 has the largest average distance, and it includes six records. In comparison, cluster 2, with six records, has a

▼ ⬙ **K Means NCluster=6**

Columns Scaled Individually

▼ **Cluster Summary**

Cluster	Count	Step	Criterion
1	1	2	0
2	6		
3	6		
4	5		
5	1		
6	3		

▼ **Cluster Means**

Cluster	Fixed	RoR	Cost	Load_factor	Demand	Sales	Nuclear	Fuel Cost
1	0.76	6.4	136	61.9	9	5714	8.3	1.92
2	1.075	11.2833333	181.333333	55.8	3.5	7087.5	36.1166667	0.90216667
3	1.185	12.4	120.833333	54.65	0.8	10456	3.75	0.8765
4	1.138	10.46	177.8	62.64	3.3	7064	0.18	1.5854
5	1.49	8.8	192	51.2	1	3300	15.6	2.044
6	1.00333333	8.86666667	223.333333	54.8333333	6.33333333	15504.6667	0	0.56566667

▼ **Cluster Standard Deviations**

Cluster	Fixed	RoR	Cost	Load_factor	Demand	Sales	Nuclear	Fuel Cost
1	0	0	0	0	0	0	0	0
2	0.10045729	1.18520978	19.694895	2.44131112	2.11108187	1523.19946	8.59794872	0.38490147
3	0.14244882	1.87705443	21.6749984	3.15052906	2.19012937	1523.20736	8.38525492	0.26527329
4	0.13332667	2.06455806	14.0057131	2.56561883	2.37402612	835.435216	0.36	0.43376196
5	0	0	0	0	0	0	0	0
6	0.18080069	1.00774776	35.7055862	2.39211668	2.4115463	1812.47719	0	0.19128397

▼ ⬙ **Parallel Coordinate Plot**

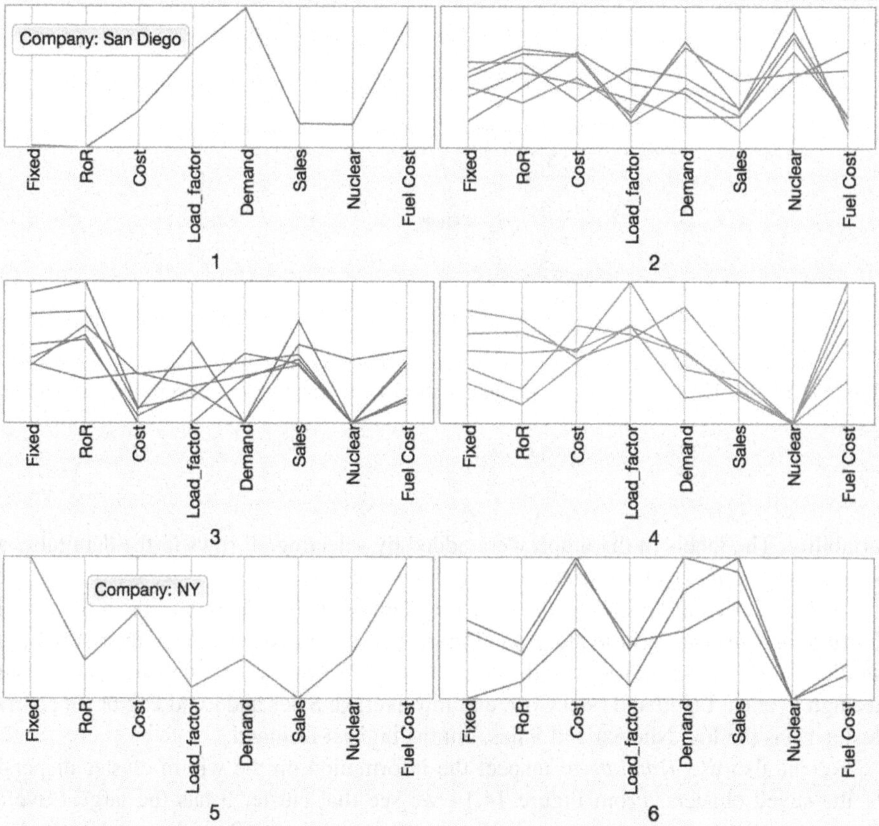

FIGURE 14.11   Output for *k*-means clustering of 22 utilities into *k=6* clusters (sorted by cluster ID)

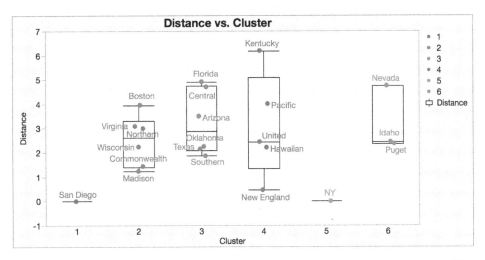

FIGURE 14.12   Graph of Normalized Distances versus Cluster Number from the Graph Builder, after saving clusters to the data table

smaller within-cluster average distance. This means that cluster 2 is more homogeneous than cluster 3.

From the distances between clusters we can learn about the separation of the different clusters. In Figure 14.15 we see two plots that are available from the cluster platform. The *biplot* (top) is a two-dimensional plot of the clusters based on a principal components analysis. The circles for each cluster in the biplot are sized in proportion to the number of observations in the cluster, and the shaded area is a *density contour*, which indicates where

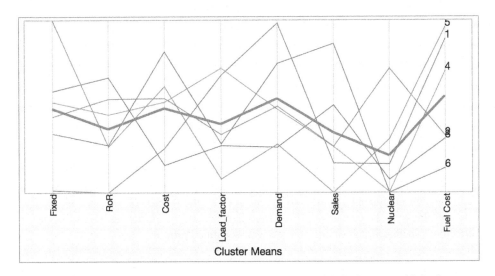

FIGURE 14.13   Parallel (profile) plot of Cluster centroids for *k*-means with *k=6*

Tabulate

Cluster	N	Distance Mean	Std Dev
1	1	0.00	.
2	6	2.49	1.05
3	6	3.23	1.35
4	5	3.06	2.15
5	1	0.00	.
6	3	3.18	1.35

FIGURE 14.14 Summary of cluster centroids, created using the saved clusters and *Tabulate*

90% of the observations in the cluster would fall. Note that the different markers were applied by selecting *Color or Mark by Column* from the *Rows* menu.

The scatterplot matrix (bottom) shows the clusters across pairs of the measurements. While its a bit difficult to see this on the printed page, from these plots we learn that cluster 6 is very different from the other clusters. However, cluster 2 is similar to clusters 3 and 4 in the biplot. From the scatterplot we learn that cluster 2 is most unique in the percent Nuclear, but is similar to clusters 3 and 4 in the other measurements. This might lead us to examine the possibility of merging two of the clusters. Cluster 1, which is a singleton cluster, appears to be very far from all the other clusters (this can be seen most clearly in the biplot).

When the number of clusters is not pre-determined by domain requirements, we can use a criterion provided by JMP for determining the number of clusters to produce, the *CCC*, or *Cubic Clustering Criterion* (Sarle, 1983). CCC is a criterion designed by SAS Institute to choose k in an automated way. "Larger positive values of the CCC indicate a better solution, as it shows a larger difference from a uniform (no clusters) distribution" (Lim, Acito, and Rusetski, 2006). CCC is reported in *Cluster Comparison* section at the top of the *K Means Cluster* output window (see Figure 14.16) when a range of values for k is used. Here the CCC identifies $k = 6$ as the best number of clusters.

We can also use a graphical approach to evaluating different numbers of clusters. An "elbow chart" is a line chart, similar to a scree plot, depicting the decline in cluster heterogeneity as we add more clusters. Figure 14.17 shows the average deviation of within-cluster distance (normalized) for different choices of k. Moving from 1 to 6 tightens clusters considerably (reduces the average within-cluster distance). Adding more than 6 clusters brings less reduction in the within-cluster homogeneity.

Finally, we can use the information on the distances between the final clusters to evaluate the cluster validity. The ratio of the sum of squared distances to cluster means for a given k to the sum of squared distances to the mean of all the records ($k = 1$) is a measure for the usefulness of the clustering. If the ratio is near 1.0, the clustering has not been very effective, whereas if it is small, we have well-separated groups.

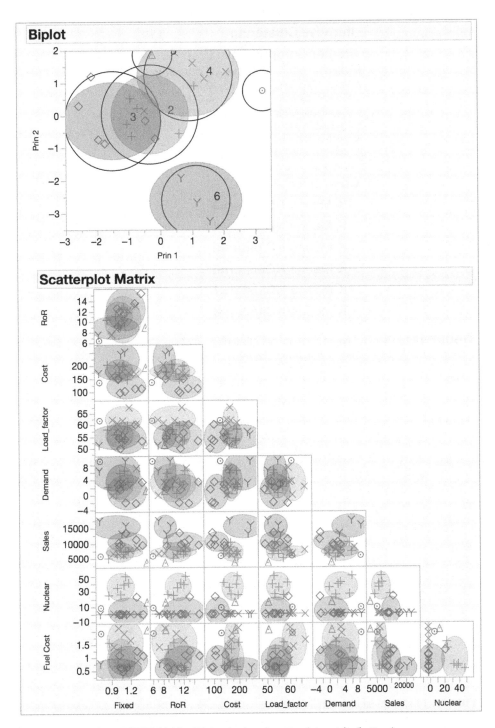

FIGURE 14.15   Biplot (top) and scatterplot matrix (bottom)

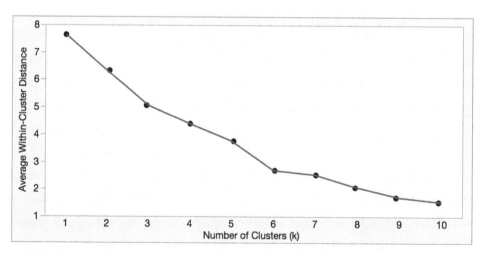

FIGURE 14.16   Initial Output for *k*-means clustering of 22 utilities, showing a comparison of clusters using the *CCC* (*Cubic Clustering Criterion*)

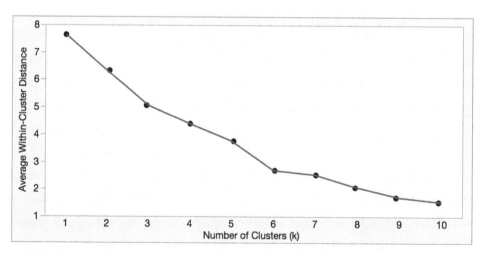

FIGURE 14.17   Comparing different choices of *k* in terms of the standard deviation of within-cluster distances by saving clusters for *k* = 1 to 10, summarizing the distances using *Tabulate*, and graphing using *Graph Builder*

PROBLEMS

14.1 Pharmaceutical Industry. An equities analyst is studying the pharmaceutical industry and would like your help in exploring and understanding the financial data
collected by her firm. Her main objective is to understand the structure of the pharmaceutical industry using some basic financial measures.

Financial data gathered on 21 firms in the pharmaceutical industry are available
in the file `Pharmaceuticals.jmp`. For each firm the following variables are
recorded:

1. Market capitalization (in billions of dollars)
2. Beta
3. Price/earnings ratio
4. Return on equity
5. Return on assets
6. Asset turnover
7. Leverage
8. Estimated revenue growth
9. Net profit margin
10. Median recommendation (across major brokerages)
11. Location of firm's headquarters
12. Stock exchange on which the firm is listed

Use cluster analysis to explore and analyze the given dataset as follows:

a. Use only the quantitative variables to cluster the 21 firms. Justify the various
choices made in conducting the cluster analysis, such as the specific clustering
algorithm/s used, and the number of clusters formed.

b. Interpret the clusters with respect to the quantitative variables that were used in
forming the clusters.

c. Is there a pattern in the clusters with respect to the qualitative variables (those not
used in forming the clusters)?

d. Provide an appropriate name for each cluster (based on variables in the dataset).

e. How do your results help the equities analyst achieve her objective?

14.2 Customer Rating of Breakfast Cereals. The dataset `Cereals.jmp` includes nutritional information, store display, and consumer ratings for 77 breakfast cereals.

Data preprocessing. Note that some cereals are missing values. These will be automatically omitted from the analysis. Use the *Columns Viewer* to identify which
variables are missing values and how many values are missing. Describe why the
missing values, in this example, might (or migh not) be a concern in this assignment.

a. Apply hierarchical clustering to the data using single linkage and complete linkage.

    i. The clustering dialog provides an option to standardize. Should this be se-
lected? Why?

ii. Look at the dendrograms and the parallel plots. Comment on the structure of the clusters and on their stability.

iii. Which method leads to the most insightful or meaningful clusters?

iv. Choose one of the methods. How many clusters would you use? Why?

b. The public elementary schools would like to choose a set of cereals to include in their daily cafeterias. Every day a different cereal is offered, but all cereals should support a healthy diet. For this goal you are requested to find a cluster of "healthy cereals."

i. Based on the variables at hand, how would you characterize "healthy cereals"?

ii. Save the clusters to the data table, and use graphical and descriptive tools to characterize and name the clusters.

iii. Which cluster of cereals is the most "healthy"? Does one cluster stand out as being the healthiest?

14.3 Marketing to Frequent Fliers. The file `EastWestAirlinesCluster.jmp` contains information on 3999 passengers who belong to an airline's frequent flier program. For each passenger the data include information on their mileage history and on different ways they accrued or spent miles in the last year. The goal is to try to identify clusters of passengers that have similar characteristics for the purpose of targeting different segments for different types of mileage offers.

a. Apply hierarchical clustering and Ward's method. Make sure to standardize the data. Use the dendrogram and the scree plot, along with practical considerations, to identify the "best" number of clusters. Use the *Color Clusters* and *Two-Way Clustering* option to help interpret the clusters. How many clusters would you select? Why?

b. What would happen if the data were not standardized?

c. Explore the clusters to try to characterize them. Try to give each cluster a label.

i. Compare the cluster centroids (select *Cluster Summary*), and click on the lines to characterize the different clusters.

ii. Save the clusters to the data table and use graphical tools and the *Column Switcher*.

d. To check the stability of the clusters, hide and exclude a random 5% of the data (use *Rows > Row Selection > Select Randomly*, and then use *Rows > Hide and Exclude*), and repeat the analysis. Does the same picture emerge?

e. Use *k*-means clustering with the number of clusters that you found above. How do these results compare to the results from hierarchical clustering? Use the built-in graphical tools to characterize the clusters.

f. Which clusters would you target for offers, and what types of offers would you target to customers in that cluster?

14.4 University Rankings. The dataset on American College and University Rankings (`Colleges.jmp`) contains information on 1302 American colleges and universities offering an undergraduate program. For each university there are several measurements. Most of these are continuous (e.g., tuition and graduation rate) while a couple are categorical (location by state and whether it is a private or public school).

a. Use the *Columns Viewer* to produce numeric summaries of all of the variables. Note that many observations are missing some measurements. Clustering methods in JMP omit records with missing values. For the purposes of this exercise, we will assume that the values are missing purely at random (not that this may in fact not be the case). So our first goal is to estimate (impute) these missing values.

b. Select all of the continuous columns, and go to *Cols > Modeling Utilities > Explore Missing Values*. Select the option *Multivariate Normal Imputation*, and click *Yes Shrinkage*. Multivariate normal imputation uses least squares regression to predict the missing values from the nonmissing variables in each row. The imputed values will be highlighted in the data table. Save the data table with imputed values under a new name.

c. Use *k*-means clustering, with only the continuous variables. Use College Name as the label. Form clusters starting with 3 and ending with 8. What is the optimal number of clusters based on the Cubic Clustering Criterion (under *Cluster Comparison*)?

d. Use the biplot, the parallel plot, and other built-in graphical and numeric summaries to explore the clusters. Save the clusters to the data table and use other graphical tools to compare and characterize the clusters. Summarize the key characteristics of each of the four clusters, and try label or name the clusters.

e. Use the categorical variables that were not used in the analysis (State and Private/Public) to characterize the different clusters. (**Hint**: Create a geographic map to explore the clusters geographically.) Is there any relationship between the clusters and the categorical information?

f. Can you think of other external information that might explain the contents of some or all of these clusters?

g. Consider Tufts University. Which cluster does Tufts belong to? Which other universities is Tufts similar to, based on the clustering and the categorical variables?

h. Return to the original data table (with the missing values). Run the same *k*-means cluster analysis using this data.

   i. Compare the results to those achieved after imputing missing values in terms of the number of clusters and the characteristics of the clusters? What are the key differences?

   ii. In the initial analysis, we assumed that the values were missing at random and imputed the missing values. Describe why this approach was, or was not, a good strategy.

PART VI

FORECASTING TIME SERIES

15

HANDLING TIME SERIES

In this chapter we describe the context of business time series forecasting and introduce the main approaches that are detailed in the next chapters, and in particular regression-based forecasting and smoothing-based methods. Our focus is on forecasting future values of a single time series. These three chapters are meant as an introduction to the general forecasting approach and methods.

In this chapter we discuss the difference between the predictive nature of time series forecasting and the descriptive or explanatory task of time series analysis. A general discussion of combining forecasting methods or results for added precision follows. Additionally, we present a time series in terms of four components (level, trend, seasonality, and noise), and present methods for visualizing the different components and for exploring time series data. We close with a discussion of data partitioning (creating training and validation sets), which is performed differently from cross-sectional data partitioning.

15.1 INTRODUCTION

Time series forecasting is performed in nearly every organization that works with quantifiable data. Retail stores use it to forecast sales. Energy companies use it to forecast reserves, production, demand, and prices. Educational institutions use it to forecast enrollment. Governments use it to forecast tax receipts and spending. International financial organizations like the World Bank and International Monetary Fund use it to forecast inflation and economic activity. Transportation companies use time series forecasting to forecast future travel. Banks and lending institutions use it (sometimes badly!) to forecast new home purchases. And venture capital firms use it to forecast market potential and to evaluate business plans.

Previous chapters in this book deal with classifying and predicting data where time is not a factor, in the sense that it is not treated differently from other variables, and

Data Mining for Business Analytics: Concepts, Techniques, and Applications with JMP Pro®, First Edition.
Galit Shmueli, Peter C. Bruce, Mia L. Stephens, and Nitin R. Patel.
© 2017 John Wiley & Sons, Inc. Published 2017 by John Wiley & Sons, Inc.

where the sequence of measurements over time does not matter. These are typically called cross-sectional data. In contrast, this chapter deals with a different type of data: time series.

With today's technology, many time series are recorded on very frequent time scales. Stock data are available at ticker level. Purchases at online and offline stores are recorded in real time. Although data might be available at a very frequent scale, for the purpose of forecasting, it is not always preferable to use this time scale. In considering the choice of time scale, one must consider the scale of the required forecasts and the level of noise in the data. For example, if the goal is to forecast next-day sales at a grocery store, using minute-by-minute sales data is likely to be less useful for forecasting than using daily aggregates. The minute-by-minute series will contain many sources of noise (e.g., variation by peak and nonpeak shopping hours) that degrade its forecasting power, and these noise errors, when the data are aggregated to a cruder level, are likely to average out.

The focus in this part of the book is on forecasting a single time series. In some cases multiple time series are to be forecasted (e.g., the monthly sales of multiple products). Even when multiple series are being forecasted, the most popular forecasting practice is to forecast each series individually. The advantage of single-series forecasting is its simplicity. The disadvantage is that it does not take into account possible relationships between series. The statistics literature contains models for multivariate time series that directly model the cross-correlations between series. Such methods tend to make restrictive assumptions about the data and the cross-series structure, and they also require statistical expertise for estimation and maintenance. Econometric models often include information from one or more series as inputs into another series. However, such models are based on assumptions of causality that are based on theoretical models. An alternative approach is to capture the associations between the series of interest and external information more heuristically. An example is using the sales of lipstick to forecast some measure of the economy, based on the observation by Leonard Lauder, chairman of Estee Lauder, that lipstick sales tend to increase before tough economic times (a phenomenon called the "leading lipstick indicator").

15.2 DESCRIPTIVE VERSUS PREDICTIVE MODELING

As with cross-sectional data, modeling time series data is done for either descriptive or predictive purposes. In descriptive modeling, or *time series analysis*, a time series is modeled to determine its components in terms of seasonal patterns, trends, relation to external factors, and so on. These can then be used for decision making and policy formulation. In contrast, *time series forecasting* uses the information in a time series (and perhaps other information) to predict future values of that series. The difference between the goals of time series analysis and time series forecasting leads to differences in the type of methods used and in the modeling process itself. For example, in selecting a method for describing a time series, priority is given to methods that produce understandable results (rather than "black box" methods) and sometimes to models based on causal arguments (explanatory models). Furthermore describing can be done in retrospect, while forecasting is prospective in nature. This means that descriptive models might use "future" information (e.g., averaging the values of yesterday, today, and tomorrow to obtain a smooth representation of today's value), whereas forecasting models cannot.

The focus in this chapter is on time series forecasting, where the goal is to predict future values of a time series. For information on time series analysis, see Chatfield (2003).

15.3   POPULAR FORECASTING METHODS IN BUSINESS

In this part of the book we focus on two main types of forecasting methods that are popular in business applications. Both are versatile and powerful, yet relatively simple to understand and deploy. One type of forecasting tool is multiple linear regression, where the user specifies a certain model and then estimates it from the time series. The other is the more data-driven tool of smoothing, where the method learns patterns from the data. Each of the two types of tools has advantages and disadvantages, as detailed in Chapters 16 and 17. We also note that data mining methods such as neural networks and others that are intended for cross-sectional data are also sometimes used for time series forecasting, especially for incorporating external information into the forecasts (see Shmueli, 2011).

Combining Methods

Before a discussion of specific forecasting methods in the following two chapters, it should be noted that a popular approach for improving predictive performance is to combine forecasting methods. Combining methods can be done via two-level (or multi-level) forecasters, where the first method uses the original time series to generate forecasts of future values, and the second method uses the residuals from the first model to generate forecasts of future forecast errors, thereby "correcting" the first level forecasts. We describe two-level forecasting in Chapter 16 (Section 16.4). Another combination approach is via "ensembles," where multiple methods are applied to the time series, and their resulting forecasts are averaged in some way to produce the final forecast. Combining methods can take advantage of the strengths of different forecasting methods to capture different aspects of the time series (also true in cross-sectional data). The averaging across multiple methods can lead to forecasts that are more robust and of higher precision.

15.4   TIME SERIES COMPONENTS

In both types of forecasting methods, regression models and smoothing, and in general, it is customary to dissect a time series into four components: level, trend, seasonality, and noise. The first three components are assumed to be invisible, as they characterize the underlying series, which we only observe with added noise. Level describes the average value of the series, trend is the change in the series from one period to the next, and seasonality describes a short-term cyclical behavior of the series that can be observed several times within the given series. Finally, noise is the random variation that results from measurement error or other causes not accounted for. It is always present in a time series to some degree.

In order to identify the components of a time series, the first step is to examine a time plot. In its simplest form, a time plot is a line chart of the series values over time, with temporal labels (e.g., calendar date) on the horizontal axis. To illustrate this, consider the following example.

Example: Ridership on Amtrak Trains

Amtrak, a US railway company, routinely collects data on ridership. Here we focus on forecasting future ridership using the series of monthly ridership between January 1991 and March 2004. These data are found in `Amtrak.jmp`, and are publicly available at www.forecastingprinciples.com. [1]

A time plot for monthly Amtrak ridership series is shown in Figure 15.1. Note that the values are in thousands of riders. (Use the *Graph Builder*, with ridership as *Y* and Month as *X*, and select the *Line* graph element. Double-click on the horizontal axis to change the label orientation.)

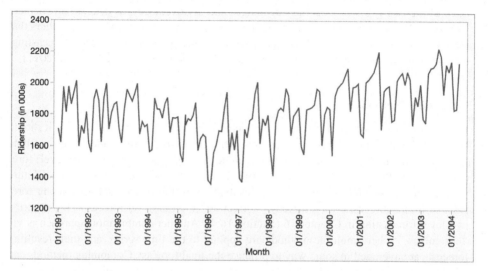

FIGURE 15.1   Monthly ridership on Amtrak trains (in thousands) from January 1991 to March 2004

Looking at the time plot reveals the nature of the series components: the overall level is around 1,800,000 passengers per month. A slight U-shaped trend is discernible during this period, with pronounced annual seasonality, with peak travel during summer (July and August).

A second step in visualizing a time series is to examine it more carefully. A few tools are useful:

Zoom in: Zooming in to a shorter period within the series can reveal patterns that are hidden when viewing the entire series. This is especially important when the time series is long. Consider a series of the daily number of vehicles passing through the Baregg tunnel in Switzerland (data are in `Baregg Daily.jmp`, and are publicly available in the same location as the Amtrak Ridership data; series D028). The series from November 1, 2003, to November 16, 2005, is shown in the top panel of Figure 15.2. Zooming in to a three-month period (bottom panel in Figure 15.2) reveals a strong day-of-week pattern that was not visible in the initial time plot of the complete time series. (In JMP, double-click on the horizontal axis to change the minimum and

---

[1] To get this series: Click on Data. Under T-Competition Data click "time-series data" (the direct URL to the data file is www.forecastingprinciples.com/files/MHcomp1.xlsas of September 2014). This file contains many time series. In the Monthly worksheet, column AI contains series M034.

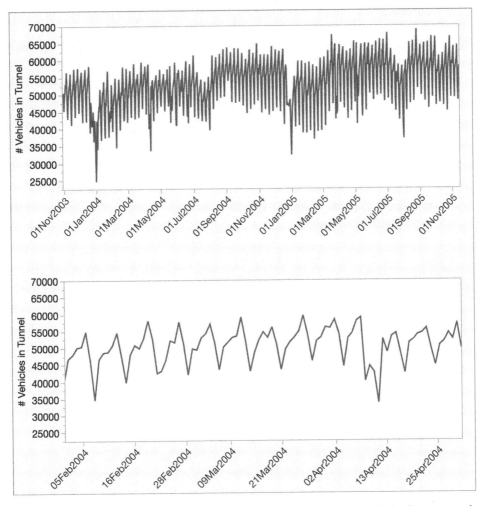

FIGURE 15.2 Time plots of the daily number of vehicles passing through the Baregg tunnel, Switzerland. The bottom panel zooms in to a 4-month period, revealing a day-of-week pattern.

maximum values, or use the *Data Filter*. The *y*-axis scale was adjusted in both graphs in Figure 15.2.)

Change scale of series: In order to better identify the shape of a trend, it is useful to change the scale of the series. One simple option is to change the vertical scale to a logarithmic scale (double-click on the vertical axis and change the scale from Linear to Log). If the trend on the new scale appears more linear, then the trend in the original series is closer to an exponential trend.

Add trend lines: Another possibility for better capturing the shape of the trend is to add a trend line. By trying different trendlines, one can see what type of trend (e.g., linear, exponential, cubic) best approximates the data. In *Graph Builder*, use the smoother icon and the slider for the smoother to explore the nature of the trend. To fit a trend line, use the *Line of Fit* graph icon, and explore different degrees (linear, quadratic, or cubic) using the *Degree* options under *Line of Fit* (on the bottom, left) A shaded confidence band for the trend line displays by default. To remove the confidence band, deselect the *Confidence, Fit* box. To explore different transformations, right-click on the column in the column selection panel, select a transformation and graph the transformed data.

Suppress seasonality: It is often easier to see trends in the data when seasonality is suppressed. Suppressing seasonality patterns can be done by plotting the series at a cruder time scale (e.g., aggregating monthly data into years). If the data are stored in JMP with a date format, then a *Date Time* transformation can be applied (in *Graph Builder*, right-click on the column and select a *Date Time* transformation). Another popular option is to use moving average charts. We will discuss these in Chapter 17 (Section 17.2).

Continuing our example of Amtrak ridership, the charts in Figure 15.3 help make the series' components more visible.

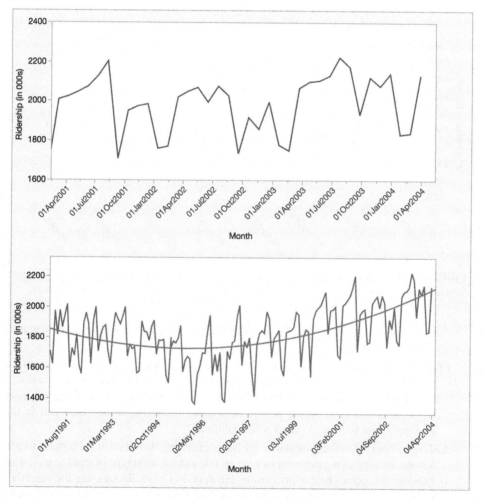

FIGURE 15.3   Plots that enhance the different components of the time series. Top: zoom-in to three years of data. Bottom: original series with overlaid quadratic trendline.

Some forecasting methods directly model these components by making assumptions about their structure. For example, a popular assumption about trend is that it is linear or exponential over the given time period or part of it. Another common assumption is about the noise structure: many statistical methods assume that the noise follows a normal distribution. The advantage of methods that rely on such assumptions is that when the assumptions are reasonably met, the resulting forecasts will be more robust and the models more understandable. Other forecasting methods, which are data-adaptive, make fewer assumptions about the structure of these components, and instead try to estimate their structure only from the data. Data-adaptive methods are advantageous when such assumptions are likely to be violated, or when the structure of the time series changes over time. Another advantage of many data-adaptive methods is their simplicity and computational efficiency.

A key criterion for deciding between model-driven and data-driven forecasting methods is to ascertain the nature of the series in terms of global and local patterns. A global pattern is one that is relatively constant throughout the series. An example is a linear trend throughout the entire series. In contrast, a local pattern is one that occurs only in a short period of the data, and then changes. An example is a trend that is approximately linear within four neighboring time points, but the trend size (slope) changes slowly over time.

Model-driven methods are generally preferable for forecasting series with global patterns, as they use all the data to estimate the global pattern. For a local pattern, a model-driven model would require specifying how and when the patterns change, which is usually impractical and often unknown. Therefore data-driven methods are preferable for local patterns. Such methods "learn" patterns from the data, and their memory length can be set to best adapt to the rate of change in the series. Patterns that change quickly warrant a "short memory," whereas patterns that change slowly warrant a "long memory." In effect, the time plot should be used not only to identify the time series component but also the global/local nature of the trend and seasonality.

15.5 DATA PARTITIONING AND PERFORMANCE EVALUATION

As in the case of cross-sectional data, in order to avoid overfitting and to be able to assess the predictive performance of the model on new data, we first partition the data into a training set and a validation set (and perhaps an additional test set). However, there is one important difference between data partitioning in cross-sectional and time series data. In cross-sectional data the partitioning is usually done randomly, with a random set of observations designated as training data and the remainder as validation data. However, in time series a random partition would create two time series with "holes!" Nearly all standard forecasting methods cannot handle time series with missing values. Therefore we partition a time series into training and validation sets differently. The series is trimmed into two periods; the earlier period is set as the training data and the later period as the validation data. Methods are then trained on the earlier training period, and their predictive performance assessed on the later validation period. Evaluation measures typically use the same metrics used in cross-sectional evaluation (see Chapter 5) with *MAPE* and *RMSE* being the most popular metrics in practice. Due to the simplicity of the calculations, *average error* and *MAE* (mean absolute error) are also used. In evaluating and comparing forecasting methods, another important tool is visualization: examining time plots of the actual and predicted series can shed light on performance and hint toward possible improvements.

Benchmark Performance: Naive Forecasts

While it is tempting to apply "sophisticated" forecasting methods, we must evaluate their added value compared to a very simple approach: the *naive forecast*. A naive forecast is simply the most recent value of the series. In other words, at time t, our forecast for any future period $t + k$ is simply the value of the series at time t. While simple, naive forecasts are sometimes surprisingly difficult to outperform with more sophisticated models. It is therefore important to benchmark against results from a naive forecasting approach.

Generating Future Forecasts

Another important difference between cross-sectional and time-series partitioning occurs when creating the actual forecasts. When forecasting future values of a time series, the method/model is run on the complete data (rather than just the training data). The three advantages of using the entire data set are: (1) The validation set, which is the most recent period, usually contains the most valuable information in terms of being the closest in time to the forecasted period; (2) With more data (the complete time series, compared to only the training set), some models can be estimated more accurately; (3) If only the training set is used to generate forecasts, then it will require forecasting farther into the future (e.g., if the validation set contains 4 time points, forecasting the next observation will require a 5-step-ahead forecast from the training set).

PARTITIONING TIME SERIES DATA IN JMP AND VALIDATING TIME SERIES MODELS

Unlike the modeling methods we've covered until now, there is no automated way to partition time series data, and validation is not built into the time series platforms in JMP. However, this can be done manually:

1. To partition a time series in JMP, hide and exclude the holdout partition (select the rows, and then select *Rows > Hide and Exclude*)
2. Fit the time series model(s) using the *Time Series* platform (from the *Analyze > Modeling* menu).
3. For the chosen model, save the prediction formula using the red triangle for the model. This produces a new data table with predicted values and residuals for the training data.
4. Calculate residuals for the validation data: Copy and paste the holdout values from the original table into the appropriate cells in this new table, create a new column, and use the *Formula Editor* to calculate the residuals *(actual - predicted)*. Note that forecasts for the validation data are automatically computed when the prediction formula is saved to the data table.

5. Summarize the residuals for the training and validation data using the *Distribution* platform, and graph the actual and forecasted time series and the residuals using the *Graph Builder*.

6. To forecast future values using both the training and validation data, return to the original data, unhide and unexclude the holdout partition, and refit the model using the *Time Series* platform. Specify the number of forecast periods (from the red triangle), and save the prediction formula for this model to the data table. Repeat steps 4 and 5 above.

These steps apply to models built using the JMP *Time Series* platform. Regression-based forecasting models can be built using the *Analyze > Fit Model* platform (see Chapter 16). Built-in model validation can be used for these models. Here we do not hide and exclude the validation data–we create a validation column, and use this column in the validation field in the Fit Model dialog.

The Time Series Validation Add-in provides a graphical approach to partitioning time series data. The Add-in, along with detailed instructions for use, can be downloaded from the JMP User Community (community.jmp.com). Additional instructions for use are built into the add-in.

Note: In the JMP *Time Series* platform (under *Analyze > Modeling*), forecasting is done automatically on 25 time periods (the number of forecast periods can be changed in the Time Series dialog). If the data contain a validation set (whose values are hidden and excluded), then forecasts are also provided for the validation data. However, the validation data are not used in creating this forecast.

PROBLEMS

15.1 Impact of September 11 on Air Travel in the United States. The Research and Innovative Technology Administration's Bureau of Transportation Statistics conducted a study to evaluate the impact of the September 11, 2001, terrorist attack on US transportation. The 2006 study report and the data can be found at http://goo.gl/w2lJPV. The goal of the study was stated as follows:

The purpose of this study is to provide a greater understanding of the passenger travel behavior patterns of persons making long distance trips before and after 9/11.

The report analyzes monthly passenger movement data between January 1990 and May 2004. Data on three monthly time series are given in file `Sept11Travel.jmp` for this period:

- Actual airline revenue passenger miles (Air RPM)
- Rail passenger miles (Rail PM)
- Vehicle miles traveled (VMT)

In order to assess the impact of September 11, BTS took the following approach: using data before September 11, they forecasted future data (under the assumption of no terrorist attack). Then they compared the forecasted series with the actual data to assess the impact of the event. Our first step therefore is to split each of the time series into two parts: pre- and post September 11. We now concentrate only on the earlier time series.

a. Is the goal of this study descriptive or predictive?

b. Plot each of the three pre-event time series (Air, Rail, Car).

    i. What time series components appear from the plot?

    ii. What type of trend appears? Change the scale of the series, add trendlines, and suppress seasonality to better visualize the trend pattern.

15.2 Performance on Training and Validation Data. Two different models were fit to the same time series. The first 100 time periods were used for the training set and the last 12 periods were treated as a hold out set. Assume that both models make sense practically and fit the data pretty well. Below are the RMSE values for each of the models:

	Training Set	Validation Set
Model A	543	690
Model B	669	675

a. Which model appears more useful for explaining the different components of this time series? Why?

b. Which model appears to be more useful for forecasting purposes? Why?

15.3 Forecasting Department Store Sales. The file `DepartmentStoreSales.jmp` contains data on the quarterly sales for a department store over a 6-year period (data courtesy of Chris Albright).

a. Create a time series plot (using the *Graph Builder*).

b. Which of the four components (level, trend, seasonality, noise) seem to be present in this series?

15.4 Shipments of Household Appliances. The file `ApplianceShipments.jmp` contains the series of quarterly shipments (in million $) of US household appliances between 1985 and 1989 (data courtesy of Ken Black).

a. Create a well-formatted time plot of the data.

b. Which of the four components (level, trend, seasonality, noise) seem to be present in this series?

15.5 Analysis of Canadian Manufacturing Workers Workhours. The time series plot below describes the average annual number of weekly hours spent by Canadian manufacturing workers (data are available in `CanadianWorkHours.jmp`–thanks to Ken Black for these data).

a. Reproduce the time plot.

b. Which of the following components (level, trend, seasonality, noise) appear to be present in this series?

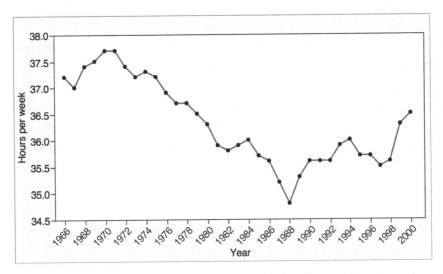

FIGURE 15.4   Average annual weekly hours spent by Canadian manufacturing workers

15.6   Souvenir Sales. The file `SouvenirSales.jmp` contains monthly sales for a souvenir shop at a beach resort town in Queensland, Australia, between 1995 and 2001. (Source: Hyndman, R.J. Time Series Data Library, http://data.is/TSDLdemo. Accessed on 05/01/14.)

   Back in 2001, the store wanted to use the data to forecast sales for the next 12 months (year 2002). They hired an analyst to generate forecasts. The analyst first partitioned the data into training and validation sets, with the validation set containing the last 12 months of data (year 2001). She then fit a regression model to sales, using the training set.

   a. Create a well-formatted time plot of the data.

   b. Change the scale on the x-axis, or on the y-axis, or on both to log-scale in order to achieve a linear relationship. Select the time plot that seems most linear.

   c. Comparing the two time plots, what can be said about the type of trend in the data?

   d. Why were the data partitioned? Partition the data into training and validation sets as explained above.

15.7   Forecasting Shampoo Sales. The file `ShampooSales.jmp` contains data on the monthly sales of a certain shampoo over a 3-year period. Source: Hyndman, R.J. Time Series Data Library, http://data.is/TSDLdemo. Accessed on 05/01/14)

   a. Create a well-formatted time plot of the data.

   b. Which of the four components (level, trend, seasonality, noise) seem to be present in this series?

   c. Do you expect to see seasonality in sales of shampoo? Why?

   d. If the goal is forecasting sales in future months, which of the following steps should be taken? Mark all that apply.
   - Partition the data into training and validation sets.
   - Tweak the model parameters to obtain good fit to the validation data.
   - Look at MAPE, RMSE, or other measures of error for the training set.
   - Look at MAPE, RMSE, or other measures of error for the validation set.

16

REGRESSION-BASED FORECASTING

A popular forecasting tool is based on multiple linear regression models, using suitable predictors to capture trend and/or seasonality. In this chapter we show how a linear regression model can be set up to capture a time series with a trend and/or seasonality. The model, which is estimated from the data, can then produce future forecasts by inserting the relevant predictor information into the estimated regression equation. We describe different types of common trends (linear, exponential, polynomial), as well as two types of seasonality (additive and multiplicative). Next we show how a regression model can be used to quantify the correlation between neighboring values in a time series (called autocorrelation). This type of model, called an autoregressive model, is useful for improving forecast precision by making use of the information contained in the autocorrelation (beyond trend and seasonality). It is also useful for evaluating the predictability of a series by evaluating whether the series is a "random walk." The various steps of fitting linear regression and autoregressive models, using them to generate forecasts, and assessing their predictive accuracy, are illustrated using the Amtrak ridership series (data are in `Amtrak.jmp`).

16.1 A MODEL WITH TREND

Linear Trend

To create a linear regression model that captures a time series with a global linear trend, the output variable (Y) is set as the time series measurement or some function of it, and the

Data Mining for Business Analytics: Concepts, Techniques, and Applications with JMP Pro®, First Edition.
Galit Shmueli, Peter C. Bruce, Mia L. Stephens, and Nitin R. Patel.
© 2017 John Wiley & Sons, Inc. Published 2017 by John Wiley & Sons, Inc.

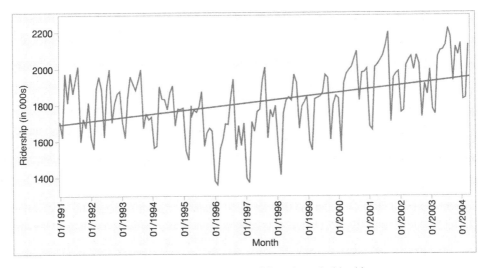

FIGURE 16.1   A linear trend fit to Amtrak ridership.

predictor (X) is set as a time index. Let us consider a simple example: fitting a linear trend to the Amtrak ridership data. This type of trend is shown in Figure 16.1.

From the time plot it is obvious that the global trend is not linear. However, we use this example to illustrate how a linear trend is fit, and later we consider more appropriate models for this series.

To obtain a linear relationship between ridership and time, we set the output variable Y as the Amtrak ridership and create a new variable that is a time index $t = 1, 2, 3, \ldots$. A snapshot of the two corresponding columns (Y_t and t) is shown in Table 16.1. This time index is then used as a single predictor in the regression model:

$$Y_t = \beta_0 + \beta_1 \times t + \epsilon$$

TABLE 16.1   Output variable (Ridership) and predictor variable (t) used to fit a linear trend

Month	Ridership	t
Jan-91	1709	1
Feb-91	1621	2
Mar-91	1973	3
Apr-91	1812	4
May-91	1975	5
Jun-91	1862	6
Jul-91	1940	7
Aug-91	2013	8
Sep-91	1596	9
Oct-91	1725	10
Nov-91	1676	11
Dec-91	1814	12
Jan-92	1615	13
Feb-92	1557	14

where Y_t is the ridership at time point t and ϵ is the standard noise term in a linear regression. Thus we are modeling three of the four time series components: level (β_0), trend (β_1), and noise (ϵ). Seasonality is not modeled.

After partitioning the data into training and validation sets, the next step is to fit a linear regression model to the training set, with t as the single predictor (using *Analyze > Fit Model*). Applying this to the Amtrak ridership data (with a validation set consisting of the last 12 months in this example) results in the estimated model shown in Figure 16.2. The actual and fitted values are plotted in Figure 16.2. Note that examining only the estimated coefficients and their statistical significance (in the *Parameter Estimates* table) can be very misleading! In this example they would indicate that the linear fit is reasonable, although it is obvious from the plot that the trend is not linear.

The difference in the magnitude of the validation statistics is also indicative of an inadequate trend shape. But an inadequate trend shape is easiest to detect by examining the plots in the Figure 16.3, which were produced using the *Graph Builder* after saving the prediction formula and the residuals for the model to the data table. The top graph is the actual and fitted series plot, and the bottom graph is the time-ordered residual plot. The inadequacy of the linear model is easy to see in these two plots.

FITTING A MODEL WITH LINEAR TREND IN JMP

First, prepare the data:

- Create the variable t: Add a new column in the data table, name the column t, and tab or enter to accept. Then, right-click on the column head and select *New Formula Column > Row > Row* to populate the column with the row number for each observation.
- Create the validation column: Add another new column, and name the column *Validation*. Enter the values "Training" for the training data and "Validation" for the validation data. In this example, the validation set consists of the last 12 months.

Next, create regression models with trends:

- Go to *Analyze > Fit Model*.
- In the dialog, select the time series data as the Y variable, select t as the model effect, and put the partition variable in the *Validation* field. Change the *Emphasis* to *Minimal Report* to turn off some default output.
- In the analysis window, graphical and statistical output can be added or removed by using *red triangle* options.

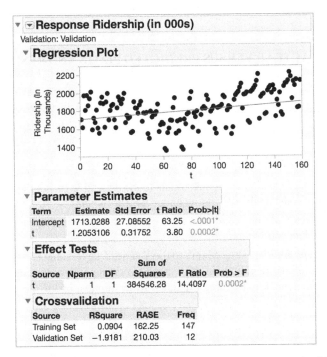

FIGURE 16.2 Fitted regression model with linear trend, regression output and validation statistics.

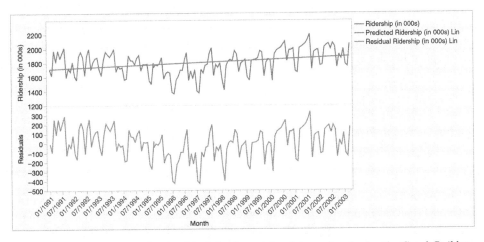

FIGURE 16.3 Fitted trend and residual plot for training data produced using the *Graph Builder*.

CREATING ACTUAL VERSUS PREDICTED PLOTS AND RESIDUAL PLOTS IN JMP

A variety of plots are available as red triangle options in the Fit Model analysis window. The actual versus predicted plot and residual plot in Figure 16.3 were produced using the *Graph Builder* for a better indication of the time-ordered behavior. To create these graphs:

- Save the predicted values and residuals to the data table (using the top red triangle in the Fit Model analysis window under *Save Columns*).
- Hide and exclude the validation data (select the rows in the data table, and use *Rows > Hide and Exclude*).
- Open *Graph Builder*. Drag Month to the *X* zone, drag Residuals to the *Y* zone, Drag Ridership above Residuals in the *Y* zone (to create a separate graph). Then click on the *Line* icon. Finally, drag Predicted Ridership just inside the *y*-axis in the top graph.

Exponential Trend

There are several alternative trend shapes that are useful and easy to fit via a linear regression model. Recall that different transformations can be explored dynamically in the JMP *Graph Builder*. One such shape is an exponential trend. An exponential trend implies a multiplicative increase/decrease of the series over time ($Y_t = ce^{\beta_1 t + \epsilon}$). To fit an exponential trend, simply replace the output variable Y with $\log Y$ where log is the natural logarithm, and fit a linear regression ($\log Y_t = \beta_0 + \beta_1 t + \epsilon$). In the Amtrak example, for instance, we would fit a linear regression of $\log(Ridership)$ on the index variable t. Exponential trends are popular in sales data, where they reflect percentage growth.

To create the variable log*Ridership* right click on *Ridership* in the data table and select *New Formula Column > Transform > Log*. Transformed variables can also be created from the *Graph Builder* and other platforms, which allow you to graphically explore a variety of transformations.

Note: As in the general case of linear regression, when comparing the predictive accuracy of models that have a different output variable, such as a linear trend model (with Y) and an exponential trend model (with $\log Y$), it is essential to compare forecasts or forecasts errors on the same scale. An exponential trend model will produce forecasts of $\log Y$, and the forecast errors reported by the software will therefore be $\log Y - \widehat{\log Y}$. To obtain forecasts in the original units, create a new column which takes an exponent of the model forecasts. Then use this column to create an additional column of forecast errors by subtracting the original Y. An example is shown in Figures 16.4 and 16.5, where an exponential trend is fit to the Amtrak ridership data. Note that the performance measures for the training and validation data are not comparable to those from the linear trend model shown in Figure 16.2. Instead, we manually compute two new columns in Figure 16.5, one that gives forecasts of ridership (in thousands) and another that gives the forecast errors in terms of ridership. To compare RMSE or average error, we would now use the new forecast errors and compute their standard deviation (for RMSE) or their average (for average error) using *Analyze > Distribution*. These would then be comparable to the numbers in Figure 16.2.

Response log(Ridership)

Validation: Validation

Parameter Estimates

Term	Estimate	Std Error	t Ratio	Prob>|t|
Intercept	7.4439865	0.015475	481.05	<.0001*
t	0.0006512	0.000181	3.59	0.0005*

Effect Tests

Source	Nparm	DF	Sum of Squares	F Ratio	Prob > F
t	1	1	0.11226500	12.8882	0.0005*

Crossvalidation

Source	RSquare	RASE	Freq
Training Set	0.0816	0.09269	147
Validation Set	−2.0907	0.10791	12

FIGURE 16.4 Output from regression model with exponential trend, fit to training data.

	Validation	Ridership (in 000s)	log(Ridership)	Pred log(Ridership)	Pred Ridership	Forecast Errors
148	Validation	2098.899	7.649168	7.540371	1882.529	216.3702
149	Validation	2104.911	7.652028	7.541023	1883.755	221.1558
150	Validation	2129.671	7.663723	7.541674	1884.982	244.6886
151	Validation	2223.349	7.70677	7.542325	1886.21	337.1386
152	Validation	2174.36	7.68449	7.542976	1887.439	286.9208
153	Validation	1931.406	7.566004	7.543628	1888.669	42.73722
154	Validation	2121.47	7.659865	7.544279	1889.899	231.5708
155	Validation	2076.054	7.638224	7.54493	1891.13	184.9236
156	Validation	2140.677	7.668877	7.545581	1892.362	248.3146
157	Validation	1831.508	7.512895	7.546233	1893.595	−62.0872
158	Validation	1838.006	7.516437	7.546884	1894.829	−56.8228
159	Validation	2132.446	7.665025	7.547535	1896.063	236.3828

FIGURE 16.5 Adjusting forecasts of log(Ridership) to the original scale and computing forecast errors in the original scale (last column).

COMPUTING FORECAST ERRORS FOR EXPONENTIAL TREND MODELS

Use *Save Columns > Prediction Formula* to save the model to the data table. The prediction formula produces forecasts for the validation data, which were not used to build the model. Use *Save Columns > Residuals* to save the residuals to the data table. If an exponential trend model was fit, save the prediction formula to the data table, and use the *Exp* function to calculate forecasts in the original units (right-click on the prediction formula column and select *New Formula Column > Transform > Exp*). Create a new column, and compute forecast errors using the *Formula Editor* (actual − predicted).

Polynomial Trend

Another nonlinear trend shape that is easy to fit via linear regression is a polynomial trend, and in particular, a quadratic relationship of the form $Y_t = \beta_0 + \beta_1 t + \beta_2 t^2 + \epsilon$. This is done by adding $t * t$ (the square of t) to the model, and fitting a multiple linear regression with the two predictors, t and $t * t$. For the Amtrak ridership data, we have already seen a U-shaped

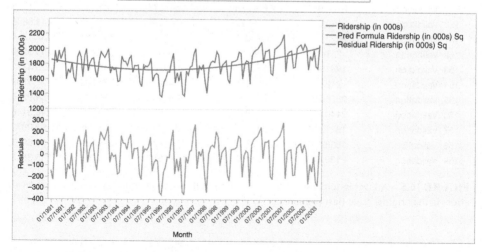

FIGURE 16.6 Regression model with quadratic trend fit to the training data

trend in the data. We therefore fit a quadratic model, concluding from the plots of model fit and residuals (Figure 16.6) that this shape adequately captures the trend. The residuals are now devoid of trend and exhibit only seasonality.

FITTING A POLYNOMIAL TREND IN JMP

To add a quadratic or squared term in the *Fit Model* dialog, select the variable (*t* in this example). Then, under *Macros*, select *Polynomial to Degree* to add both *t* and *t * t* to the model. Note that a default setting in the *Fit Model* platform is to *center polynomials*. In this example (Figure 16.6), we have deselected this option (from the red triangle in the *Fit Model* dialog). Note that models with and without centering will lead to the same predictions.

In general, any type of trend shape can be fit as long as it has a mathematical representation. However, the underlying assumption is that this shape is applicable throughout the period of data that we have and also during the period that we are going to forecast. Do not choose an overly complex shape. Although it will fit the training data well, it will in fact be overfitting them. To avoid overfitting, always examine performance on the validation set and refrain from choosing overly complex trend patterns.

16.2   A MODEL WITH SEASONALITY

A seasonal pattern in a time series means that observations that fall in some seasons have consistently higher or lower values than those that fall in other seasons. Examples are day-of-week patterns, monthly patterns, and quarterly patterns. The Amtrak ridership monthly time series, as can be seen in the time plot, exhibits strong monthly seasonality (with highest traffic during summer months).

Seasonality is modeled in a regression model by creating a new categorical variable that denotes the season for each observation. This categorical variable is then turned into indicator variables when the regression model is fit in JMP. To illustrate this, we created a new "Season" column for the Amtrak ridership data, as shown in Figure 16.2. This column contains the month names.

For *m* seasons JMP creates $m - 1$ indicator variables in the Fit Model platform, holding out the highest level. (For a discussion of indicator coding in JMP and coefficient interpretation, see Chapter 6.[1]) We then partition the data into training and validation sets (see Section 13.5) and fit the regression model to the training data. For this example, we again use a validation set consisting of the last 12 months.

---

[1] JMP displays only $m - 1$ indicator variables in the *Parameter Estimates* table because the coefficients for the *m* seasons sum to zero. If the coefficients for $m - 1$ variables are added together, the negative of this total is the coefficient for the *m*th season. Including the *m*th variable would cause redundant information and multicollinearity errors.

TABLE 16.2 New categorical variable (Season) to be used in a linear regression model

Month	Ridership	Season
Jan-91	1709	Jan
Feb-91	1621	Feb
Mar-91	1973	Mar
Apr-91	1812	Apr
May-91	1975	May
Jun-91	1862	Jun
Jul-91	1940	Jul
Aug-91	2013	Aug
Sep-91	1596	Sep
Oct-91	1725	Oct
Nov-91	1676	Nov
Dec-91	1814	Dec
Jan-92	1615	Jan
Feb-92	1557	Feb
Mar-92	1891	Mar
Apr-92	1956	Apr
May-92	1885	May

Since the *Month* column has a date format, we can right-click on the *Month* column in the data table and use *New Formula Column* and the *Month Abbr.* transformation under *Date Time*. Here we've renamed this new column *Season*. Since this column is based on the *Month* column, the values are automatically ordered January to December in all analyses, and December will be held out of regression models. To re-order the values in analyses (if desired), use the *Value Ordering* column property.

The top panel of Figure 16.7 shows the output of a linear regression fit to Ridership (Y) on 11 month indicator variables using the training data. To show the fitted model and residuals (bottom, Figure 16.7), we again use the *Graph Builder*. The model appears to capture the seasonality in the data. However, since we have not included a trend component in the model (as shown in Section 16.1), the fitted values do not capture the existing trend. Therefore the residuals, which are the difference between the actual and fitted values, clearly display the remaining U-shaped trend.

When seasonality is added as described above, the model captures *additive seasonality*. This means that the average value of Y in a given season is a certain amount more or less than the overall average.[2] Specifically, the coefficient for a month tells us the average amount by which that month's average value is higher or lower than the overall average. For example, the coefficient for February (-238.5) indicates that the average number of

---

[2]If we have balanced data (an equal number of observations per month), the intercept will equal the overall mean for the training data. Here we have one more observation for Jan, Feb, and Mar than we do for the other months, so the intercept is not equal to the overall mean.

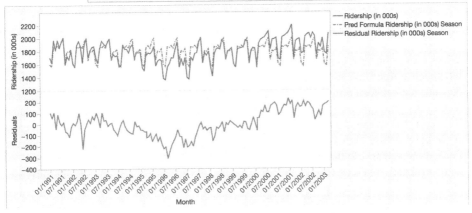

▼ ⬛ **Response Ridership (in 000s)**

Validation: Validation

▼ **Parameter Estimates**

Term	Estimate	Std Error	t Ratio	Prob>|t|
Intercept	1804.7887	9.706055	185.94	<.0001*
Season[Jan]	−200.8382	31.31879	−6.41	<.0001*
Season[Feb]	−238.5159	31.31879	−7.62	<.0001*
Season[Mar]	62.020342	31.31879	1.98	0.0497*
Season[Apr]	50.447259	32.47699	1.55	0.1227
Season[May]	81.687592	32.47699	2.52	0.0131*
Season[Jun]	40.336342	32.47699	1.24	0.2164
Season[Jul]	144.79151	32.47699	4.46	<.0001*
Season[Aug]	189.83759	32.47699	5.85	<.0001*
Season[Sep]	−143.1899	32.47699	−4.41	<.0001*
Season[Oct]	−3.681575	32.47699	−0.11	0.9099
Season[Nov]	−13.51924	32.47699	−0.42	0.6779

▼ **Effect Tests**

Source	Nparm	DF	Sum of Squares	F Ratio	Prob > F
Season	11	11	2386794.3	15.6870	<.0001*

▼ **Crossvalidation**

Source	RSquare	RASE	Freq
Training Set	0.5611	112.71	147
Validation Set	−3.6374	264.77	12

FIGURE 16.7 Fitted regression model with seasonality. Regression output, residuals by row, and validation statistics (top), and plot of fitted and actual training data (bottom).

passengers in February is lower than the overall average monthly number of passengers by 238.5 thousand passengers. Similarly, the coefficient for August indicates that the average ridership in August is higher than the overall average monthly ridership by 189.8 thousand passengers.

The coefficient for December, the holdout level, is not shown. To display all of the coefficients, including the holdout level, select *Estimates > Expanded Estimates* from the top red triangle. The regression equation can be displayed by selecting *Estimates > Show*

Prediction Expression from the top red triangle, and can be seen graphically using the *Prediction Profiler*.

Using regression models we can also capture *multiplicative seasonality*, where values on a certain season are on average higher or lower by a certain *percentage* compared to the average. To fit multiplicative seasonality, we use the same model as above, except that we use log Y as the output variable.

16.3   A MODEL WITH TREND AND SEASONALITY

Now we can create models that capture both trend and seasonality by including predictors of both types. For example, from our exploration of the Amtrak ridership data, it appears that a quadratic trend and monthly seasonality are both warranted. We therefore fit a model to the training data with Season, t and $t * t$. The output and fit from this final model are shown in Figure 16.8. If we are satisfied with this model, after evaluating its predictive performance on the validation data and comparing it against alternatives, we can re-fit it to the entire un-partitioned series. This re-fitted model can then be saved to the data table and used to generate k-step-ahead forecasts (denoted by F_{t+k}). To do this, we add rows to the data table and plug in the appropriate month and index terms.

(*Note*: A version of the Amtrak ridership data with the validation column, the variables t, and *Season*, along with the saved prediction formula and residuals after fitting the model with quadratic trend and seasonality, are found in AmtrakTS.jmp.)

16.4   AUTOCORRELATION AND ARIMA MODELS

When we use linear regression for time series forecasting, we are able to account for patterns such as trend and seasonality. However, ordinary regression models do not account for dependence between observations, which in cross-sectional data is assumed to be absent. Yet, in the time series context, observations in neighboring periods tend to be correlated. Such correlation, called *autocorrelation*, is informative and can help in improving forecasts. If we know that a high value tends to be followed by high values (positive autocorrelation), then we can use that to adjust forecasts. We will now discuss how to compute the autocorrelation of a series, and how best to utilize the information for improving forecasts.

Computing Autocorrelation

Correlation between values of a time series in neighboring periods is called *autocorrelation* because it describes a relationship between the series and itself. To compute autocorrelation, we compute the correlation between the series and a *lagged* version of the series. A lagged series is a "copy" of the original series that is moved forward one or more time periods. A lagged series with lag-1 is the original series moved forward one time period; a lagged series with lag-2 is the original series moved forward two time periods, etc. Table 16.3 shows the first 24 months of the Amtrak ridership series, the lag-1 series and the lag-2 series. (To create a lagged variable in JMP, create a new column, and then use the formula editor with the *Row, Lag* function. In JMP 13, use the *New Formula Column* shortcut to create lag columns using *Row > Lag or Row > Multiple Lag*.)

Response Ridership (in 000s)

Validation: Validation

Parameter Estimates

Term	Estimate	Std Error	t Ratio	Prob>|t|
Intercept	1874.8259	18.80734	99.69	<.0001*
Season[Jan]	−209.2717	19.92373	−10.50	<.0001*
Season[Feb]	−248.135	19.92303	−12.45	<.0001*
Season[Mar]	51.12797	19.92373	2.57	0.0114*
Season[Apr]	58.172797	20.65092	2.82	0.0056*
Season[May]	88.489971	20.64767	4.29	<.0001*
Season[Jun]	46.128048	20.64537	2.23	0.0271*
Season[Jul]	149.48503	20.644	7.24	<.0001*
Season[Aug]	193.34541	20.64354	9.37	<.0001*
Season[Sep]	−140.9553	20.644	−6.83	<.0001*
Season[Oct]	−2.807692	20.64537	−0.14	0.8920
Season[Nov]	−14.0936	20.64767	−0.68	0.4961
t	−5.246521	0.586749	−8.94	<.0001*
t*t	0.0437566	0.003841	11.39	<.0001*

Effect Tests

Source	Nparm	DF	Sum of Squares	F Ratio	Prob > F
Season	11	11	2523859.0	41.0649	<.0001*
t	1	1	446724.2	79.9536	<.0001*
t*t	1	1	725213.9	129.7970	<.0001*

Crossvalidation

Source	RSquare	RASE	Freq
Training Set	0.8253	71.100	147
Validation Set	0.8306	50.599	12

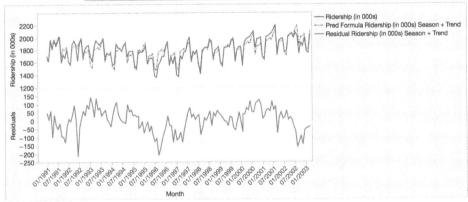

FIGURE 16.8 Regression model with monthly (additive) seasonality and quadratic trend, fit to Amtrak Ridership data. Regression output, and validation statistics (top), and plot of fitted and actual training data and residuals (bottom).

To compute the lag−1 autocorrelation, which measures the linear relationship between values in consecutive time periods, we compute the correlation between the original series and the lag-1 series (e.g., via the *Multivariate Methods > Multivariate* platform) to be

0.08. Note that although the original series shown above has 24 time periods, the lag-1 autocorrelation will only be based on 23 pairs (because the lag-1 series does not have a value for Jan-91). Similarly, the lag-2 autocorrelation, measuring the relationship between values that are two time periods apart, is the correlation between the original series and the lag-2 series (yielding −0.15).

We can use the JMP *Time Series* platform (under the *Analyze > Modeling* menu), to directly compute the autocorrelations of a series at different lags. For example, the output from the *Time Series* platform for the 24-month ridership data is shown in Figure 16.9 (all but the first 24 months in the series have been hidden and excluded). The output from the *Time Series* platform is shown. A bar chart of the autocorrelations at different lags (the *Autocorrelation Function Plot*, or *ACF Plot*) is provided (bottom left).

A few typical autocorrelation behaviors are useful to explore:

Strong autocorrelation (positive or negative) at a lag k larger than 1 and its multiples $(2k, 3k, \ldots)$ typically reflects a cyclical pattern. For example, strong positive autocorrelation at lag 12 in monthly data will reflect an annual seasonality (where values during a given month each year are positively correlated).

Positive lag-1 autocorrelation (called "stickiness") describes a series where consecutive values move generally in the same direction. In the presence of a strong linear trend, we would expect to see a strong and positive lag-1 autocorrelation.

TABLE 16.3   First 24 months of Amtrak ridership series

Month	Ridership	Lag−1 Series	Lag−2 Series
Jan-91	1709		
Feb-91	1621	1709	
Mar-91	1973	1621	1709
Apr-91	1812	1973	1621
May-91	1975	1812	1973
Jun-91	1862	1975	1812
Jul-91	1940	1862	1975
Aug-91	2013	1940	1862
Sep-91	1596	2013	1940
Oct-91	1725	1596	2013
Nov-91	1676	1725	1596
Dec-91	1814	1676	1725
Jan-92	1615	1814	1676
Feb-92	1557	1615	1814
Mar-92	1891	1557	1615
Apr-92	1956	1891	1557
May-92	1885	1956	1891
Jun-92	1623	1885	1956
Jul-92	1903	1623	1885
Aug-92	1997	1903	1623
Sep-92	1704	1997	1903
Oct-92	1810	1704	1997
Nov-92	1862	1810	1704
Dec-92	1875	1862	1810

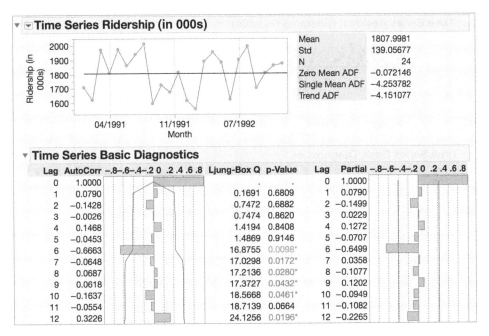

FIGURE 16.9   JMP Time Series platform output first 24 months of the Amtrak ridership data (partial output with only the first 12 lags AER displayed).

Negative lag-1 autocorrelation reflects swings in the series, where high values are immediately followed by low values, and vice versa.

Examining the autocorrelation of a series can therefore help detect seasonality patterns. In this example, the ACF plot (bottom left in Figure 16.9), we see the strongest autocorrelation is at lag-6, and the autocorrelation is negative. This indicates a biannual pattern in ridership, with 6-month swings from high to low ridership. A look at the time plot (top, in Figure 16.9) confirms the high-summer low-winter pattern (the x-axis has been rescaled to better show this pattern).

Note that a partial autocorrelation function plot, or *PACF* (bottom right in Figure 16.9), is produced by default in the JMP Time Series platform. This plot shows the autocorrelation at different lags after adjusting for (or, removing the influence of) all other other points. Throughout this chapter, we will limit our discussion to the ACF plot.

In addition to looking at autocorrelations of the raw series, it is very useful to look at autocorrelations of *residual series*. For example, after fitting a time series (or regression) model, we can save the residuals to the data table. Then we can examine the autocorrelation of the series of *residuals* using the Time Series platform. If we have adequately modeled the seasonal pattern, then the residual series should show no autocorrelation at the season's lag.

Figure 16.10 displays the autocorrelations for the residuals from the regression model with seasonality and quadratic trend shown in Figure 16.8 (note that the rows in the validation set were first hidden and excluded). It is clear that the 6-month (and 12-month)

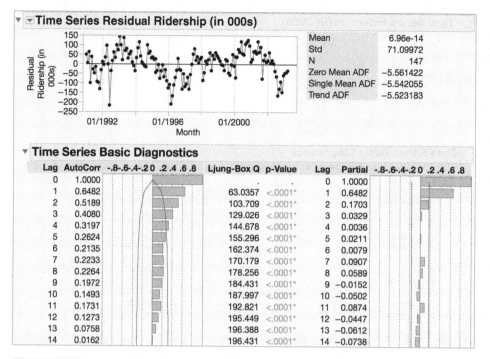

FIGURE 16.10 Amtrak data: time series output showing autocorrelation of residuals from regression model with seasonality and quadratic trend.

cyclical behavior no longer dominates the series of residuals, indicating that the regression model captured them adequately. However, we can also see a strong positive autocorrelation from lag-1 on, indicating a positive relationship between neighboring residuals. This is valuable information, which can be used to improve forecasts.

Improving Forecasts by Integrating Autocorrelation Information

In general, there are two approaches to taking advantage of autocorrelation. One is by directly building the autocorrelation into the regression model, and the other is by constructing a second-level forecasting model on the residual series.

Among regression-type models that directly account for autocorrelation are autoregressive models, or the more general class of models called ARIMA models (autoregressive integrated moving average models). Autoregression models are similar to linear regression models, except that the predictors are the past values of the series. For example, an autoregression model of order 2, denoted AR(2), can be written as

$$Y_t = \beta_0 + \beta_1 Y_{t-1} + \beta_2 Y_{t-2} + \epsilon \qquad (16.4.1)$$

Estimating such models is roughly equivalent to fitting a linear regression model with the series as the output, and the two lagged series (at lag-1 and 2 in this example) as the predictors. However, using the designated ARIMA estimation methods (e.g., those available in the JMP *Time Series > ARIMA*) platform will produce more accurate results than ordinary linear regression estimation.[3]

FITTING AR (AUTOREGRESSION) MODELS IN THE JMP TIME SERIES PLATFORM

To fit an AR model to training data in JMP, first hide and exclude the validation data. Then select *ARIMA* from the red triangle in the *Time Series* platform, and set *p, Autoregressive Order* to the required order. The *Model Comparison* window will display a summary of the fit of the model (drag the scroll bar to view fit statistics). Select the *Graph* box to show the fitted values and ACF and PACF residual plots at lags 0–25. To view the model results, select the *Report* box (this is selected by default). The red triangle for the model can be used to change the results displayed or to save the prediction formula to the data table.

Specifying ARIMA models in JMP is facilitated by the *ARIMA Model Group* option within the *Time Series* platform. This allows you to specify the range of orders for the model, and to also address seasonality. The *Time Series* platform will be discussed in greater detail in Chapter 17. For more information on fitting ARIMA and Seasonal ARIMA models in JMP, search for *Time Series* in the book *Fitting Specialized Models* (under the *Help* menu).

Moving from AR to ARIMA models creates a larger set of more flexible forecasting models, and can include a seasonal component (using *Seasonal ARIMA* or *ARIMA Model Group*). But fitting these models requires much more statistical expertise. Even with the simpler AR models, fitting AR models to time series that contain patterns such as trends and seasonality requires you to perform operations that remove the trend and seasonality before fitting the AR model, and then to choose the order of the model, and this can be challenging.

Because ARIMA modeling is less robust and requires more experience and statistical expertise than other methods, the use of such models for forecasting raw series is generally less popular in practical forecasting. We therefore direct the interested reader to classic time series textbooks (e.g., see Chapter 4 in Chatfield, 2003).

Fitting AR Models to Residuals

We now discuss a particular use of AR models that is straightforward in the context of forecasting, and can provide significant improvement to short-term forecasts. This relates to the second approach for utilizing autocorrelation, which requires constructing a second-level forecasting model for the residuals, as follows:

1. Generate a k-step-ahead forecast of the series (F_{t+k}), using any forecasting method.

---

[3] ARIMA model estimation differs from ordinary regression estimation by accounting for the dependence between observations.

2. Generate a k-step-ahead forecast of the forecast error (residual) (E_{t+k}), using an AR (or other) model.

3. Improve the initial k-step-ahead forecast of the series by adjusting it according to its forecasted error: *Improved* $F^*_{t+k} = F_{t+k} + E_{t+k}$.

In particular, we can fit low-order AR models to series of residuals (or *forecast errors*) that can then be used to forecast future forecast errors. To fit an AR model to the series of residuals, we first examine the autocorrelations of the residual series. We then choose the order of the AR model according to the lags in which autocorrelation appears. Often, when autocorrelation exists at lag-1 and higher, it is sufficient to fit an AR(1) model of the form

$$E_t = \beta_0 + \beta_1 E_{t-1} + \epsilon \qquad (16.4.2)$$

where E_t denotes the residual (or *forecast error*) at time t.

For example, although the autocorrelations in Figure 16.10 appear large from lags 1 to 5 or so, it is likely that an $AR(1)$ would capture all these relationships. The reason is that if immediate neighboring values are correlated, then the relationship propagates to values that are two periods away, then three periods away, and so forth. The result of fitting an AR(1) model to the Amtrak ridership residual series is shown in Figure 16.11. The AR(1) coefficient (0.65) is close to the lag-1 autocorrelation that we found earlier (Figure 16.10).

The forecasted residual for April 2003, highlighted in Figure 16.12, is computed by plugging in the most recent residual from March 2003 (equal to -33.786) into the AR(1) model: $0.063 + (0.647)(-33.786) = -21.8$. The negative value tells us that the regression model will produce a ridership forecast for April 2003 that is too high, and that we should adjust it down by subtracting 21.8 thousand riders. In this example, the regression model (with quadratic trend and seasonality) produced a forecast of 2115 thousand riders, and the improved two-stage model (regression + AR(1) correction) corrected it by reducing it to 2093.2 thousand riders. The actual value for April 2003 turned out to be 2099 thousand riders–much closer to the improved forecast.

Last, to examine whether we have indeed accounted for the autocorrelation in the series, and that no more information remains in the series, we examine the autocorrelations of the series of residuals-of-residuals (the residuals obtained after the AR(1) was applied to the regression residuals). This can be seen in Figure 16.13. It is clear that no more autocorrelation remains, and that the addition of the AR(1) model has captured the autocorrelation information adequately.

From the plot of the actual versus forecasted residual series (under *Forecast* in Figure 16.11), we can see that the AR(1) model fits the residual series quite well. Note, however, that the plot is based on the training data (until March 2003), and that the forecast after that time is also shown. To evaluate predictive performance of the two-level model (regression + AR(1)), we would have to examine performance (e.g., via MAPE, RMSE, or MAE metrics) on the validation data, in a similar way to the calculation that we performed for April 2003 above.

We mentioned earlier that improving forecasts via an additional AR layer is useful for short-term forecasting. The reason is that an AR model of order k will usually only provide useful forecasts for the next k periods, and after that forecasts will rely on earlier forecasts rather than on actual data. For example, to forecast the residual of May 2003, when the

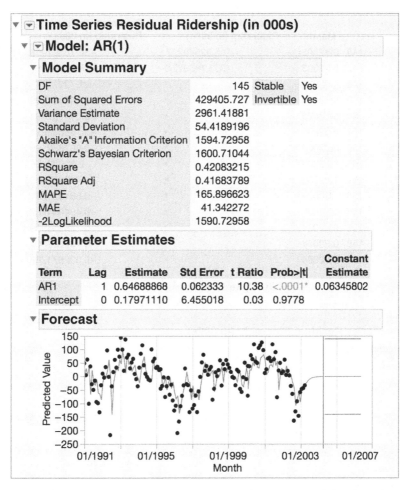

FIGURE 16.11   Fitting an AR(1) model to the residual series (training data only).

time of prediction is March 2003, we would need the residual for April 2003. However, because that value is not available, it would be replaced by its forecast. Hence the forecast for May 2003 would be based on the forecast for April 2003.

Evaluating Predictability

Before attempting to forecast a time series, it is important to determine whether it is predictable, in the sense that its past can be used to predict its future beyond the naive forecast. One useful way to assess predictability is to test whether the series is a "random walk." A random walk is a series in which changes from one time period to the next are random. According to the Efficient Market Hypothesis in economics, asset prices are random walks and therefore predicting stock prices is a game of chance.[4]

---

[4]There is some controversy surrounding the Efficient Market Hypothesis, with claims that there is slight autocorrelation in asset prices, which does make them predictable to some extent. However, transaction costs and bid-ask spreads tend to offset any prediction benefits.

	Month	Actual Residual Ridership (in 000s)	Predicted Residual Ridership (in 000s)
146	02/2003	−43.92549227	−30.18391458
147	03/2003	−33.7856734	−28.35144587
148	04/2003	•	−21.79211178
149	05/2003	•	−14.03361249
150	06/2003	•	−9.014727094
151	07/2003	•	−5.768066924
152	08/2003	•	−3.667839199
153	09/2003	•	−2.30922565
154	10/2003	•	−1.430353919
155	11/2003	•	−0.861821741
156	12/2003	•	−0.494044709
157	01/2004	•	−0.256133909
158	02/2004	•	−0.102232104
159	03/2004	•	−0.002674768

FIGURE 16.12   Forecasted residual.

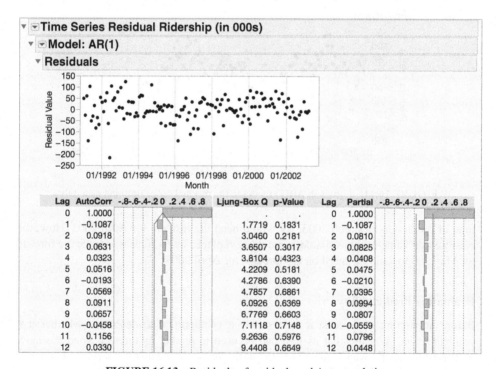

FIGURE 16.13   Residuals-of-residuals and Autocorrelations

A random walk is a special case of an AR(1) model, where the slope coefficient is equal to 1:

$$Y_t = \beta_0 + Y_{t-1} + \epsilon_t \tag{16.4.3}$$

We can also write this formula as

$$Y_t - Y_{t-1} = \beta_0 + \epsilon_t \qquad (16.4.4)$$

We see from this equation that the difference between the values at periods $t - 1$ and t is random, hence the term "random walk." Forecasts from such a model are basically equal to the most recent observed value (the naive forecast), reflecting the lack of any other information.

To test whether a series is a random walk, we fit an AR(1) model and test the hypothesis that the slope coefficient is equal to 1 ($H_0 : \beta_1 = 1$ vs. $H_1 : \beta_1 \neq 1$). If the null hypothesis is rejected (reflected by a small p-value), then the series is not a random walk, and we can attempt to predict it.

As an example, consider the AR(1) model shown in Figure 16.11. The slope coefficient (0.647) is more than 3 standard errors away from 1, indicating that this is not a random walk. In contrast, consider the AR(1) model fitted to the series of S&P500 monthly closing prices between May 1995 and August 2003 (available in SP500.jmp, shown in Figure 16.14). Here the slope coefficient is 0.983, with a standard error of 0.0145. The coefficient is sufficiently close to 1 (around one standard error away), indicating that this is a random walk. Forecasting this series using any of the methods described earlier (aside from the naive forecast) is therefore futile.

SUMMARY: FITTING REGRESSION-BASED TIME SERIES MODELS IN JMP

To fit models with trend and/or seasonality, we use the standard *Fit Model* platform and the built-in validation option.

- In this chapter, we have changed the default output by using the *red triangle* options. For example, we have turned off some of the standard graphical and statistical output. To show all available options, click the *Alt* or *Option* button on your keyboard before using the red triangle. Some graphs have been resized (using the *cursor*) and repositioned on the screen (using the *selection tool*, and other graphs have been created using the *Graph Builder*, as described earlier.

- To fit a regression model with an exponential trend, take the log of the time series and use the *Fit Model* platform. Save the prediction formula to the data table. Use the *Exp* transformation to adjust the forecasts to the original scale. Use a formula to calculate the residuals.

- To fit an autoregressive (AR) model, partition the data and hide and exclude the validation set. Then fit the model using the *ARIMA* option within *Time Series* platform. A time plot, ACF, and residual plots are provided, but forecasts and forecast errors can be saved to the data table (using *Save Prediction Formula*) and graphed using the *Graph Builder*.

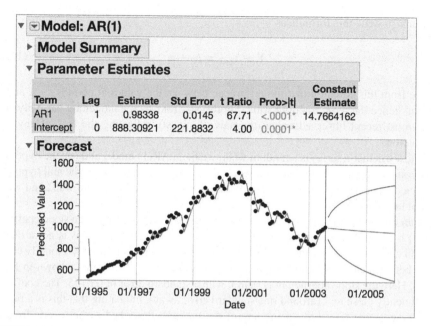

FIGURE 16.14 AR(1) model fitted to S&P500 monthly closing prices (May 1995–Aug 2003). (*Note:* Ignore the *p*-values reported in the output; they test a different hypothesis than the one discussed in the text.)

PROBLEMS

16.1 Impact of September 11 on Air Travel in the United States. The Research and Innovative Technology Administration's Bureau of Transportation Statistics conducted a study to evaluate the impact of the September 11, 2001, terrorist attack on US transportation. The 2006 study report and the data can be found at http://goo.gl/w2lJPV. The goal of the study was stated as follows:

> The purpose of this study is to provide a greater understanding of the passenger travel behavior patterns of persons making long distance trips before and after 9/11.

The report analyzes monthly passenger movement data between January 1990 and May 2004. Data on three monthly time series are given in file Sept11Travel.jmp for this period: Actual (1) airline revenue passenger miles (Air RPM), (2) rail passenger miles (Rail PM), and (3) auto miles traveled (VMT).

In order to assess the impact of September 11, BTS took the following approach: using data before September 11, they forecasted future data (under the assumption of no terrorist attack). Then they compared the forecasted series with the actual data to assess the impact of the event. Our first step therefore is to split each of the time series into two parts: pre- and post-September 11. We now concentrate only on the earlier time series.

a. Plot the pre-event AIR RPM time series.

  i. What time series components appear from the plot?

b. In Figure 16.15 we show is a time plot of the seasonally adjusted pre-September-11 AIR RPM series.

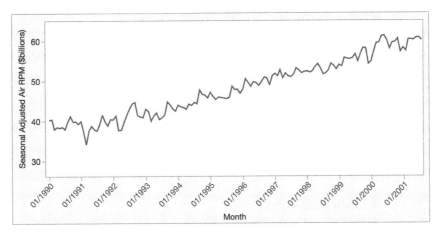

FIGURE 16.15   Seasonally adjusted pre-September-11 AIR RPM series

Which of the following methods would be adequate for forecasting this series?

- Linear regression model with seasonality
- Linear regression model with trend
- Linear regression model with seasonality and trend

c. In part (d) you will run a linear regression model for the AIR RPM series that will produce a seasonally adjusted series similar to the one shown in part (b), with multiplicative seasonality. What is the output variable? What are the predictors? Prepare the data for this regression model. (HINT: Use a *Date Time* transformation to create a new column with the month, and make sure the modeling type for this column is nominal.)

d. Run the regression model from part (c) using only the pre-event data (using the *Fit Model* platform).

    i. What can we learn from the statistical insignificance of the coefficients for September and October?

    ii. Save the residuals to the data table. The actual value of AIR RPM (air revenue passenger miles) in January 1990 was 35.153577 billion. What is the residual for this month, using the regression model? Report the residual in terms of air revenue passenger miles.

e. Create an ACF (autocorrelation) plot of the regression residuals using the *Time Series* platform.

    i. What does the ACF plot tell us about the regression model's forecasts?

    ii. How can this information be used to improve the model?

f. Fit linear regression models to Air, Rail, and to Auto with additive seasonality and an appropriate trend. For Air and Rail, fit a linear trend. For Rail, use a quadratic trend. Remember to use only pre-event data. Once the models are estimated, use them to forecast each of the three post-event series.

i. For each series (Air, Rail, Auto), plot the complete pre-event and post-event actual series overlayed with the predicted series.

ii. What can be said about the effect of the September 11 terrorist attack on the three modes of transportation? Discuss the magnitude of the effect, its time span, and any other relevant aspects.

16.2 Analysis of Canadian Manufacturing Workers Workhours.

The time series plot in Figure 16.16 describes the average annual number of weekly hours spent by Canadian manufacturing workers. (Data are available in `Canadi-anWorkHours.jmp`, data courtesy of Ken Black.)

a. Which one model of the following regression-based models would fit the series best?

- Linear trend model
- Linear trend model with seasonality
- Quadratic trend model
- Quadratic trend model with seasonality

b. If we computed the autocorrelation of this series, would the lag-1 autocorrelation exhibit negative, positive, or no autocorrelation?

i. How can you see this from the plot in Figure 16.16?

ii. Create a time series plot using the *time series* platform. Use the ACF plot to verify your answer to the previous question.

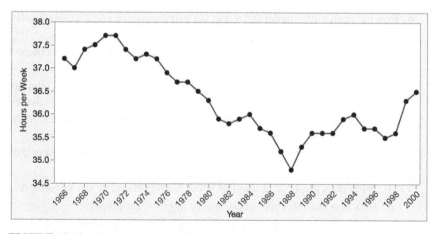

FIGURE 16.16   Average annual weekly hours spent by Canadian manufacturing workers

16.3 Regression Modeling of Toys "R" Us Revenues. Figure 16.17 is a time series plot of the quarterly revenues of Toys "R" Us between 1992 and 1995. (Thanks to Chris Albright for suggesting the use of these data, which are available in `ToysRUsRevenues.jmp`.)

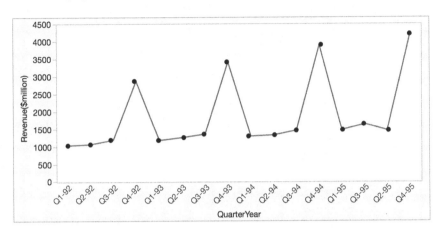

FIGURE 16.17   Quarterly Revenues of Toys "R" Us, 1992–1995

a. Fit a regression model with a linear trend and quarter (you'll need to create a new nominal column, "Quarter"). Use the first 14 quarters as the training set, and the last two quarters as the validation set (create a new "Validation" column to indicate the partition).

b. A partial output of the regression model is shown in Figure 16.18. Use this output to answer the following questions:

   i. Name two statistics (and their values) that measure how well this model fits the training data?

   ii. Name statistic (and its value) that measures the predictive accuracy of this model?

   iii. What is the coefficient for the trend? Interpret this value.

   iv. After adjusting for trend, which quarter ($Q1, Q2, Q3$, or $Q4$) has the highest average sales?

16.4 Forecasting Walmart Stock. Figure 16.19 shows the series of Walmart daily closing prices between February 2001 and February 2002. (Thanks to Chris Albright for suggesting the use of these data, which are publicly available, e.g., at http://finance.yahoo.comand are in the file `WalMartStock.jmp`.)

a. Run an AR(1) model for the close price series. Report the parameter estimates table.

b. Which of the following is/are relevant for testing whether this stock is a random walk?

   • The autocorrelations of the close prices series
   • The AR(1) slope coefficient
   • The AR(1) constant coefficient

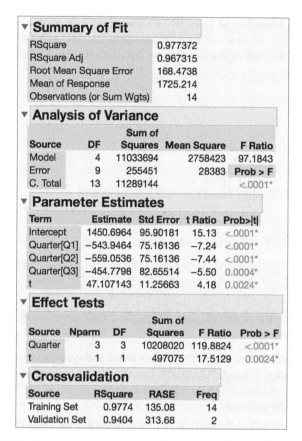

FIGURE 16.18 Output for regression model fitted to Toys "R" Us time series

c. Does the AR model indicate that this is a random walk? Explain how you reached your conclusion.

d. What are the implications of finding that a time series is a random walk? Choose the correct statement(s) below:

- It is impossible to obtain forecasts that are more accurate than naive forecasts for the series.
- The series is random.
- The changes in the series from one period to the other are random.

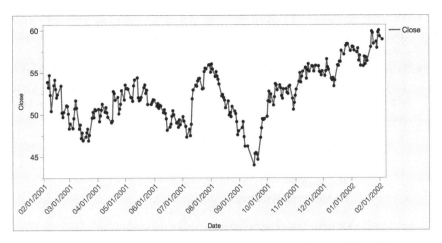

FIGURE 16.19   Daily close price of Walmart stock, February 2001–2002

16.5 Forecasting Department Store Sales. The time-series plot in Figure 16.20 describes actual quarterly sales for a department store over a 6-year period. (Data are available in DepartmentStoreSales.jmp, data courtesy of Chris Albright.) Assume that the first record is Q1.

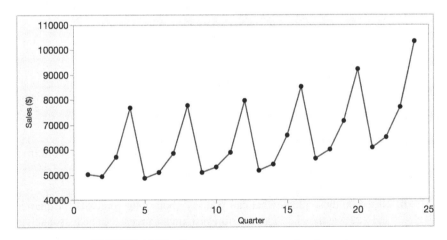

FIGURE 16.20   Department store quarterly sales series

a. The forecaster decided that there is an exponential trend in the series. The forecaster needs to fit a regression-based model that accounts for this trend. Which of the following operations must be performed?

- Take log of Quarter index
- Take log of sales
- Take an exponent of sales
- Take an exponent of Quarter index

b. Fit a regression model with an exponential trend and seasonality using *Fit Model*, and only first 20 quarters as the training data (remember to first partition the series into training and validation series).

c. A regression model with trend (Quarter) and seasonality (Quarter 2) was fit. A partial output is shown in Figure 16.21. From the output, after adjusting for trend, are Q2 average sales higher, lower, or approximately equal to the average Q1 sales?

Summary of Fit

RSquare	0.979125
RSquare Adj	0.973558
Root Mean Square Error	0.032766
Mean of Response	11.02138
Observations (or Sum Wgts)	20

Analysis of Variance

Source	DF	Sum of Squares	Mean Square	F Ratio
Model	4	0.75537034	0.188843	175.8917
Error	15	0.01610445	0.001074	**Prob > F**
C. Total	19	0.77147479		<.0001*

Parameter Estimates

Term	Estimate	Std Error	t Ratio	Prob>|t|
Intercept	10.904956	0.015448	705.93	<.0001*
Quarter	0.0110879	0.001295	8.56	<.0001*
Quarter 2[Q1]	−0.156011	0.012838	−12.15	<.0001*
Quarter 2[Q2]	−0.131055	0.012707	−10.31	<.0001*
Quarter 2[Q3]	0.009332	0.012707	0.73	0.4740

FIGURE 16.21   Output from regression model fit to department store sales training series

d. Use this model to forecast sales in quarters 21 and 22.

e. The plots in Figure 16.22 describe the fit (top) and forecast errors (bottom) from this regression model.

   i. Recreate these plots.

   ii. Based on these plots, what can you say about your forecasts for quarters 21 and 22? Are they likely to overforecast, underforecast, or be reasonably close to the real sales values?

f. Looking at the residual plot, which of the following statements appear true?

   • Seasonality is not captured well.
   • The regression model fits the data well.
   • The trend in the data is not captured well by the model.

g. Which of the following solutions is adequate *and* a parsimonious solution for improving model fit?

   • Fit a quadratic trend model to the residuals (with Quarter and Quarter^2).

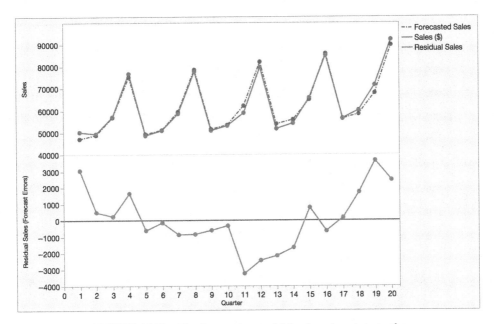

FIGURE 16.22   Fit of regression model for department store sales

- Fit an AR model to the residuals.
- Fit a quadratic trend model to sales (with Quarter and Quarter^2).

16.6 Souvenir Sales. Figure 16.23 shows a time plot of monthly sales for a souvenir shop at a beach resort town in Queensland, Australia, between 1995 and 2001. (Data are available in `SouvenirSales.jmp`.) Source: Hyndman, R.J. Time Series Data Library http://data.is/TSDLdemo. Accessed on 05/01/14.) The series is presented twice, in Australian dollars and in log-scale. Back in 2001, the store wanted to use the data to forecast sales for the next 12 months (year 2002). They hired an analyst to generate forecasts. The analyst first partitioned the data into training and validation sets, with the validation set containing the last 12 months of data (year 2001). She then fit a regression model to sales.

a. Based on the two time plots, which predictors should be included in the regression model? What is the total number of predictors in the model?

b. Run a regression model on the training data with Sales (in Australian dollars) as the output variable, and with a linear trend and monthly predictors. Call this model A.

   i. Examine the estimated coefficients. Which month tends to have the highest average sales during the year? Why is this reasonable?

   ii. What is the estimated coefficient for trend? What does this mean?

c. Run a regression model with log(Sales) as the output variable, and with a linear trend and monthly predictors. Again, remember to use the validation column. Call this model B.

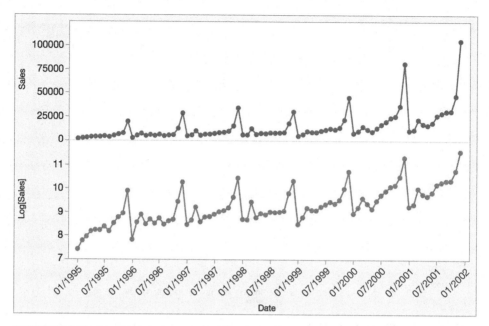

FIGURE 16.23 Monthly sales at Australian souvenir shop in dollars (top) and in log-scale (bottom).

    i. Fitting a model to log(Sales) with a linear trend is equivalent to fitting a model to Sales (in dollars) with what type of trend?

    ii. What is the estimated coefficient for trend? What does this mean?

    iii. Use this model to forecast the sales in February 2002. What is the extra step needed?

  d. Compare the two regression models (A and B) in terms of forecast performance. Which model is preferable for forecasting? Mention at least two reasons based on the information in the outputs.

  e. Continuing with Model B (with log(Sales) as output), fit an AR model with lag 2 [ARIMA(2,0,0)] to the forecast errors.

    i. Examining the ACF plot and the estimated coefficients of the AR(2) model (and their statistical significance), what can we learn about the forecasts that result from Model B?

    ii. Use the autocorrelation information to compute an improved forecast for January 2002, using model B and the AR(2) model above.

  f. How would you model these data differently if the goal was to understand the different components of sales in the souvenir shop between 1995 and 2001? Mention two differences.

16.7 **Shipments of Household Appliances.** The time plot in Figure 16.24 shows the series of quarterly shipments (in million $) of US household appliances between 1985 and 1989. (Data are available in `ApplianceShipments.jmp`, data courtesy of Ken Black.)

  a. If we compute the autocorrelation of the series, which lag (>0) is most likely to have the largest coefficient (in absolute value)?

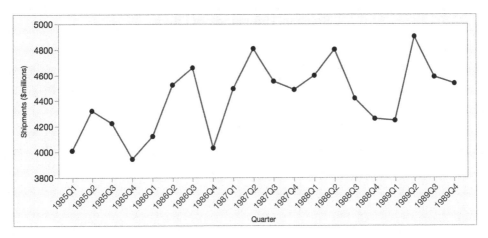

FIGURE 16.24 Quarterly shipments of US household appliances over 5 years.

b. Use the *Time Series* platform to create a time plot. Examine the ACF plot, and compare with your answer in part (a).

16.8 Forecasting Australian Wine Sales. Figure shows time plots of monthly sales of six types of Australian wines (red, rose, sweet white, dry white, sparkling, and forti-fied) for 1980 to 1994. (Data are available in `AustralianWines.jmp`.) Source: Hyndman, R.J. Time Series Data Library, http://data.is/TSDLdemo. Accessed on 07/25/15.) The units are thousands of liters. You are hired to obtain short term fore-casts (2–3 months ahead) for each of the six series, and this task will be repeated every month.

a. Which forecasting method would you choose if you had to choose the same method for all series? Why?

b. Fortified wine has the largest market share of the these six types of wine. You are asked to focus on fortified wine sales alone, and to produce as accurate as possible forecast for the next 2 months.

- Start by partitioning the data using the period until December 1993 as the training set.
- Fit a regression model to sales with a linear trend and seasonality.

i. Create the "actual vs. forecast" and forecast errors plots. What can you say about the model fit?

ii. Use the regression model to forecast sales in January and February 1994.

c. Create an ACF plot for the residuals from the model until lag 12. Examining this plot, which of the following statements are reasonable?

- Decembers (month 12) are not captured well by the model.
- There is a strong correlation between sales on the same calendar month.
- The model does not capture the seasonality well.
- We should try to fit an autoregressive model with lag 12 to the residuals.

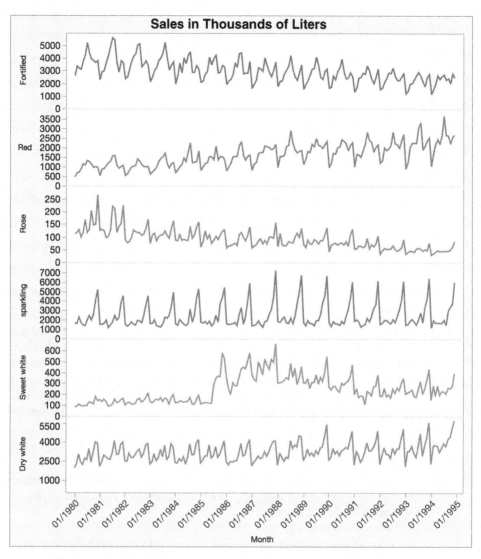

FIGURE 16.25 Monthly sales of six types of Australian wines between 1980-1994

17

SMOOTHING METHODS

In this chapter we describe a set of popular and flexible methods for forecasting time series that rely on smoothing. Smoothing is based on averaging over multiple periods in order to reduce the noise. We start with two simple smoothers, the moving average and simple exponential smoother, which are suitable for forecasting series that contain no trend or seasonality. In both cases forecasts are averages of previous values of the series (the length of the series history that is considered and the weights that are used in the averaging differ between the methods). We also show how a moving average can be used, with a slight adaptation, for data visualization. We then proceed to describe smoothing methods that are suitable for forecasting series with a trend and/or seasonality. Smoothing methods are data driven, and are able to adapt to changes in the series over time. Although highly automated, the user must specify smoothing constants, which determine how fast the method adapts to new data. We discuss the choice of such constants, and their meaning. The different methods are illustrated using the Amtrak ridership series.

17.1 INTRODUCTION

A second class of methods for time series forecasting is smoothing methods. Unlike regression models, which rely on an underlying theoretical model for the components of a time series (e.g., linear model or multiplicative seasonality), smoothing methods are data driven, in the sense that they estimate time-series components directly from the data without a predetermined structure. Data-driven methods are especially useful in series where patterns change over time. Smoothing methods "smooth" out the noise in a series in an attempt to uncover the patterns. Smoothing is done by averaging the series over multiple periods, where different smoothers differ by the number of periods averaged, how the average is computed, how many times averaging is performed, and so on. We now describe two types

Data Mining for Business Analytics: Concepts, Techniques, and Applications with JMP Pro®, First Edition.
Galit Shmueli, Peter C. Bruce, Mia L. Stephens, and Nitin R. Patel.

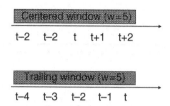

FIGURE 17.1   Schematic of centered moving average (top) and trailing moving average (bottom), both with window width $w = 5$

of smoothing methods that are popular in business applications due to their simplicity and adaptivity. These are the moving average method and exponential smoothing.

17.2   MOVING AVERAGE

The moving average is a simple smoother: it consists of averaging across a window of consecutive observations, thereby generating a series of averages. A moving average with window width w means averaging across each set of w consecutive values, where w is determined by the user.

In general, there are two types of moving averages: a *centered moving average* and a *trailing moving average*. Centered moving averages are powerful for visualizing trends because the averaging operation can suppress seasonality and noise, thereby making the trend more visible. In contrast, trailing moving averages are useful for forecasting. The difference between the two is in terms of placing the window over the time series.

Centered Moving Average for Visualization

In a centered moving average, the value of the moving average at time t (MA_t) is computed by centering the window around time t and averaging across the w values within the window:

$$MA_t = \left(Y_{t-(w-1)/2} + \cdots + Y_{t-1} + Y_t + Y_{t+1} + \cdots + Y_{t+(w-1)/2} \right) / w$$

For example, with a window of width $w = 5$, the moving average at time point $t = 3$ means averaging the values of the series at time points $1, 2, 3, 4, 5$; at time point $t = 4$ the moving average is the average of the series at time points $2, 3, 4, 5, 6$, and so on. This is illustrated in the top panel of Figure 17.1.

Choosing the window width in a seasonal series is straightforward: since the goal is to suppress seasonality for better visualizing the trend, the default choice should be the length of a season. Returning to the Amtrak ridership data, the annual seasonality indicates a choice of $w = 12$. Figure 17.2 shows a centered moving average line overlaid on the original series. We can see a global U-shape, but unlike the regression model that fits a strict U-shape, the moving average shows some deviation, such as the slight dip during the last year.

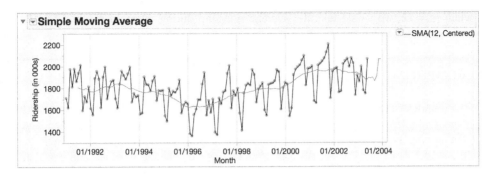

FIGURE 17.2   Centered moving average with window $w = 12$, overlaid on Amtrak ridership series. This helps visualize trends.

This plot is obtained by using the *Time Series* platform, and choosing *Simple Moving Average* from the list of available *Smoothing Models* (see Figure 17.3). This allows you to set the length of the window, and to specify whether to use a centered moving average (*Centered*) or a trailing moving average (*No Centering*).[1]

Simple Smoothing Average Specification		
Enter smoothing window width.		12
⚪ No Centering		
⚫ Centered		
⚪ Centered and Double Smoothed for Even Number of Terms		
	Cancel	OK

FIGURE 17.3   JMP time series simple moving average specification window.

Trailing Moving Average for Forecasting

Centered moving averages are computed by averaging across data in the past and the future of a given time point. In this sense, they cannot be used for forecasting, since at the time of forecasting the future is typically unknown. Hence, for purposes of forecasting, we use *trailing moving averages*, where the window of width w is set on the most recent available w values of the series. The k-step-ahead forecast F_{t+k} ($k = 1, 2, 3, \ldots$) is then the average of these w values (see also bottom plot in Figure 17.1):

$$F_{t+k} = \left(Y_t + Y_{t-1} + \cdots + Y_{t-w+1} \right) / w$$

For example, in the Amtrak ridership series, to forecast ridership in February 1992 or later months, given information until January 1992 and using a moving average with window

---

[1]For an even window width, such as $w = 4$, obtaining the centered moving average at time point $t = 3$ requires averaging across two windows: across time points $1, 2, 3, 4$; across time points $2, 3, 4, 5$; and finally the average of the two averages is the final moving average. In JMP, use the *Centered and Double Smoothed* option (see Figure 17.3).

width $w = 12$, we would take the average ridership during the most recent 12 months (February 1991 to January 1992).

COMPUTING A TRAILING MOVING AVERAGE FORECAST IN JMP

To compute a trailing moving average in JMP, use *Simple Moving Average* under *Smoothing Models* in the *Time Series* platform, and select *No Centering*. Then save the moving averages to a data table using *Save to Data Table* from the red triangle for the model. Note that this creates a new data table. To generate forecasts and calculate residuals:

- Copy the last moving average in the training data and paste over the moving averages in the validation set (since information is only known until this value).
- Copy the values in the validation set (from the original data table) into the corresponding cells in the new data table.
- Create a new column (*Forecast*) and use the formula editor to create a *Lag(1)* of the moving average column using the *Row > Lag* function.
- Create a new column (*Residual*) and use a formula to calculate the errors *(actual-forecast)*.

Using JMP, we illustrate a 12-month moving average forecaster for the Amtrak ridership. We again partition the series, leaving the last 12 months as the validation set (using *Hide and Exclude*). Applying a moving average forecaster with window $w = 12$ and *No Centering*, we save the moving averages to a data table and apply the steps outlined in the text box to obtain the output partially shown in Figure 17.4. Note that for the first 11 records of the training set, there is no forecast because there are fewer than 12 past values to average. Also note that the forecasts for all months in the validation period should be identical (1942.73) because the method assumes that information is known only until March 2003.

From the graph in Figure 17.5 it is clear that the moving average forecaster is inadequate for generating monthly forecasts, since it does not capture the seasonality in the data. Hence seasons with high ridership are underforecasted, and seasons with low ridership are overforecasted. A similar issue arises when forecasting a series with a trend: the moving average "lags behind," thereby underforecasting in the presence of an increasing trend, and overforecasting in the presence of a decreasing trend.

In general, the moving average can be used for forecasting *only in series that lack seasonality and trend*. Such a limitation might seem impractical. However, there are a few popular methods for removing trends (de-trending) and removing seasonality (de-seasonalizing) from a series, such as regression models. The moving average can be used to forecast such de-trended and de-seasonalized series, and then the trend and seasonality can be added back to the forecast. For example, consider the regression model shown in Figure 16.8 in Chapter 16, which yields residuals devoid of seasonality and trend. We can apply a moving average forecaster to that series of residuals (also called forecast errors), thereby creating a forecast for the next *forecast error*.

	Time Ridership (in 000s)	Actual Ridership (in 000s)	SMA(12)	Forecast	Residual
1	01/1991	1708.917	•	•	•
2	02/1991	1620.586	•	•	•
3	03/1991	1972.715	•	•	•
4	04/1991	1811.665	•	•	•
5	05/1991	1974.964	•	•	•
6	06/1991	1862.356	•	•	•
7	07/1991	1939.86	•	•	•
8	08/1991	2013.264	•	•	•
9	09/1991	1595.657	•	•	•
10	10/1991	1724.924	•	•	•
11	11/1991	1675.667	•	•	•
12	12/1991	1813.863	1809.5365		•
13	01/1992	1614.827	1801.6956667	1809.5365	−194.7095
14	02/1992	1557.088	1796.4041667	1801.6956667	−244.6076667
15	03/1992	1891.223	1789.6131667	1796.4041667	94.818833333
16	04/1992	1955.981	1801.6395	1789.6131667	166.36783333
17	05/1992	1884.714	1794.1186667	1801.6395	83.0745
18	06/1992	1623.042	1774.1758333	1794.1186667	−171.0766667
19	07/1992	1903.309	1771.1299167	1774.1758333	129.13316667
20	08/1992	1996.712	1769.7505833	1771.1299167	225.58208333

FIGURE 17.4   Partial output for moving average forecaster with $w = 12$ applied to Amtrak ridership series.

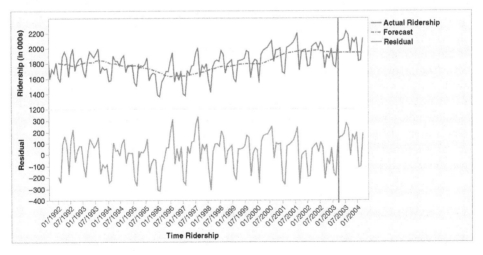

FIGURE 17.5   Actual vs. forecasted ridership (top) and residuals (bottom) from moving average forecaster (created using the *Graph Builder*). The vertical line (double-click on the axis to add a reference line) separates the training and validation data.

For example, to forecast ridership in April 2003 (the first period in the validation set), assuming that we have information until March 2003, we use the regression model in Figure 16.8 to generate a forecast for April 2003 (which yields 2115 thousand riders).

	Time Residual Ridership	Actual Residual Ridership (in 000s)	SMA(12)	Forecast	Residual
136	04/2002	19.604064209	36.184632072	40.264327008	−20.6602628
137	05/2002	2.9528597099	31.779840604	36.184632072	−33.23177236
138	06/2002	−35.96776146	18.920785936	31.779840604	−67.74760206
139	07/2002	−65.20779929	8.2398014014	18.920785936	−84.12858522
140	08/2002	−164.7277538	−12.913613	8.2398014014	−172.9675552
141	09/2002	−129.881125	−17.27620727	−12.913613	−116.967512
142	10/2002	−92.54932945	−26.7602314	−17.27620727	−75.27312219
143	11/2002	−146.913534	−43.10876873	−26.7602314	−120.1533026
144	12/2002	−58.82582178	−48.97048593	−43.10876873	−15.71705305
145	01/2003	−46.75823422	−54.35463299	−48.97048593	2.2122517133
146	02/2003	−43.92549227	−63.11879326	−54.35463299	10.429140723
147	03/2003	−33.7856734	−66.33213339	−63.11879326	29.333119854
148	04/2003	•	−66.33213339	−66.33213339	•

FIGURE 17.6   Applying MA to the residuals from the regression model (which lack trend and seasonality), to forecast the April 2003 residual.

We then use a 12-month moving average (using the period April 2002 to March 2003) to forecast the *forecast error* for April 2003, which yields −66.33 (as shown in Figure 17.6). The negative value implies that the regression model's forecast for April 2003 is too high, and therefore we should adjust it by reducing approximately 66 thousand riders from the regression model's forecast of 2,115,000 riders.

Choosing Window Width (w)

With moving average forecasting or visualization, the only choice that the user must make is the width of the window (w). As with other methods such as k-nearest neighbors, the choice of the smoothing parameter is a balance between undersmoothing and oversmoothing. For visualization (using a centered window), wider windows will expose more global trends, while narrow windows will reveal local trends. Hence examining several window widths is useful for exploring trends of differing local/global nature. For forecasting (using a trailing window), the choice should incorporate domain knowledge in terms of relevance of past observations and how fast the series changes. Empirical predictive evaluation can also be done by experimenting with different values of w and comparing performance. However, care should be taken not to overfit!

17.3 SIMPLE EXPONENTIAL SMOOTHING

A popular forecasting method in business is exponential smoothing. Its popularity derives from its flexibility, ease of automation, cheap computation, and good performance. Simple exponential smoothing is similar to forecasting with a moving average, except that instead of taking a simple average over the w most recent values, we take a *weighted average* of *all* past values, such that the weights decrease exponentially into the past. The idea is to give more weight to recent information, yet not to completely ignore older information.

Like the moving average, simple exponential smoothing should only be used for forecasting *series that have no trend or seasonality*. As mentioned earlier, such series can be obtained by removing trend and/or seasonality from raw series, and then applying exponential smoothing to the series of residuals (which are assumed to contain no trend or seasonality).

The exponential smoother generates a forecast at time $t + 1$ (F_{t+1}) as follows:

$$F_{t+1} = \alpha Y_t + \alpha(1 - \alpha)Y_{t-1} + \alpha(1 - \alpha)^2 Y_{t-2} + \cdots \qquad (17.1)$$

where α is a constant between 0 and 1 called the *smoothing parameter* or *smoothing weight*. The formulation above displays the exponential smoother as a weighted average of all past observations, with exponentially decaying weights.

It turns out that we can write the exponential forecaster in another way, which is very useful in practice:

$$F_{t+1} = F_t + \alpha E_t \qquad (17.2)$$

where E_t is the forecast error at time t. This formulation presents the exponential forecaster as an "active learner": it looks at the previous forecast (F_t) and how far it was from the actual value (E_t), and then corrects the next forecast based on that information. If last period the forecast was too high, the next period is adjusted down. The amount of correction depends on the value of the smoothing parameter α. The formulation in (17.2) is also advantageous in terms of data storage and computation time: it means that we need to store and use only the forecast and forecast error from the most recent period, rather than the entire series. In applications where real-time forecasting is done, or many series are being forecasted in parallel and continuously, such savings are critical.

Note that forecasting further into the future yields the same as a one-step-ahead forecast. Because the series is assumed to lack trend and seasonality, forecasts into the future rely only on information until the time of prediction. Hence the k-step-ahead forecast is equal to

$$F_{t+k} = F_{t+1}$$

Choosing Smoothing Parameter α

The smoothing weight, α, determines the rate of learning. A value close to 1 indicates fast learning (i.e., only the most recent observations have influence on forecasts), whereas a value close to 0 indicates slow learning (past observations have a large influence on forecasts). This can be seen by plugging 0 or 1 into equation (17.1) or (17.2). Hence the choice of α depends on the required amount of smoothing, and on how relevant the history is for generating forecasts. Default values that have been shown to work well are around 0.1 to 0.2. Some trial and error can also help in the choice of α: examine the time plot of the actual and predicted series, as well as the predictive accuracy (e.g., MAPE, RMSE, or MAE of the validation set).

By default, JMP will search for the optimal weight (the value of α), but you can also specify a fixed value or range of values to search for the optimal weight. Finding the α value that optimizes predictive accuracy on the validation set can be used to deter-

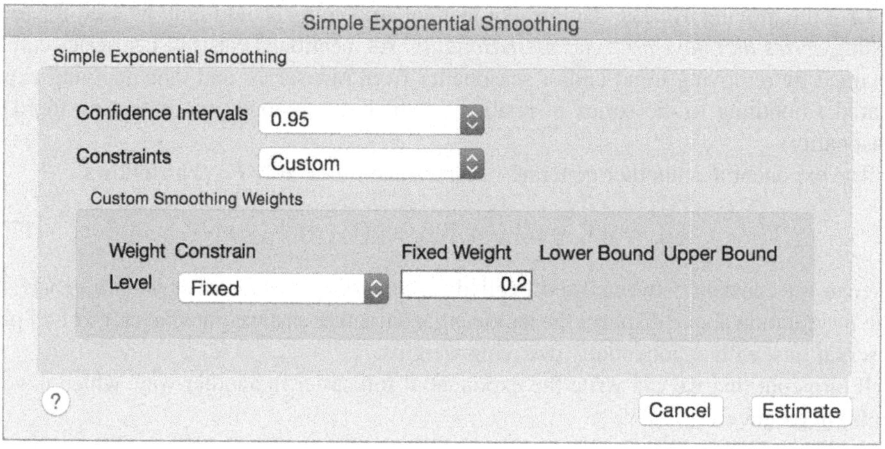

FIGURE 17.7   The Simple Exponential Smoothing dialog: Change the constraint to *Custom*, and then use *Fixed* to specify a value of the smoothing weight α.

mine the degree of local versus global nature of the trend. *Note*: Beware of choosing the "best α" based on training data for forecasting purposes, as this will most likely lead to model overfitting and low predictive accuracy on future data. Therefore, although JMP provides by default the best alpha (the optimal weight), the model is being fit to the training data and the user should consider trying different values it to avoid overfitting.

FITTING SIMPLE EXPONENTIAL SMOOTHING MODELS IN JMP

Simple Exponential Smoothing is one of many models available from the *Time Series > Smoothing Model* menu. This smoothing model allows you to specify the *Constraint* set on the smoothing parameter α. By default, JMP will search for the optimal value of α. The *Custom* constraint option allows you to enter a specific *Fixed* value or a *Bounded* range of values to search for the optimal smoothing weight (as shown in Figure 17.7).

All smoothing models in JMP produce summary statistics, parameter estimates for the model, forecasts (using a specified number of forecast periods), ACF and residual ACF plots, and more. Check the *Report* and *Graph* boxes in the *Model Comparison* section to display model results. Use the red triangle for the model to save the prediction formula and residuals to the data table.

Note that smoothing can also be applied to a column in the data table using the formula editor or by right-clicking on the column head and selecting *New Formula Column > Row > Moving Average*, and then specifying the smoothing parameter.

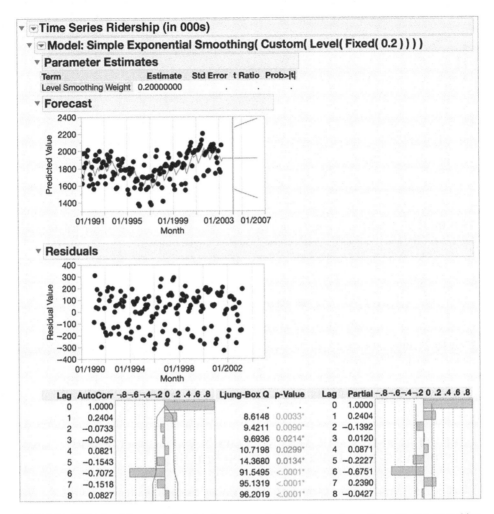

FIGURE 17.8 Partial output for simple exponential smoothing forecaster with the smoothing weight $\alpha = 0.2$, applied to the series of residuals from the regression model (which lacks trend and seasonality).

To illustrate forecasting with simple exponential smoothing, we return to the residuals from the regression model, as these residuals are assumed to contain no trend or seasonality (the data are in AmtrakTS.jmp). To forecast the residual on April 2003, we apply exponential smoothing to the entire period until March 2003, and use the $\alpha = 0.2$ value. The partial output is shown in Figure 17.8, and the actual versus forecasted time series and residuals are shown in Figure 17.9. The forecast for the residual for April 2003 is -58.220 (see Figure 17.10). This implies that we should adjust the regression's forecast by reducing 58,220 riders from that forecast.

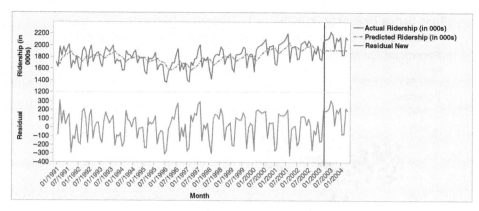

FIGURE 17.9   Graphs of the actual vs. forecasted ridership from the simple exponential smoothing forecaster (top) and residuals (bottom), created using the Graph Builder.

	Month	Predocted Residual Ridership (in 000s)
148	04/2003	−58.22022654

FIGURE 17.10   Forecast for April 2003 residual (partial output from saving the probability formula for the model to a data table).

CREATING PLOTS FOR ACTUAL VERSUS FORECASTED SERIES AND RESIDUALS SERIES USING THE GRAPH BUILDER

Follow the following steps to produce charts similar to those in Figure 17.9:

1. Save the prediction formula for the model to the data table.
2. Copy the values in the validation set (from the original data table) into the corresponding cells in the new data table.
3. Create a new column (*Residual New*) and use a formula to calculate the errors *(actual-forecast)*.
4. Graph using the *Graph Builder* as discussed in Chapter 16 and customize as needed. For example, double-click on the *x*-axis to add the reference line for the validation data and change the scaling, double-click on an axis label to change, add different graph elements, and right-click on the legend to change the line style.

Relation between Moving Average and Simple Exponential Smoothing

In both smoothing methods the user must specify a single parameter: in moving averages, the window width (w) must be set; in exponential smoothing, the smoothing parameter (or smoothing weight), α, must be set (or, the software determines an optimal value). In both cases the parameter determines the importance of fresh information over older information. In fact, the two smoothers are approximately equal if the window width of the moving average is equal to $w = 2/\alpha - 1$.

17.4 ADVANCED EXPONENTIAL SMOOTHING

As mentioned earlier, both moving average and simple exponential smoothing should only be used for forecasting series with no trend or seasonality—meaning that series that have only a level and noise. One solution for forecasting series with trend and/or seasonality is first to remove those components (e.g., via regression models). Another solution is to use a more sophisticated version of exponential smoothing, one that can capture trend and/or seasonality.

Series with a Trend

For series that contain a trend, we can use "double exponential smoothing." Unlike in regression models, the trend shape is not assumed to be global, but rather it can change over time. In double exponential smoothing, the local trend is estimated from the data and is updated as more data arrive. As in simple exponential smoothing, the level of the series is estimated from the data and is updated as more data arrive.

JMP provides two methods for double exponential smoothing—*linear exponential smoothing* (the *Holt* method) and *double exponential smoothing* (the *Brown* method). The k-step-ahead forecast for both methods is given by combining the level estimate at time t (L_t) and the trend estimate at time t (T_t):

$$F_{t+k} = L_t + kT_t \tag{17.3}$$

Note that in the presence of a trend, one-, two-, three-step-ahead (etc.) forecasts are no longer identical. Using the Holt method, the level and trend are updated through a pair of updating equations:

$$L_t = \alpha Y_t + (1 - \alpha)(L_{t-1} + T_{t-1}) \tag{17.4}$$

$$T_t = \beta \left(L_t - L_{t-1} \right) + (1 - \beta)T_{t-1} \tag{17.5}$$

The first equation means that the level at time t is a weighted average of the actual value at time t and the level in the previous period, adjusted for trend (in the presence of a trend, moving from one period to the next requires factoring in the trend). The second equation means that the trend at time t is a weighted average of the trend in the previous period and the more recent information on the change in level.[2] Here there are two smoothing parameters, α and β, which determine the rate of learning. As in simple exponential smoothing, they are both constants between 0 and 1, with higher values leading to faster learning (more weight to most recent information). Optimal values for the constants are determined by JMP, or custom values or ranges of values can be specified by the user.

Brown's double exponential smoothing is a special case of Holt's exponential smoothing, where $\alpha = \beta$ (Hyndman et al., 2008). In Brown's double exponential smoothing, the trend values are not smoothed directly. Instead, the Brown method applies simple exponential smoothing twice: it applies simple exponential smoothing to a series that has already been

---

[2]There are various ways to estimate the initial values L_1 and T_1, but the differences among these ways usually disappear after a few periods.

simple exponentially smoothed. The equations for Brown are similar to Holt's, but the equation for the level estimate in the Brown approach is not adjusted for trend:

$$L_t = \alpha Y_t + (1 - \alpha)(L_{t-1}) \tag{17.6}$$

$$T_t = \beta \left(L_t - L_{t-1} \right) + (1 - \beta)T_{t-1} \tag{17.7}$$

Series with a Trend and Seasonality

For series that contain both trend and seasonality, the "Holt–Winter exponential smoothing" method can be used. This is a further extension of double exponential smoothing, where the k-step-ahead forecast also takes into account the season at period $t + k$. Assuming seasonality with M seasons (e.g., for weekly seasonality $M = 7$), the forecast is given by

$$F_{t+k} = \left(L_t + kT_t \right) S_{t+k-M} \tag{17.8}$$

(Note that by the time of forecasting t, the series must have included at least one full cycle of seasons in order to produce forecasts using this formula, i.e., $t > M$.)

Being an adaptive method, Holt–Winter's exponential smoothing allows the level, trend, and seasonality patterns to change over time. These three components are estimated and updated as more information arrives. The three updating equations are given by

$$L_t = \alpha Y_t / S_{t-M} + (1 - \alpha)(L_{t-1} + T_{t-1}) \tag{17.9}$$

$$T_t = \beta \left(L_t - L_{t-1} \right) + (1 - \beta)T_{t-1} \tag{17.10}$$

$$S_t = \gamma Y_t / L_t + (1 - \gamma)S_{t-M} \tag{17.11}$$

The first equation is similar to that in Holt's double exponential smoothing, except that it uses the seasonally adjusted value at time t rather than the raw value. This is done by dividing Y_t by its seasonal index, as estimated in the last cycle. The second equation is identical to double exponential smoothing (both Holt and Brown methods). The third equation means that the seasonal index is updated by taking a weighted average of the seasonal index from the previous cycle and the current trend-adjusted value. Note that this formulation describes a multiplicative seasonal relationship where values on different seasons differ by percentage amounts.

There is also an additive version of Holt–Winter's exponential smoothing method, which is referred to as Winter's Method (Additive) in JMP. In the additive model, seasons differ by a constant amount rather than by a percentage amount (see more in Shmueli, 2012). The three equations for the additive model are very similar to those for the multiplicative model, but the seasonal index is *subtracted* from Y_t rather than being divided by it:

$$L_t = \alpha \left(Y_t - S_{t-M} \right) + (1 - \alpha)(L_{t-1} + T_{t-1}) \tag{17.12}$$

$$T_t = \beta \left(L_t - L_{t-1} \right) + (1 - \beta)T_{t-1} \tag{17.13}$$

$$S_t = \gamma \left(Y_t - L_t \right) + (1 - \gamma)S_{t-M} \tag{17.14}$$

The Holt–Winter multiplicative method is not directly available in JMP. For an approximation to the multiplicative method, apply the additive Winter's method in JMP to the log-transformed series, such as *log(Ridership)*.

To illustrate forecasting a series with Winter's exponential smoothing, consider the raw Amtrak ridership data. As we observed earlier, the data contain both a trend and monthly seasonality. Figure 17.11 shows part of the JMP output using a 12-month cycle

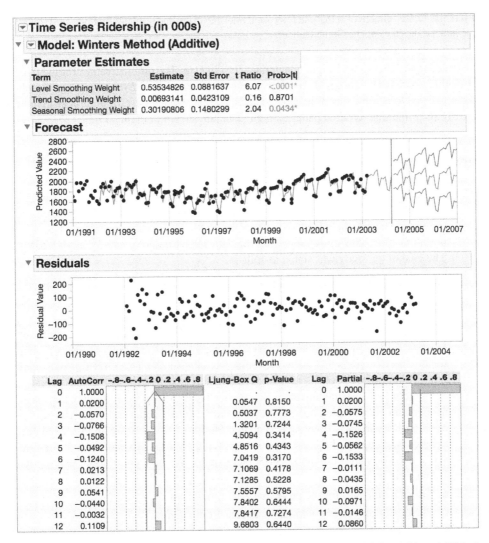

FIGURE 17.11 Partial output for Winters Method (Winters Exponential Smoothing–Additive) applied to Amtrak ridership series

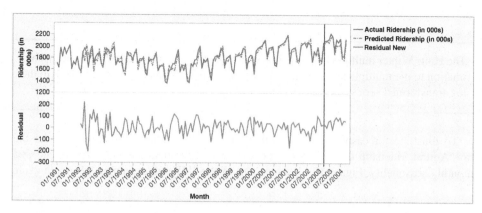

FIGURE 17.12   Graphs of actual and forecasted ridership (Top) and residuals (bottom) for Winter's method, produced using the Graph Guilder.

and the default smoothing weight constraint, *Zero to One*. The optimal values for the three smoothing parameters are given under Estimate in the *Parameter Estimates* table. Graphs for actual and forecasted ridership and residuals, recreated using the *Graph Builder* to more clearly show the time-ordered patterns in the data, are shown in Figure 17.12.

PROBLEMS

17.1 Impact of September 11 on Air Travel in the United States. The Research and Innovative Technology Administration's Bureau of Transportation Statistics conducted a study to evaluate the impact of the September 11, 2001, terrorist attack on US transportation. The 2006 study report and the data can be found at http://goo.gl/w2lJPV. The goal of the study was stated as follows:

> The purpose of this study is to provide a greater understanding of the passenger travel behavior patterns of persons making long distance trips before and after 9/11.

The report analyzes monthly passenger movement data between January 1990 and May 2004. Data on three monthly time series are given in file Sept11Travel.jmp for this period: (1) actual airline revenue passenger miles (Air), (2) rail passenger miles (Rail), and (3) vehicle miles traveled (Car).

In order to assess the impact of September 11, BTS took the following approach: using data before September 11, they forecasted future data (under the assumption of no terrorist attack). Then they compared the forecasted series with the actual data to assess the impact of the event. Our first step therefore is to split each of the time series into two parts: pre- and post-September 11. We now concentrate only on the earlier time series.

a. Create a time plot for the pre-event AIR time series.

   i. What time series components appear from the plot?

b. A time plot of the seasonally adjusted pre-September 11 AIR series is shown in Figure 17.13. Which of the following smoothing methods would be adequate for forecasting this series?

- Moving average (with what window width?)
- Simple exponential smoothing
- Holt exponential smoothing (linear exponential smoothing)
- Winter's exponential smoothing (Winter's method, additive)

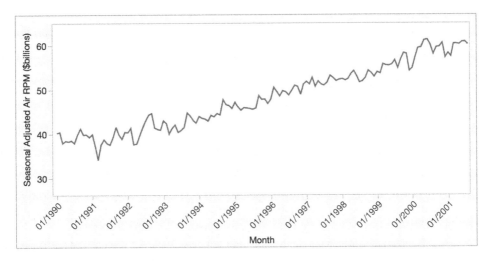

FIGURE 17.13   Seasonally adjusted pre-September 11 AIR series.

17.2 Relation between Moving Average and Exponential Smoothing. Suppose that we apply a moving average to a series, using a very short window span. If we wanted to achieve an equivalent result using simple exponential smoothing, what value should the smoothing coefficient take?

17.3 Forecasting with a Moving Average: For a given time series of sales, the training set consists of 50 months. The first 5 months' data are shown below:

Month	Sales
Sept 98	27
Oct 98	31
Nov 98	58
Dec 98	63
Jan 99	59

a. Compute the sales forecast for January '99 based on a moving average with $w = 4$.

b. Compute the forecast error for this forecast.

17.4 Optimizing Winter's Exponential Smoothing. Figure 17.14 shows an output from applying Winter's exponential smoothing to data, using "optimal" smoothing constants.

	Term	Estimate
1	Level Smoothing Weight	1.000
2	Trend Smoothing Weight	0.000
3	Seasonal Smoothing Weight	0.246

FIGURE 17.14   Optimized smoothing constants.

a. The value of zero that is obtained for the trend smoothing constant means that (choose one of the following):

- There is no trend.
- The trend is estimated only from the first 2 points.
- The trend is updated throughout the data.
- The trend is statistically insignificant.

b. What is the interpretation of the level smoothing constant value?

17.5 Forecasting Department Store Sales. The time series plot in Figure 17.15 describes actual quarterly sales for a department store over a 6-year period. (Data are available in DepartmentStoreSales.jmp, data courtesy of Chris Albright.)

a. Which of the following methods would not be suitable for forecasting this series?

- Moving average of raw series
- Moving average of deseasonalized series
- Simple exponential smoothing of the raw series
- Holt's exponential smoothing of the raw series

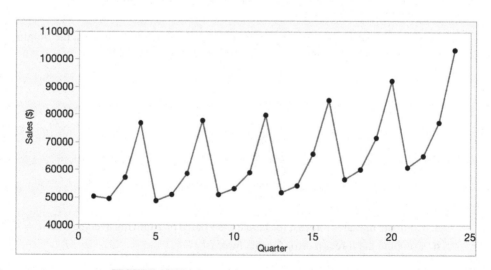

FIGURE 17.15   Department store quarterly sales series.

- Winter's exponential smoothing of the raw series (additive)
- Winter's exponential smoothing of the log(Sales) (multiplicative)
- Regression model fit to the raw series
- Random walk model fit to the raw series

b. The forecaster was tasked to generate forecasts for 4 quarters ahead. He therefore partitioned the data so that the last 4 quarters were designated as the validation period. The forecaster approached the task by using two models: a regression model with trend and seasonality and Winter's exponential smoothing.

  i. Create the necessary columns, and run the regression model on these data. Save the residuals and the predicted values to the data table, and create the top plots in Figure 17.16.

  ii. Run Winter's method on these data. Report the values of the smoothing parameters. Save the prediction formula to the data table, and follow the steps outlined in this chapter to create the graphs for the actual versus forecasted values and residuals shown at the bottom of Figure 17.16.

  iii. The forecasts and errors for the validation data are given in Figure 17.17. Compute the average errors and the mean absolute errors (MAE) for the

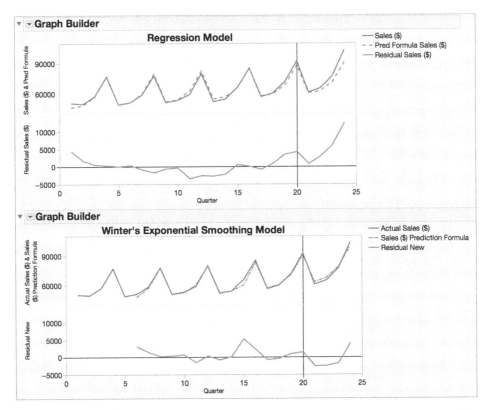

FIGURE 17.16 Forecasts and forecast errors using regression (top) and exponential smoothing (bottom).

forecasts of quarters 21–24 for each of the two models. (Hint: To calculate mean absolute errors, first take the absolute value of each error, then use either *Tabulate* or *Distribution*).

Forecasts and Errors for Validation Data from Regression Model

Quarter	Sales ($)	Pred Sales ($)	Residual Sales ($)
21	60800	60121.7	678.3
22	64900	62067.7	2832.3
23	76997	70903.7	6093.3
24	103337	90885.5	12451.5

Forecasts and Errors for Validation Data from Winters Method (Additive)

Quarter	Sales ($)	Pred Sales ($)	Residual Sales ($)
21	60800	63426.3	−2626.3
22	64900	67423.4	−2523.4
23	76997	78752.1	−1755.1
24	103337	99228.7	4108.3

FIGURE 17.17 Forecasts for validation sets.

c. Compare the fit and residuals from the exponential smoothing and regression models shown in Figure 17.16. Using all the information thus far, which model is more suitable for forecasting quarters 21–24?

d. Run Winter's exponential smoothing on log(Sales) using the training data.

   i. Compute forecasts and forecast errors, and create graphs for actual versus forecasted values and residuals.

   ii. Calculate the average error and MAE for the forecasts of quarters 21–24.

   iii. Based on the graphs and forecast errors, how does this model perform relative to the regression model and the Winter's (additive) model?

   iv. Of the three models, which does the best job of forecasting quarters 21–24? Explain.

17.6 **Shipments of Household Appliances.** The time plot in Figure 17.18 shows the series of quarterly shipments (in $ millions) of US household appliances between 1985 and 1989. (Data are available in `ApplianceShipments.jmp`, data courtesy of Ken Black.)

   a. Which of the following methods would be suitable for forecasting this series if applied to the raw data?

   • Moving average
   • Simple exponential smoothing
   • Holt's exponential smoothing
   • Winter's exponential smoothing

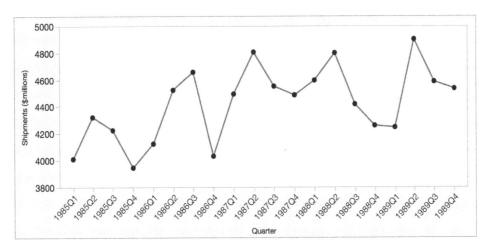

FIGURE 17.18   Quarterly shipments of US household appliances over 5 years.

b. Apply a moving average with window span $w = 4$ to the data. Use all but that last year as the training set. Create a time plot of the moving average series.

   i. What does the MA(4) chart reveal?

   ii. Use the MA(4) model to forecast appliance sales in Q1-1990.

   iii. Use the MA(4) model to forecast appliance sales in Q1-1991.

   iv. Is the forecast for Q1-1990 most likely to underestimate, overestimate, or accurately estimate the actual sales on Q1-1990? Explain.

   v. Management feels most comfortable with moving averages. The analyst therefore plans to use this method for forecasting future quarters. What else should be considered before using the MA(4) to forecast future quarterly shipments of household appliances?

c. We now focus on forecasting beyond 1989. In the following, continue to use all but the last year as the training set, and the last four quarters as the validation set. First, fit a regression model to sales with a linear trend and quarterly seasonality to the training data. Next, apply Winter's exponential smoothing to the training data. Choose an adequate "season length."

   i. Compute the average error and MAE for the validation data using the regression model.

   ii. Compute the average error and MAE for the validation data using Winter's exponential smoothing.

   iii. Which model would you prefer to use for forecasting Q1-1990? Give three reasons.

17.7 Forecasting Shampoo Sales. The time series in in Figure 17.19 describes monthly sales of a certain shampoo over a 3-year period. (Data are available

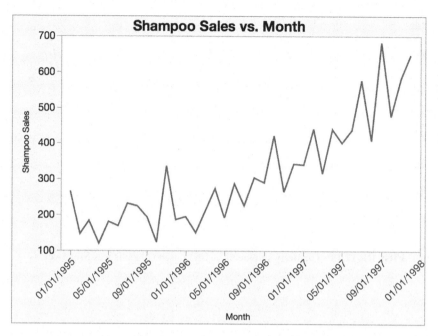

FIGURE 17.19 Monthly sales of a certain shampoo

in `ShampooSales.jmp`. Source: Hyndman, R.J. Time Series Data Library, http://data.is/TSDLdemo. Accessed on 05/01/14.)

a. Which of the following methods would be suitable for forecasting this series if applied to the raw data?

- Moving average
- Simple exponential smoothing
- Holt's exponential smoothing
- Winter's exponential smoothing

b. Fit the four models in the Time Series platform, and compare the built in statistics, graphs, and residual plots.

c. Of these models, which is the best at forecasting future time periods? Explain why.

17.8 Natural Gas Sales. Figure 17.20 shows a time plot of quarterly natural gas sales (in billions of BTU) of a certain company, over a period of 4 years. (Data courtesy of George McCabe. Data are in `Natural Gas Sales.jmp`.) The company's analyst is asked to use a moving average model to forecast sales in Winter 2005.

a. Reproduce the time plot with the overlaying SMA(4) line (use the *Smoothing Model > Simple Moving Average* option in the *Time Series* platform).

b. What can we learn about the series from the MA line?

c. Run a moving average forecaster with adequate season length (refer to the text box in this chapter for computing a trailing moving average forecast in JMP).

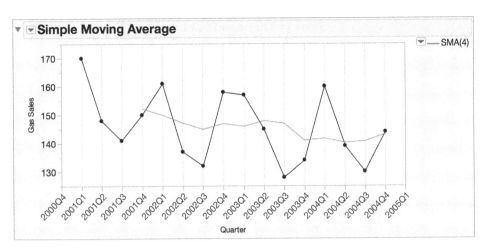

FIGURE 17.20   Quarterly sales of natural gas over 4 years.

d. Are forecasts generated by this method expected to overforecast, underforecast, or accurately forecast actual sales? Why?

17.9 Forecasting Australian Wine Sales. Figure 17.21 shows time plots of monthly sales of six types of Australian wines (red, rose, sweet white, dry white, sparkling, and fortified) for 1980–1994 (Data are available in `AustralianWines.jmp`, source: Hyndman, R.J. Time Series Data Library, http://data.is/TSDLdemo. Accessed on 05/01/14.) The units are thousands of liters. You are hired to obtain short-term forecasts (2–3 months ahead) for each of the six series, and this task will be repeated every month.

a. Which forecasting method would you choose if you had to choose the same method for all series? Why?

b. Fortified wine has the largest market share of the above six types of wine. You are asked to focus on fortified wine sales alone, and produce an accurate as possible forecast for the next 2 months.

- Start by partitioning the data, using the period until December 1993 as the training set.
- Apply Winter's exponential smoothing to sales with an appropriate season length (use the default values for the smoothing constants).

c. Examine ACF plot for the residuals from the model until lag 12.

i. Examining this plot, which of the following statements are reasonable?

- Decembers (month 12) are not captured well by the model.
- There is a strong correlation between sales on the same calendar month.
- The model does not capture the seasonality well.
- We should try to fit an autoregressive model with lag 12 to the residuals.

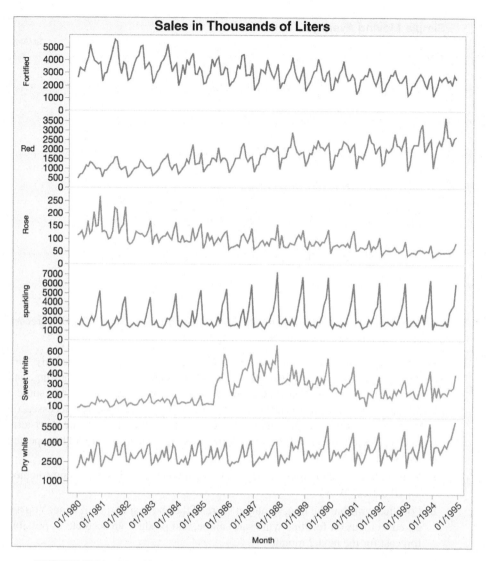

FIGURE 17.21   Monthly sales of six types of Australian wines between 1980 and 1994.

- We should first deseasonalize the data and then apply Winter's exponential smoothing.
- We should fit a Winter's exponential smoothing model to the series of log(sales).

   ii. How can you handle the effect above without adding another layer to your model?

PART VII

CASES

18

CASES

18.1 CHARLES BOOK CLUB

`CharlesBookClub.jmp` is the dataset for this case study.

The Book Industry

Approximately 50,000 new titles, including new editions, are published each year in the United States, giving rise to a \$25 billion industry in 2001.[1] In terms of percentage of sales, this industry may be segmented as follows:

16%	Textbooks
16%	Trade books sold in bookstores
21%	Technical, scientific, and professional books
10%	Book clubs and other mail-order books
17%	Mass-market paperbound books
20%	All other books

Book retailing in the United States in the 1970s was characterized by the growth of bookstore chains located in shopping malls. The 1980s saw increased purchases in bookstores stimulated through the widespread practice of discounting. By the 1990s the superstore concept of book retailing gained acceptance and contributed to double-digit growth of the book industry. Conveniently situated near large shopping centers, superstores maintain large inventories of 30,000 to 80,000 titles and employ well-informed sales personnel. Book

[1] The Charles Book Club case was derived, with the assistance of Ms. Vinni Bhandari, from *The Bookbinders Club, a Case Study in Database Marketing*, prepared by Nissan Levin and Jacob Zahavi, Tel Aviv University; used with permission.

Data Mining for Business Analytics: Concepts, Techniques, and Applications with JMP Pro[®], First Edition.
Galit Shmueli, Peter C. Bruce, Mia L. Stephens, and Nitin R. Patel.
© 2017 John Wiley & Sons, Inc. Published 2017 by John Wiley & Sons, Inc.

retailing changed fundamentally with the arrival of Amazon, which started out as an online bookseller and, as of 2015, has been the world's largest online retailer of any kind. Amazon margins were small and the convenience factor high, putting intense competitive pressure on all other book retailers. Borders, one of the two major superstore chains, discontinued operations in 2011.

Subscription-based book clubs have offer an alternative model that has persisted, though it too has suffered from the dominance of Amazon.

Historically, book clubs offered their readers different types of membership programs. Two common membership programs are the continuity and negative option programs, extended contractual relationships between the club and its members. Under a *continuity program*, a reader signs up by accepting an offer of several books for just a few dollars (plus shipping and handling) and an agreement to receive a shipment of one or two books each month thereafter at more-standard pricing. The continuity program has been most common in the children's book market, where parents are willing to delegate the rights to the book club to make a selection, and much of the club's prestige depends on the quality of its selections.

In a *negative option program*, readers get to select which and how many additional books they would like to receive. However, the club's selection of the month is delivered to them automatically unless they specifically mark "no" on their order form by a deadline date. Negative option programs sometimes result in customer dissatisfaction and always give rise to significant mailing and processing costs.

In an attempt to combat these trends, some book clubs have begun to offer books on a *positive option basis*, but only to specific segments of their customer base that are likely to be receptive to specific offers. Rather than expanding the volume and coverage of mailings, some book clubs are beginning to use database-marketing techniques to target customers more accurately. Information contained in their databases is used to identify who is most likely to be interested in a specific offer. This information enables clubs to design special programs carefully tailored to meet their customer segments' varying needs.

Database Marketing at Charles

The Club   The Charles Book Club (CBC) was established in December 1986 on the premise that a book club could differentiate itself through a deep understanding of its customer base and by delivering uniquely tailored offerings. CBC focused on selling specialty books by direct marketing through a variety of channels, including media advertising (TV, magazines, newspapers) and mailing. CBC is strictly a distributor and does not publish any of the books that it sells. In line with its commitment to understanding its customer base, CBC has built and maintained a detailed database about its club members. Upon enrollment, readers are required to fill out an insert and mail it to CBC. Through this process, CBC has created an active database of 500,000 readers; most are acquired through advertising in specialty magazines.

The Problem   CBC sends mailings to its club members each month containing the latest offerings. On the surface, CBC appears very successful: mailing volume are increasing, book selection is diversifying and growing, and their customer database is increasing. However, their bottom-line profits are falling. The decreasing profits has led CBC to re-visit their original plan of using database marketing to improve mailing yields and to stay profitable.

A Possible Solution   CBC embraces the idea of deriving intelligence from their data to allow them to know their customers better and enable multiple targeted campaigns where each target audience receives appropriate mailings. CBC's management have decided to focus its efforts on the most profitable customers and prospects, and to design targeted marketing strategies to best reach them. They have two processes in place:

1. **Customer acquisition**

   o New members are acquired by advertising in specialty magazines, newspapers, and on TV.

   o Direct mailing and telemarketing contact existing club members.

   o Every new book offered to club members before general advertising.

2. **Data collection**

   o All customer responses are recorded and maintained in the database.

   o Any information not being collected that is critical is requested from the customer.

For each new title, they use a two-step approach:

1. Conduct a market test involving a random sample of 7000 customers from the database to enable analysis of customer responses. The analysis would create and calibrate response models for the current book offering.

2. Based on the response models, compute a score for each customer in the database. Use this score and a cutoff value to extract a target customer list for direct mail promotion.

Targeting promotions is considered to be of prime importance. Other opportunities to create successful marketing campaigns based on customer behavior data (returns, inactivity, complaints, compliments, etc.) will be addressed by CBC at a later stage.

Art History of Florence   A new title, *The Art History of Florence*, is ready for release. CBC sent a test mailing to a random sample of 4000 customers from its customer base. The customer responses have been collated with past purchase data. The dataset has been randomly partitioned into three parts: *Training Data* (1800 customers), initial data to be used to fit response models; *Validation Data* (1400 customers), holdout data used to compare the performance of different response models, and *Test Data* (800 customers): data to be used only after a final model has been selected to estimate the probable performance of the model when it is deployed. Each row (or case) in the data table (other than the header) corresponds to one market test customer. Each column is a variable, with the header row giving the name of the variable. The variable names and descriptions are given in Table 18.1.

Data Mining Techniques

Various data mining techniques can be used to mine the data collected from the market test. No one technique is universally better than another. The particular context and the particular characteristics of the data are the major factors in determining which techniques perform better in an application. For this assignment we focus on two fundamental techniques:

TABLE 18.1   List of Variables in Charles Book Club Dataset

Variable Name	Description
Seq#	Sequence number in the partition
ID#	Identification number in the full (unpartitioned) market test dataset
Gender	0 = Male 1 = Female
M	Monetary–Total money spent on books
R	Recency–Months since last purchase
F	Frequency–Total number of purchases
FirstPurch	Months since first purchase
ChildBks	Number of purchases from the category child books
YouthBks	Number of purchases from the category youth books
CookBks	Number of purchases from the category cookbooks
DoItYBks	Number of purchases from the category do-it-yourself books
RefBks	Number of purchases from the category reference books (atlases, encyclopedias, dictionaries)
ArtBks	Number of purchases from the category art books
GeoBks	Number of purchases from the category geography books
ItalCook	Number of purchases of book title *Secrets of Italian Cooking*
ItalAtlas	Number of purchases of book title *Historical Atlas of Italy*
ItalArt	Number of purchases of book title *Italian Art*
Florence	= 1 if *The Art History of Florence* was bought; = 0 if not
Related Purchase	Number of related books purchased

k-nearest neighbor and logistic regression. We compare them with each other as well as with a standard industry practice known as *RFM (recency, frequency, monetary) segmentation.*

RFM Segmentation   The segmentation process in database marketing aims to partition customers in a list of prospects into homogeneous groups (segments) that are similar with respect to buying behavior. The homogeneity criterion we need for segmentation is the propensity to purchase the offering. But since we cannot measure this attribute, we use variables that are plausible indicators of this propensity.

In the direct marketing business the most commonly used variables are the *RFM variables*:

> *R recency*, time since last purchase
> *F frequency*, number of previous purchases from the company over a period
> *M monetary*, amount of money spent on the company's products over a period

The assumption is that the more recent the last purchase, the more products bought from the company in the past, and the more money spent in the past buying the company's products, the more likely the customer is to purchase the product offered.

The 1800 observations in the training data and the 1400 observations in the validation data have been divided into recency, frequency and monetary categories as follows:

Recency

0–2 months (Rcode = 1)
3–6 months (Rcode = 2)

7–12 months (Rcode = 3)
13 months and up (Rcode = 4)

Frequency

1 book (Fcode = 1)
2 books (Fcode = 2)
3 books and up (Fcode = 3)

Monetary

$0–$25 (Mcode = 1)
$26–$50 (Mcode = 2)
$51–$100 (Mcode = 3)
$101–$200 (Mcode = 4)
$201 and up (Mcode = 5)

Figures 18.1 and 18.2 display the 1800 customers in the training data cross-tabulated by these categories using the *Categorical* platform in JMP (under *Analyze > Consumer Research*). To generate the summary for buyers in Figure 18.1, run the saved script *Data Filter Training Buyers* (in the top left corner of the data table), then run the two categorical scripts. To create the summary for all customers, first run the *Data Filter Training* saved script before running the categorical scripts.

Assignment

1. What is the response rate for the training data customers taken as a whole? What is the response rate for each of the $4 \times 5 \times 3 = 60$ combinations of RFM categories? Which combinations have response rates in the training data that are above the overall response in the training data?
2. Suppose that we decide to send promotional mail only to the "above-average" RFM combinations identified in part (a). Compute the response rate in the validation data using these combinations.
3. Rework parts 1 and 2 with three segments:

   Segment 1:   consisting of RFM combinations that have response rates that exceed twice the overall response rate
   Segment 2:   consisting of RFM combinations that exceed the overall response rate but do not exceed twice that rate
   Segment 3:   consisting of the remaining RFM combinations

   Which RFM combination has the highest lift?

The *k*-Nearest Neighbor   The *k*-nearest neighbor technique can be used to create segments based on product proximity to similar products of the products offered as well as the propensity to purchase (as measured by the RFM variables).

▼ ▼ Categorical

Validation = Training Data and Florence = 1

▼ Fcode by Mcode

Freq		Mcode				
		1	2	3	4	5
Fcode	1	2	2	10	7	17
	2	0	3	5	9	17
	3	0	1	1	15	62
	Total Responses	2	6	16	31	96

▼ ▼ Categorical

Validation = Training Data and Florence = 1

▼ Rcode*Fcode by Mcode

			Freq	Mcode				
				1	2	3	4	5
Rcode	1	Fcode	1	0	0	0	2	1
			2	0	1	0	0	1
			3	0	1	0	0	5
			Total Responses	0	2	0	2	7
	2	Fcode	1	1	0	1	1	5
			2	0	0	3	5	5
			3	0	0	0	4	10
			Total Responses	1	0	4	10	20
	3	Fcode	1	1	0	1	2	5
			2	0	1	1	2	4
			3	0	0	0	4	31
			Total Responses	1	1	2	8	40
	4	Fcode	1	0	2	8	2	6
			2	0	1	1	2	7
			3	0	0	1	7	16
			Total Responses	0	3	10	11	29

FIGURE 18.1 RFM counts for buyers.

▼ ▾ Categorical

Validation = Training Data

▼ Fcode by Mcode

Freq		Mcode				
		1	2	3	4	5
	1	20	40	93	166	219
	2	0	32	91	180	247
Fcode	3	0	2	33	179	498
	Total Responses	20	74	217	525	964

▼ ▾ Categorical

Validation = Training Data

▼ Rcode*Fcode by Mcode

			Freq	Mcode				
				1	2	3	4	5
Rcode	1	Fcode	1	2	2	6	10	15
			2	0	3	4	12	16
			3	0	1	2	11	45
			Total Responses	2	6	12	33	76
	2	Fcode	1	3	5	17	28	26
			2	0	2	17	30	31
			3	0	0	3	34	66
			Total Responses	3	7	37	92	123
	3	Fcode	1	7	15	24	51	86
			2	0	12	29	55	85
			3	0	1	17	53	165
			Total Responses	7	28	70	159	336
	4	Fcode	1	8	18	46	77	92
			2	0	15	41	83	115
			3	0	0	11	81	222
			Total Responses	8	33	98	241	429

FIGURE 18.2 RFM counts for all customers (buyers and non buyers).

For *The Art History of Florence*, a possible segmentation by product proximity could be created using the following variables:

M: monetary–total money (in dollars) spent on books

R: recency–months since last purchase

F: frequency–total number of past purchases

FirstPurch: months since first purchase

RelatedPurch: total number of past purchases of related books (i.e., sum of purchases from the art and geography categories and of titles *Secrets of Italian Cooking*, *Historical Atlas of Italy*, and *Italian Art*).

4. Clear the row states (use *Rows > Clear Row States*), and then use the *K Nearest Neighbor* option under the *Analyze > Modeling > Partition* platform to classify cases. The *Y, Response* is Florence. Use validation, and $k = 15$. Determine the best value for k. How good is the model at correctly predicting Florence = 1 (purchased)?

Logistic Regression   The logistic regression model offers a powerful method for modeling response because it yields well-defined purchase probabilities. (The model is especially attractive in consumer-choice settings because it can be derived from the random utility theory of consumer behavior under the assumption that the error term in the customer's utility function follows a type I extreme value distribution.)

Use the training set data of 1800 observations to construct three logistic regression models with:

- The full set of 15 predictors in the dataset as independent variables and "Florence" as the dependent variable (do not include Rcode, Mcode, or Fcode)
- A subset that you judge to be the best
- Only the *R*, *F*, and *M* variables
  Save the probability formula for each model to the data table.

5. Use the *Model Comparison* platform from *Analyze > Modeling* to compare these models on the validation set. (Hint: Use the *Validation* column as a *By* or *Group* variable in the *Model Comparison* dialog.) Select the *Lift Curve* from the top red triangle, and compare the lift curves for buyers for the three models. Which model has the highest lift at portion = 0.1?

6. Create a subset of the dataset with just the validation data, and sort the data in descending order of the probability of buying (*Prob[1]*) for the best model. If the cutoff criterion for a campaign is a 30% likelihood of a purchase, find the customers in the validation data that would be targeted and count the number of buyers in this set.

18.2   GERMAN CREDIT

GermanCredit.jmp is the dataset for this case study.

Background

Money-lending has been around since the advent of money; it is perhaps the world's second-oldest profession. The systematic evaluation of credit risk, though, is a relatively recent arrival, and lending was largely based on reputation and very incomplete data. Thomas Jefferson, the third President of the United States, was in debt throughout his life and unreliable in his debt payments, yet people continued to lend him money. It wasn't until the beginning of the 20th century that the Retail Credit Company was founded to share information about credit. That company is now Equifax, one of the big three credit scoring agencies (the others are Transunion and Experion).

Individual and local human judgment are now largely irrelevant to the credit reporting process. Credit agencies and other big financial institutions extending credit at the retail level collect huge amounts of data to predict whether defaults or other adverse events will occur, based on numerous customer and transaction information.

Data

This case deals with an early stage of the historical transition to predictive modeling, in which humans were employed to label records as good or poor credit. The German Credit dataset[2] has 30 variables and 1000 records, each record being a prior applicant for credit. Each applicant was rated as "good credit" (700 cases) or "bad credit" (300 cases).

All the variables are explained in Table 18.2. New applicants for credit can also be evaluated on these 30 predictor variables and classified as a good or a bad credit risk based on the predictor variables.

The consequences of misclassification have been assessed as follows: The costs of a false positive (incorrectly saying that an applicant is a good credit risk) outweigh the benefits of a true positive (correctly saying that an applicant is a good credit risk) by a factor of 5. This is summarized in the Average Net Profit Table in Figure 18.3.

Because decision makers are used to thinking of their decision in terms of net profits, we use this table in assessing the performance of the various models.

Assignment

1. Review the list of predictor variables and guess what their role in a credit decision might be.
2. Use the *Columns Viewer* and *Distribution* to become familiar with the data. Are there any surprises in the data? Are there any potential issues with data quality?
3. Use the *Profit Matrix* column property to enter the average net profits for the response. Recall that the labels (Actual and Predicted) are reversed in the *Profix Matrix*.

---

[2]This dataset is available from ftp.ics.uci.edu/pub/machine-learning-databases/statlog/.

4. Divide the data into training and validation sets, and develop classification models using the following data mining techniques: logistic regression, classification trees, and neural networks. (Recall that the random seed can be set prior to creating the validation column to ensure reproducibility.)

5. Save the prediction formulas for each model to the data table. Several columns are saved to the data table, including an expected profit column for each response. Group the columns for each model (select the columns and select *Cols > Group Columns*), and rename the groupings with the model name. Which technique had the highest net (expected) profits for the validation data?

6. Use the *Confusion Matrix* option from the *Model Comparison* platform to compare the misclassifications for these models on the validation set. Since a profit matrix was specified, the confusion matrix is reported, along with a decision matrix based on the profix matrix. Which technique had the lowest misclassification rate? Compare the confusion matrix to the decision matrix for your best model. How do these differ?

7. Let us try and improve our performance for your best model. Rather than accept the initial classification of all applicants' credit status, use the "probability of success" (saved as PROB[1]) for your model as a basis for selecting the best credit risks first, followed by poorer risk applicants.

   a. Create a subset of your data with just the validation data. Sort the validation data on "probability of success" for your model.

   b. Calculate the cumulative sum of the expected profits for this model (right-click on the expected profits column and select *New Formula Column > Row > Cumulative Sum*).

   c. How far into the validation data do you go to get maximum net profit? (Express this is as a percentile.)

   d. If this model is scored to future applicants, what "probability of success" cutoff should be used in extending credit?

18.3   TAYKO SOFTWARE CATALOGER

`TaykoAllData.jmp` is the dataset for this case study.

Background

Tayko is a software catalog firm that sells games and educational software.[3] It started out as a software manufacturer and later added third-party titles to its offerings. It has recently put together a revised collection of items in a new catalog, which it is preparing to roll out in a mailing.

In addition to its own software titles, Tayko's customer list is a key asset. In an attempt to expand its customer base, it has recently joined a consortium of catalog firms that specialize in computer and software products. The consortium affords members the opportunity to mail catalogs to names drawn from a pooled list of customers. Members supply their own customer lists to the pool, and can "withdraw" an equivalent number of names each quarter. Members are allowed to do predictive modeling on the records in the pool so that they can do a better job of selecting names from the pool.

---

[3]©Resampling Stats, Inc. 2006; used with permission.

TABLE 18.2   Variables for the German Credit Dataset

Variable	Variable Name	Description	Variable Type	Code Description
1	OBS#	Observation numbers	Categorical	Sequence number in dataset
2	CHK_ACCT	Checking account status	Categorical	0: < 0 DM
				1: 0 − 200 DM
				2 : > 200 DM
				3: no checking account
3	DURATION	Duration of credit in months	Numerical	
4	HISTORY	Credit history	Categorical	0: no credits taken
				1: all credits at this bank paid back duly
				2: existing credits paid back duly until now
				3: delay in paying off in the past
				4: critical account
5	NEW_CAR	Purpose of credit	Binary	car (new) 0: no, 1: yes
6	USED_CAR	Purpose of credit	Binary	car (used) 0: no, 1: yes
7	FURNITURE	Purpose of credit	Binary	furniture/equipment 0: no, 1: yes
8	RADIO/TV	Purpose of credit	Binary	radio/television 0: no, 1: yes
9	EDUCATION	Purpose of credit	Binary	education 0: no, 1: yes
10	RETRAINING	Purpose of credit	Binary	retraining 0: no, 1: yes
11	AMOUNT	Credit amount	Numerical	
12	SAV_ACCT	Average balance in savings account	Categorical	0: < 100 DM
				1 : 101 − 500 DM
				2 : 501 − 1000 DM
				3 : > 1000 DM
				4 : unknown/ no savings account
13	EMPLOYMENT	Present employment since	Categorical	0 : unemployed
				1: < 1 year
				2: 1 − 3 years
				3: 4 − 6 years
				4: ≥ 7 years
14	INSTALL_RATE	Installment rate as % of disposable income	Numerical	
15	MALE_DIV	Applicant is male and divorced	Binary	0: no, 1:yes

(continued)

TABLE 18.2   *(Continued)*

Var.	Variable Name	Description	Variable Type	Code Description
16	MALE_SINGLE	Applicant is male and single	Binary	0: no, 1:yes
17	MALE_MAR_WID	Applicant is male and married or a widower	Binary	0: no, 1:yes
18	CO-APPLICANT	Application has a coapplicant	Binary	0: no, 1:yes
19	GUARANTOR	Applicant has a guarantor	Binary	0: no, 1:yes
20	PRESENT_RESIDENT	Present resident since (years)	Categorical	0: $\leq$ 1 year 1: 1 − 2 years 2: 2 − 3 years 3: $\geq$ 3 years
21	REAL_ESTATE	Applicant owns real estate	Binary	0: no, 1:yes
22	PROP_UNKN_NONE	Applicant owns no property (or unknown)	Binary	0: no, 1:yes
23	AGE	Age in years	Numerical	
24	OTHER_INSTALL	Applicant has other installment plan credit	Binary	0: no, 1:yes
25	RENT	Applicant rents	Binary	0: no, 1:yes
26	OWN_RES	Applicant owns residence	Binary	0: no, 1:yes
27	NUM_CREDITS	Number of existing credits at this bank	Numerical	
28	JOB	Nature of job	Categorical	0 : unemployed/ unskilled— nonresident 1 : unskilled— resident 2 : skilled employee/ official 3 : management/ self-employed/ highly qualified employee/officer
29	NUM_DEPENDENTS	Number of people for whom liable to provide maintenance	Numerical	
30	TELEPHONE	Applicant has phone in his or her name	Binary	0: no, 1:yes
31	FOREIGN	Foreign worker	Binary	0: no, 1:yes
32	RESPONSE	Credit rating is good	Binary	0: no, 1:yes

Note: The original dataset had a number of categorical variables, some of which have been transformed into a series of binary variables. (Several ordered categorical variables have been left as is.)

TABLE 18.3   Average Net Profit (Deutsche Marks)

	Predicted (Decision)	
Actual	No (Bad, Reject)	Yes (Good, Accept)
No	0	−500
Yes	0	100

The Mailing Experiment

Tayko has supplied its customer list of 200,000 names to the pool, which totals over 5,000,000 names, so it is now entitled to draw 200,000 names for a mailing. Tayko would like to select the names that have the best chance of performing well, so it conducts a test—it draws 20,000 names from the pool and does a test mailing of the new catalog.

This mailing yielded 1065 purchasers, a response rate of 0.053. Average spending was $103 for each of the purchasers, or $5.46 per catalog mailed. To optimize the performance of the data mining techniques, it was decided to work with a stratified sample that contained equal numbers of purchasers and nonpurchasers. For ease of presentation, the dataset for this case includes just 1000 purchasers and 1000 nonpurchasers, an apparent response rate of 0.5. Therefore, after using the dataset to predict who will be a purchaser, we must adjust the purchase rate back down by multiplying each case's "probability of purchase" by 0.053/0.5, or 0.107.

Data

There are two response variables in this case. *Purchase* indicates whether a prospect responded to the test mailing and purchased something. *Spending* indicates, for those who made a purchase, how much they spent. The overall procedure in this case will be to develop two models. One will be used to classify records as *purchase* or *no purchase*. The second will be used for those cases that are classified as *purchase* and will predict the amount they will spend.

Table 18.4 provides a description of the variables available in this case. A validation column is used because we will be developing two different models in this case and want to use the same training, validation, and test sets for both models.

Assignment

1. Each catalog costs approximately $2 to mail (including printing, postage, and mailing costs). Estimate the gross profit that the firm could expect from the remaining 180,000 names if it selects them randomly from the pool.
2. Develop a model for classification of a customer as a purchaser or nonpurchaser.
   a. Use the validation column, and develop models to classify customers using:

   i. Stepwise logistic regression
   ii. Classification trees

   b. Choose one model on the basis of its performance with the validation data.
3. Develop a model for predicting spending among the purchasers.

TABLE 18.4   Description of Variables for Tayko Dataset

Variable	Variable Name	Description	Variable Type	Code Description
1	US	Is it a US address?	Binary	1: yes
				0: no
2–16	Source_*	Source catalog for the record (15 possible sources)	Binary	1: yes 0: no
17	Freq.	Number of transactions in last year at source catalog	Numerical	
18	last_update_days_ago	How many days ago last update was made to customer record	Numerical	
19	1st_update_days_ago	How many days ago first update to customer record was made	Numerical	
20	RFM%	Recency–frequency–monetary percentile, as reported by source catalog (see Section 18.1)	Numerical	
21	Web_order	Customer placed at least one order via web	Binary	1: yes 0: no
22	Gender=mal	Customer is male	Binary	1: yes 0: no
23	Address_is_res	Address is a residence	Binary	1: yes 0: no
24	Purchase	Person made purchase in test mailing	Binary	1: yes 0: no
25	Spending	Amount (dollars) spent by customer in test mailing	Numerical	
26	Validation	Variable indicating which partition the record will be assigned to	Nominal	Training Validation Test

  a. Use the data filter to select purchasers, and then make a subset (a new data table) that includes just the purchasers.
  b. Use the validation column, and develop models for predicting spending, using:

    i. Multiple linear regression (use best subset selection)
    ii. Regression trees

  c. Choose one model on the basis of its performance with the validation data.
  4. Return to the original data, with data on both purchasers and nonpurchasers. Now focus on the test set. It has not been used in our analysis up to this point, so it will

give an unbiased estimate of the performance of our models. Create a new data table (a subset) with just the test portion of the data and all of the columns (including all of the columns from the saved models). Save this new data table as *Score Analysis*.

a. Arrange the following columns so that they are adjacent:
   i.   Predicted probability of purchase (*success*)
   ii.   Actual spending (dollars)
   iii. Predicted spending (dollars)

b. Add a column for "adjusted probability of purchase" by multiplying "predicted probability of purchase" by 0.107. *This is to adjust for oversampling the purchasers* (see above).

c. Add a column and calculate the expected spending using the *Formula Editor* [adjusted probability of purchase × predicted spending].

d. Sort all records on the "expected spending" column.

e. Calculate cumulative lift [= cumulative "actual spending" divided by the average spending that would result from random selection (each adjusted by 0.107)].

h. Using this cumulative lift, estimate the gross profit that would result from mailing to the 180,000 names on the basis of your data mining models.

Note: Although Tayko is a hypothetical company, the data in this case (modified slightly for illustrative purposes) were supplied by a real company that sells software through direct sales. The concept of a catalog consortium is based on the Abacus Catalog Alliance. Details can be found at www.abacusalliance.com/faqs/membership-provide/.

18.4 POLITICAL PERSUASION

VoterPersuasion.jmp is the dataset for this case study[4].

Background

At first, when you think of political persuasion, you may think of the efforts that political campaigns undertake to persuade you that their candidate is better than the other candidate. In truth, campaigns are less about persuading people to change their minds and more about persuading those who agree with you to actually go out and vote. Predictive analytics now plays a big role in this effort, but in 2004 it was a new arrival in the political toolbox.

Predictive Analytics Arrives in US Politics

In January of 2004, candidates in the US presidential campaign were competing in the Iowa caucuses, part of a lengthy state-by-state primary campaign that culminates in the selection of the Republican and Democratic candidates for president. Among the Democrats, Howard

[4]©Statistics.com, Inc. 2015; used with permission. Thanks to Ken Strasma, President of HaystaqDNA and director of targeting for the 2004 Kerry campaign and the 2008 Obama campaign, for the data used in this case, and for sharing the information in the following writeup.

Dean was leading in national polls. The Iowa caucuses, however, are a complex and intensive process attracting only the most committed and interested voters. Those participating are not a representative sample of voters nationwide. Surveys of those planning to take part showed a close race between Dean and three other candidates, including John Kerry.

Kerry ended up winning by a surprisingly large margin, and the better than expected performance was due to his campaign's innovative and successful use of predictive analytics to learn more about the likely actions of individual voters. This allowed the campaign to target voters in such a way as to optimize performance in the caucuses. For example, once the model showed sufficient support in a precinct to win that precinct's delegate to the caucus, money and time could be redirected to other precincts where the race was closer.

Political Targeting

Targeting of voters is not new in politics. It has traditionally taken three forms:

- Geographic
- Demographic
- Individual

In geographic targeting, resources are directed to a geographic unit–state, city, county, and the like–on the basis of prior voting patterns or surveys that reveal the political tendency in that geographic unit. It has significant limitations, though. If a county is only, say, 52% in your favor, it may be in the greatest need of attention, but if messaging is directed to everyone in the county, nearly half of it is reaching the wrong people.

In demographic targeting, the messaging is intended for demographic groups–older voters, younger women voters, Hispanic voters, and so forth. The limitation of this method is that it is often not easy to implement–messaging is hard to deliver just to single demographic groups.

Traditional individual targeting, the most effective form of targeting, was done on the basis of surveys asking voters how they plan to vote. The big limitation of this method is, of course, the cost. The expense of reaching all voters in a phone or door-to-door survey can be prohibitive.

The use of predictive analytics adds power to the individual targeting method, and reduces cost. A model allows prediction to be rolled out to the entire voter base, not just those surveyed, and brings to bear a wealth of information. Geographic and demographic data remain part of the picture, but they are used at an individual level.

Uplift

In a classical predictive modeling application for marketing, a sample of data is selected and an offer is made (e.g., on the web) or a message is sent (e.g., by mail), and a predictive model is developed to classify individuals as responding or not responding. The model is then applied to new data, propensities to respond are calculated, individuals are ranked by their propensity to respond, and the marketer can then select those most likely to respond to mailings or offers.

Some key information is missing from this classical approach: How would the individual respond in the absence of the offer or mailing? Might a high propensity customer be inclined to purchase regardless of the offer? Might a person's propensity to buy actually be diminished by the offer? Uplift modeling (see Chapter 13) allows us to estimate the effect of "offer versus no offer" or "mailing versus no mailing" at the individual level.

In this case we will apply uplift modeling to actual voter data that have been augmented with the results of a hypothetical experiment. The experiment consisted of the following steps:

1. Conduct a pre-survey of the voters to determine their inclination to vote democratic.
2. Randomly split the voters into two samples–control and treatment.
3. Send a flyer promoting the Democratic candidate to the treatment group.
4. Conduct another survey of the voters to determine their inclination to vote Democratic.

Data

The data in this case are in the file `VoterPersuasion.jmp`. The target variable is MOVED_AD, where a 1 = "opinion moved in favor of the Democratic candidate" and 0 = "opinion did not move in favor of the Democratic candidate." This variable encapsulates the information from the pre- and post-surveys. The important predictor variable is *Flyer*, a binary variable that indicates whether or not a voter received the flyer. In addition there are numerous other predictor variables from these sources:

1. Government voter files
2. Political party data
3. Commercial consumer and demographic data
4. Census neighborhood data

Government voter files are maintained, and made public, to ensure the integrity of the voting process. They contain essential data for identification purposes such as name, address, and date of birth. The file used in this case also contains party identification (needed if a state limits participation in party primaries to voters in that party). Parties also staff elections with their own poll-watchers, who record whether an individual votes in an election. These data (termed "derived" in the case data) are maintained and curated by each party, and can be readily matched to the voter data by name. Demographic data at the neighborhood level are available from the census, and can be appended to the voter data by address matching. Consumer and additional demographic data (buying habits, education) can be purchased from marketing firms and appended to the voter data (matching by name and address).

Assignment

The task in this case is to develop an uplift model that predicts the uplift for each voter, first using the *Two Model* approach and then using the built-in JMP utility for uplift modeling. Uplift is defined as the increase in propensity to move one's opinion in a Democratic direction. First, review the variables in the dataset to understand which data source they are probably coming from. Then:

1. Overall, how well did the flyer do in moving voters in a Democratic direction? (Look at the target variable among those who got the flyer, compared to those who did not.)

2. Explore the data to learn more about the relationships between predictor variables and MOVED_AD using data visualization. Which of the predictors seem to have good predictive potential? Show supporting charts and/or tables.

3. Use the validation column that is in the dataset, make decisions about predictor inclusion, and fit three predictive models accordingly. For each model, give sufficient detail about the method used, its parameters, and the predictors used, so that your results can be replicated, and save the scripts for your models to the data table. Save the prediction formulas to the data table.

4. Among your three models, choose the best one in terms of predictive power. Which one is it? Why did you choose it?

5. Using your chosen model, report the propensities for the first three records in the validation set.

6. Create a derived variable that is the opposite of *Flyer*. Call it *Flyer-reversed*. Using your chosen model, refit the same model using the *Flyer-reversed* variable as a predictor, instead of *Flyer*. Report the propensities for the first 3 records in the validation set.

7. For each record, uplift is computed based on the following difference:

$$P(success|Flyer = 1) - P(success|Flyer = 0)$$

Compute the uplift for each of the voters in the validation set, and report the uplift for the first 3 records.

8. If a campaign has the resources to mail the flyer only to 10% of the voters, what uplift cutoff should be used?

9. Use the Uplift platform in JMP Pro (under *Analyze > Consumer Research*) to fit an uplift model for MOVED_AD. Recall that the Uplift platform in JMP is based on the *Partition* platform.

   • What are the overall rates (propensities) for Flyer = 0 and Flyer = 1 (before splitting)?

   • What is the uplift before splitting? Interpret this value.

   • Click *Go* to automatically build and prune the model. Using the leaf report and the uplift graph, determine which node has the highest uplift for the training set? What is the uplift for this node? Provide an interpretation of this uplift value.

   • Save the difference formula to the data table. Create a subset of this table containing only the validation set, and sort the differences (the uplift) in descending order. What is the highest uplift for the validation set?

   • Based on this model, if a campaign has the resources to mail the flyer only to 10% of the voters, what uplift cutoff should be used? Characterize these voters in terms of the predictors in your model.

18.5   TAXI CANCELLATIONS

`Taxi-cancellation-case.xlsx` (an Excel file) is the dataset for this case study.

Business Situation

In late 2013, the taxi company Yourcabs.com in Bangalore, India was facing a problem with the drivers using their platform–not all drivers were showing up for their scheduled calls. Drivers would cancel their acceptance of a call, and, if the cancellation did not occur with adequate notice, the customer would be delayed or even left high and dry.

Bangalore is a key tech center in India, and technology was transforming the taxi industry. Yourcabs.com featured an online booking system (though customers could phone in as well), and presented itself as a taxi booking portal. The Uber ride sharing service would start its Bangalore operations in mid-2014.

Yourcabs.com had collected data on its bookings from 2011 to through 2013, and posted a contest on Kaggle, in coordination with the Indian School of Business, to see what it could learn about the problem of cab cancellations.

The data presented for this case are a randomly selected subset of the original data, with 10,000 rows, one row for each booking. There are 17 input variables, including user (customer) ID, vehicle model, whether the booking was made online or via a mobile app, type of travel, type of booking package, geographic information, and the date and time of the scheduled trip. The target variable of interest is the binary indicator of whether a ride was canceled. The overall cancellation rate is between 7% and 8%. The data are in an Excel file.

Assignment

1. How can a predictive model based on these data be used by Yourcabs.com?

2. How can a profiling model (identifying predictors that distinguish canceled/uncanceled trips) be used by Yourcabs.com?

3. Import the data into JMP, and explore, prepare, and transform the data to facilitate predictive modeling. The Excel file contains a tab with descriptions of the variables. Depending on your familiarity with Excel you may choose to do some data preparation in Excel before importing the data into JMP. Or, the data preparation can be done directly within JMP. Here are some hints:

   - In exploratory modeling it is useful to move fairly soon to at least an initial model without solving *all* data preparation issues. One example is the GPS information–other geographic information is available, so you could defer the challenge of how to interpret/use the GPS information.

   - How will you deal with missing data, such as cases where "NULL" is indicated? (Consider using a column property or a column utility in JMP.)

   - Think about what useful information might be held within the date and time fields (the booking timestamp and the trip timestamp). The Excel file has a tab with hints

[4]©Statistics.com, Inc. and Galit Shmueli 2015; used with permission.

on how to transform the date variables. Recall that the JMP formula editor can be used to extract information from variables and derive new variables.

- Think also about the categorical variables, and how to deal with them. Should we recode some of them? Use only some of the variables?

4. Fit several predictive models of your choice. Do they provide information on how the predictor variables relate to cancellations? Compare the models, and select one best-performing model.

5. Report the predictive performance of your model in terms of error rates (the confusion matrix). How well does the model perform? Can the model be used in practice?

6. Examine the predictive performance of your model in terms of ranking (lift). How well does the model perform? Can the model be used in practice?

18.6 SEGMENTING CONSUMERS OF BATH SOAP

`BathSoap.jmp` is the dataset for this case study.

Business Situation

CRISA is an Asian market research agency that specializes in tracking consumer purchase behavior in consumer goods (both durable and nondurable).[5] In one major research project, CRISA tracks numerous consumer product categories (e.g., "detergents") and, within each category, perhaps dozens of brands. To track purchase behavior, CRISA constituted household panels in over 100 cities and towns in India, covering most of the Indian urban market. The households were carefully selected using stratified sampling to ensure a representative sample; a subset of 600 records is analyzed here. The strata were defined on the basis of socioeconomic status and the market (a collection of cities).

CRISA has both transaction data (each row is a transaction) and household data (each row is a household), and for the household data CRISA maintains the following information:

- Demographics of the households (updated annually)
- Possession of durable goods (car, washing machine, etc., updated annually; an "affluence index" is computed from this information)
- Purchase data of product categories and brands (updated monthly)

CRISA has two categories of clients: (1) advertising agencies that subscribe to the database services, obtain updated data every month, and use the data to advise their clients on advertising and promotion strategies, and (2) consumer goods manufacturers that monitor their market share using the CRISA database.

---

[5]©Cytel, Inc. and Resampling Stats, Inc. 2006; used with permission.

Key Problems

CRISA has traditionally segmented markets on the basis of purchaser demographics. They would now like to segment the market based on two key sets of variables more directly related to the purchase process and to brand loyalty:

1. Purchase behavior (volume, frequency, susceptibility to discounts, and brand loyalty)
2. Basis of purchase (price, selling proposition)

Doing so would allow CRISA to gain information about what demographic attributes are associated with different purchase behaviors and degrees of brand loyalty, and thus deploy promotion budgets more effectively. More effective market segmentation would enable CRISA's clients (in this case, a firm called IMRB) to design more cost-effective promotions targeted at appropriate segments. Thus multiple promotions could be launched, each targeted at different market segments at different times of the year. This would result in a more cost-effective allocation of the promotion budget to different market segments. It would also enable IMRB to design more effective customer reward systems and thereby increase brand loyalty.

Data

The data in Table 18.5 profile each household, each row containing the data for one household.

Measuring Brand Loyalty

Several variables in this case measure aspects of brand loyalty. The number of different brands purchased by the customer is one measure. However, a consumer who purchases one or two brands in quick succession, then settles on a third for a long streak, is different from a consumer who constantly switches back and forth among three brands. How often customers switch from one brand to another is another measure of loyalty. Yet a third perspective on the same issue is the proportion of purchases that go to different brands—a consumer who spends 90% of his or her purchase money on one brand is more loyal than a consumer who spends more equally among several brands.

All three of these components can be measured with the data in the purchase summary worksheet.

Assignment

1. Use k-means clustering to identify clusters of households based on the following:
   a. Variables that describe purchase behavior (including brand loyalty)
   b. Variables that describe the basis for purchase
   c. Variables that describe both purchase behavior and basis of purchase
   Note 1: How should k be chosen? Think about how the clusters would be used. It is likely that the marketing efforts would support two to five different promotional approaches.

TABLE 18.5 Description of variables for each household

Variable Type	Variable Name	Description
Member ID	Member id	Unique identifier for each household
Demographics	SEC	Socioeconomic class (1 = high, 5 = low)
	FEH	Eating habits(1 = vegetarian, 2 = vegetarian but eat eggs, 3 = nonvegetarian, 0 = not specified)
	MT	Native language (see table in worksheet)
	SEX	Gender of homemaker (1 = male, 2 = female)
	AGE	Age of homemaker
	EDU	Education of homemaker (1 = minimum, 9 = maximum)
	HS	Number of members in household
	CHILD	Presence of children in household (4 categories)
	CS	Television availability (1 = available, 2 = unavailable)
	Affluence Index	Weighted value of durables possessed
Purchase summary over the period	No. of Brands	Number of brands purchased
	Brand Runs	Number of instances of consecutive purchase of brands
	Total Volume	Sum of volume
	No. of Trans	Number of purchase transactions (multiple brands purchased in a month are counted as separate transactions
	Value	Sum of value
	Trans/ Brand Runs	Average transactions per brand run
	Vol/Trans	Average volume per transaction
	Avg. Price	Average price of purchase
Purchase within promotion	Pur Vol	Percent of volume purchased
	No Promo - %	Percent of volume purchased under no promotion
	Pur Vol Promo 6%	Percent of volume purchased under promotion code 6
	Pur Vol OthePromo %	Percent of volume purchased under other promotions
Brandwise purchase	Br. Cd. (**57, 144**), **55, 272, 286, 24, 481, 352, 5**, and 999 (others)	Percent of volume purchased of the brand
Price categorywise purchase	Price Cat 1 to 4	Percent of volume purchased under the price category
Selling proposition-wise purchase	Proposition Cat 5 to 15	Percent of volume purchased under the product proposition category

Note 2: How should the percentages of total purchases comprised by various brands be treated? Isn't a customer who buys all brand A just as loyal as a customer who buys all brand B? What will be the effect on any distance measure of using the brand share variables as is? Consider using a single derived variable.

2. Select what you think is the best segmentation and comment on the characteristics (demographic, brand loyalty, and basis for purchase) of these clusters. (This information would be used to guide the development of advertising and promotional campaigns.)

3. Develop a model that classifies the data into these segments. Since this information would most likely be used in targeting direct mail promotions, it would be useful to select a market segment that would be defined as a *success* in the classification model.

Although not used in the assignment, two additional datasets were used in the derivation of the summary data. CRISA_Purchase_Data is a transaction database in which each row is a transaction. Multiple rows in this dataset corresponding to a single household were consolidated into a single household row in CRISA_Summary_Data.

The *Durables* sheet (in the `BathSoap.xls` file) contains information used to calculate the affluence index. Each row is a household, and each column represents a durable consumer good. A 1 in the column indicates that the durable is possessed by the household; a 0 indicates that it is not possessed. This value is multiplied by the weight assigned to the durable item. For example, a 5 indicates the weighted value of possessing the durable. The sum of all the weighted values of the durables possessed equals the affluence index.

18.7 DIRECT-MAIL FUNDRAISING

`Fundraising.jmp` and `FutureFundraising.jmp` are the datasets used for this case study.

Background

Note: Be sure to read the information about oversampling and adjustment in Chapter 5 before starting to work on this case.

A national veterans' organization wishes to develop a predictive model to improve the cost-effectiveness of their direct marketing campaign. The organization, with its in-house database of over 13 million donors, is one of the largest direct-mail fundraisers in the United States. According to their recent mailing records, the overall response rate is 5.1%. Out of those who responded (donated), the average donation is $13.00. Each mailing, which includes a gift of personalized address labels and assortments of cards and envelopes, costs $0.68 to produce and send. Using these facts, we take a sample of this dataset to develop a classification model that can effectively capture donors so that the expected net profit is maximized. Weighted sampling is used, underrepresenting the nonresponders so that the sample has equal numbers of donors and nondonors.

Data

The file Fundraising.jmp contains 3120 records with 50% donors (TARGET_B = 1) and 50% nondonors (TARGET_B = 0). The amount of donation (TARGET_D) is also included but is not used in this case. The descriptions for the 22 variables (including two target variables) are listed in Table 18.6.

TABLE 18.6 Description of Variables for the Fundraising Dataset

Variable	Description
ZIP	Zip code group (Zip codes were grouped into five groups: 1 = the potential donor belongs to this zip group.)
	00000–19999 ← zipconvert_1
	20000–39999 ← zipconvert_2
	40000–59999 ← zipconvert_3
	60000–79999 ← zipconvert_4
	80000–99999 ← zipconvert_5
HOMEOWNER	1 = homeowner, 0 = not a homeowner
NUMCHLD	Number of children
INCOME	Household income
GENDER	0 = male, 1 = female
WEALTH	Wealth rating uses median family income and population statistics from each area to index relative wealth within each state. The segments are denoted 0 to 9, with 9 being the highest wealth group and zero the lowest. Each rating has a different meaning within each state.
HV	Average home value in potential donor's neighborhood in hundreds of dollars
ICmed	Median family income in potential donor's neighborhood in hundreds of dollars
ICavg	Average family income in potential donor's neighborhood in hundreds
IC15	Percent earning less than $15K in potential donor's neighborhood
NUMPROM	Lifetime number of promotions received to date
RAMNTALL	Dollar amount of lifetime gifts to date
MAXRAMNT	Dollar amount of largest gift to date
LASTGIFT	Dollar amount of most recent gift
TOTALMONTHS	Number of months from last donation to July 1998 (the last time the case was updated)
TIMELAG	Number of months between first and second gift
AVGGIFT	Average dollar amount of gifts to date
TARGET_B	Target variable: binary indicator for response 1 = donor, 0 = nondonor
TARGET_D	Target variable: donation amount (in dollars). We will NOT be using this variable for this case.

Assignment

Step 1: Prepare the data. Partition the dataset into 60% training and 40% validation (set the seed to 12345). Specify a profit matrix for TARGET_*B* (Again, the expected donation, given that they are donors, is $13.00, and the total cost of each mailing is $0.68.)

Step 2: Build the model. Follow the next series of steps to build, evaluate, and choose a model.

1. *Select classification tool and parameters.* Run at least two classification models of your choosing. Be sure NOT to use TARGET_*D* in your analysis. Describe the two models that you chose, with sufficient detail (method, parameters, variables, etc.) so that they can be replicated. Save the scripts for the models to the data table, and save the probability formulas to the data table.

2. *Classification under asymmetric response and cost.* What is the reasoning behind using weighted sampling to produce a training set with equal numbers of donors and nondonors? Why not use a simple random sample from the original dataset?

3. *Calculate net profit.* For each method, calculate the lift of net profit for both the training and validation set based on the actual response rate (5.1%). (Hint: To calculate estimated net profit, we will need to undo the effects of the weighted sampling and calculate the net profit that would reflect the actual response distribution of 5.1% donors and 94.9% nondonors. To do this, divide each row's net profit by the oversampling weights applicable to the actual status of that row (create a weight formula column). The oversampling weight for actual donors is 50%/5.1% = 9.8. The oversampling weight for actual non donors is 50%/94.9% = 0.53.)

4. *Draw lift curves.* Draw each model's net profit lift curve for the validation set onto a single graph (net profit on the *y*-axis, proportion of list or number mailed on the *x*-axis). Is there a model that dominates?

5. *Select best model.* From your answer in 2, what do you think is the "best" model?

Step 3: Testing. The file `FutureFundraising.jmp` contains the attributes for future mailing candidates.

6. Using your "best" model from step 2 (number 5), which of these candidates do you predict as donors and nondonors? (Hint: Copy and paste the formula(s) for you final model into new columns in this table.) Sort them in descending order of the probability of being a donor. Starting at the top of this sorted list, roughly how far down would you go in a mailing campaign?

18.8 PREDICTING BANKRUPTCY

`Bankruptcy.jmp` is the dataset for this case study.

Predicting Corporate Bankruptcy[6]

Just as doctors check blood pressure and pulse rate as vital indicators of the health of a patient, so business analysts scour the financial statements of a corporation to monitor its financial health. Whereas blood pressure, pulse rate, and most medical vital signs, however, are measured through precisely defined procedures, financial variables are recorded under much less specific general principles of accounting. A primary issue in financial analysis, then, is how predictable is the health of a company?

One difficulty in analyzing financial report information is the lack of disclosure of actual cash receipts and disbursements. Users of financial statements have had to rely on proxies for cash flow, perhaps the simplest of which is income (INC) or earnings per share. Attempts to improve INC as a proxy for cash flow include using income plus depreciation (INCDEP), working capital from operations (WCFO), and cash flow from operations (CFFO). CFFO is obtained by adjusting income from operations for all noncash expenditures and revenues and for changes in the current asset and current liabilities accounts.

A further difficulty in interpreting historical financial disclosure information is caused whenever major changes are made in accounting standards. For example, the Financial Accounting Standards Board issued several promulgations in the middle 1970s that changed the requirements for reporting accruals pertaining to such things as equity earnings, foreign currency gain and losses, and deferred taxes. One effect of changes of this sort was that earnings figures became less reliable indicators of cash flow.

In the light of these difficulties in interpreting accounting information, just what are the important vital signs of corporate health? Is cash flow an important signal? If not, what is? If so, what is the best way to approximate cash flow? How can we predict the impending demise of a company?

To begin to answer some of these important questions, we conducted a study of the financial vital signs of bankrupt and healthy companies. We first identified 66 failed firms from a list provided by Dun and Bradstreet. These firms were in manufacturing or retailing and had financial data available on the Compustat Research tape. Bankruptcy occurred somewhere between 1970 and 1982.

For each of these 66 failed firms, we selected a healthy firm of approximately the same size (as measured by the book value of the firm's assets) from the same industry (3 digit SIC code) as a basis of comparison. This matched sample technique was used to minimize the impact of any extraneous factors (e.g., industry) on the conclusions of the study.

The study was designed to see how well bankruptcy can be predicted two years in advance. A total of 24 financial ratios were computed for each of the 132 firms using data from the Compustat tapes and from Moody's Industrial Manual for the year that was two years prior to the year of bankruptcy. Table **??** lists the 24 ratios together with an explanation of the abbreviations used for the fundamental financial variables. All these variables are contained in a firm's annual report with the exception of CFFO. Ratios were used to facilitate comparisons across firms of various sizes.

The first four ratios using CASH in the numerator might be thought of as measures of a firm's cash reservoir with which to pay debts. The three ratios with CURASS in the numerator capture the firm's generation of current assets with which to pay debts. Two ratios, CURDEBT/DEBT and ASSETS/DEBTS, measure the firm's debt structure. Inventory and receivables turnover are measured by COGS/INV and SALES/REC, and SALES/ASSETS measures the firm's ability to generate sales. The final 12 ratios are asset flow measures.

TABLE 18.7   PREDICTING CORPORATE BANKRUPTCY: FINANCIAL VARIABLES AND RATIOS

Abbreviation	financial Variable	Ratio	Definition
ASSETS	Total assets	R_1	CASH/CURDEBT
CASH	Cash	R_2	CASH/SALES
CFFO	Cash flow from operations	R_3	CASH/ASSETS
COGS	Cost of goods sold	R_4	CASH/DEBTS
CURASS	Current assets	R_5	CFFO/SALES
CURDEBT	Current debt	R_6	CFFO/ASSETS
DEBTS	Total debt	R_7	CFFO/DEBTS
INC	Income	R_8	COGS/INV
INCDEP	Income plus depreciation	R_9	CURASS/CURDEBT
INV	Inventory	R_{10}	CURASS/SALES
REC	Receivables	R_{11}	CURASS/ASSETS
SALES	Sales	R_{12}	CURDEBT/DEBTS
WCFO	Working capital from operations	R_{13}	INC/SALES
		R_{14}	INC/ASSETS
		R_{15}	INC/DEBTS
		R_{16}	INCDEP/SALES
		R_{17}	INCDEP/ASSETS
		R_{18}	INCDEP/DEBTS
		R_{19}	SALES/REC
		R_{20}	SALES/ASSETS
		R_{21}	ASSETS/DEBTS
		R_{22}	WCFO/SALES
		R_{23}	WCFO/ASSETS
		R_{24}	WCFO/DEBTS

Assignment

1. What data mining technique(s) would be appropriate in assessing whether there are groups of variables that convey the same information and how important that information is? Conduct such an analysis.

2. Comment on the distinct goals of profiling the characteristics of bankrupt firms versus simply predicting (black-box style) whether a firm will go bankrupt and whether both goals, or only one, might be useful. Also comment on the classification methods that would be appropriate in each circumstance.

3. Explore the data to gain a preliminary understanding of which variables might be important in distinguishing bankrupt from nonbankrupt firms.

4. Using your choice of classifiers, use JMP Pro to produce several models to predict whether or not a firm goes bankrupt, assessing model performance on a validation partition.

5. Based on the above, comment on which variables are important in classification, and discuss their effect.

18.9 TIME SERIES CASE: FORECASTING PUBLIC TRANSPORTATION DEMAND

`bicup2006.xls` is the dataset for this case study.

Background

Forecasting transportation demand is important for multiple purposes such as staffing, planning, and inventory control. The public transportation system in Santiago de Chile has gone through a major effort of reconstruction. In this context, a business intelligence competition took place in October 2006 that focused on forecasting demand for public transportation. This case is based on the competition, with some modifications.

Problem Description

A public transportation company is expecting an increase demand for its services and is planning to acquire new buses and to extend its terminals. These investments require a reliable forecast of future demand. To create such forecasts, one can use data on historic demand. The company's data warehouse has data on each 15-minute interval between 6:30 AM and 22:00, on the number of passengers arriving at the terminal. As a forecasting consultant you have been asked to create a forecasting method that can generate forecasts for the number of passengers arriving at the terminal.

Available Data

Part of the historic information is available in the file `bicup2006.xls`. The file contains the worksheet "Historic Information" with known demand for a 3-week period, separated into 15-minute intervals. The second worksheet ("Future") contains dates and times for a future 3-day period, for which forecasts should be generated (as part of the 2006 competition).

Import the data from the two worksheets into one JMP data table using the JMP *Excel Wizard*.

Assignment Goal

Your goal is to create a model/method that produces accurate forecasts. To evaluate your accuracy, partition the given historic data into two periods: a training period (the first two weeks) and a validation period (the last week). Models should be fitted only to the training data and evaluated on the validation data.

Although the competition winning criterion was the lowest mean absolute error (MAE) on the future 3-day data, this is *not* the goal for this assignment. Instead, if we consider a more realistic business context, our goal is to create a model that generates reasonably good forecasts on any time/day of the week. Consider not only predictive metrics such as MAE and MAPE, but also look at actual and forecasted values, overlaid on a time plot.

Assignment

For your final model, present the following summary:

1. Name of the method/combination of methods.
2. A brief description of the method/combination.
3. All estimated equations associated with constructing forecasts from this method.
4. Two of the following statistics: MAPE, MAE, average error, and RMSE (or RASE) for the training period and the validation period.
5. Forecasts for the future period (March 22–24), in 15-min bins.
6. A single chart showing the fit of the final version of the model to the entire period (including training, validation, and future). Note that this model should be fitted using the combined training and validation data.

Tips and Suggested Steps

1. Use exploratory analysis to identify the components of this time series. Is there a trend? Is there seasonality? If so, how many "seasons" are there? Are there any other visible patterns? Are the patterns global (the same throughout the series) or local?
2. Consider the frequency of the data from a practical and technical point of view. What are some options?
3. Compare the weekdays and weekends. How do they differ? Consider how these differences can be captured by different methods.
4. Examine the series for missing values or unusual values. Think of solutions.
5. Based on the patterns that you found in the data, which models or methods should be considered?

REFERENCES

Agrawal, R., Imielinski, T., and Swami, A. (1993). "Mining associations between sets of items in massive databases." In *Proceedings of the 1993 ACM-SIGMOD International Conference on Management of Data* (pp. 207–216). New York: ACM Press.

Berry, M. J. A., and Linoff, G. S. (1997). *Data Mining Techniques*. New York: John Wiley & Sons, Inc.

Berry, M. J. A., and Linoff, G. S. (2000). *Mastering Data Mining*. New York: John Wiley & Sons, Inc.

Breiman, L., Friedman, J., Olshen, R., and Stone, C. (1984). *Classification and Regression Trees*. Boca Raton, FL: Chapman & Hall/CRC (orig. published by Wadsworth).

Chatfield, C. (2003). *The Analysis of Time Series: An Introduction*. 6th ed. Boca Raton, FL: Chapman & Hall/CRC.

Delmaster, R., and Hancock, M. (2001). *Data Mining Explained*. Boston: Digital Press.

Efron, B. (1975). The Efficiency of Logistic Regression Compared to Normal Discriminant Analysis. *Journal of the American Statistical Association*, **70**(352), pp. 892–898.

Few, S. (2004). *Show Me the Numbers*. Burlingame, CA: Analytics Press.

Few, S. (2009). *Now You See It*. Burlingame, CA: Analytics Press.

Han, J., and Kamber, M. (2001). *Data Mining: Concepts and Techniques*. San Diego, CA: Academic Press.

Hand, D., Mannila, H. and Smyth, P. (2001). *Principles of Data Mining*. Cambridge, MA: MIT Press.

Hastie, T., Tibshirani, R., and Friedman, J. (2001). *The Elements of Statistical Learning*. New York: Springer.

Hosmer, D. W., and Lemeshow, S. (2000). *Applied Logistic Regression*, 2nd ed. New York: Wiley-Interscience.

Hyndman, R., Koehler, A. B., Ord, J.K., and Snyder, J.K. (2008). *Forecasting with Exponential Smoothing: The State Space Approach*. New York: Springer Science & Business Media.

Jank, W., and Yahav, I. (2010). E-Loyalty Networks in Online Auctions. *Annals of Applied Statistics*, **4**(1), pp. 151–178.

Johnson, W., and Wichern, D. (2002). *Applied Multivariate Statistics*. Upper Saddle River, NJ: Prentice Hall.

Larsen, K. (2005). "Generalized naive Bayes classifiers," *SIGKDD Explorations*, **7**(1), pp. 76–81.

Lyman, P., and Varian, H. R. (2003). "How much information." Retrieved from http://www.sims.berkeley.edu/how-much-info-2003on Nov. 29, 2005.

Manski, C. F. (1977). "The structure of random utility models," *Theory and Decision*, **8**, pp. 229–254.

McCullugh, C. E., Paal, B., and Ashdown, S. P. (1998). "An optimisation approach to apparel sizing." *Journal of the Operational Research Society,* **49**(5), pp. 492–499.

Pregibon, D. (1999). "2001: A statistical odyssey." Invited talk at The Fifth ACM SIGKDD International Conference on Knowledge Discovery and Data Mining, ACM Press, NY, p. 4.

Sall, John (2002). "Monte Carlo calibration of distributions of partition statistics." *White Paper* retreived from http://www.jmp.com/content/dam/jmp/documents/en/white-papers/montecarlocal.pdf.
SAS Institute Inc. 2015. Discovering JMP® 12. Cary, NC: SAS Institute Inc.
SAS Institute Inc. 2015. JMP® 12 Fitting Linear Models. Cary, NC: SAS Institute Inc.
SAS Institute Inc. 2015. JMP® 12 Specialized Models. Cary, NC: SAS Institute Inc.

Shmueli, G. (2012). *Practical Time Series Forecasting: A Hands-On Guide*, 2nd ed. CreateSpace.

Trippi, R., and Turban, E. (eds.) (1996). *Neural Networks in Finance and Investing*. New York: McGraw-Hill.

Veenhoven, R., World Database of Happiness, Erasmus University Rotterdam. Available at http://worlddatabaseofhappiness.eur.nl.

DATA FILES USED IN THE BOOK

1. Accidents.jmp
2. Accidents NN.jmp
3. Accidents1000 DA.jmp
4. Airfares.jmp
5. Amtrak.jmp
6. ApplianceShipments.jmp
7. AustralianWines.jmp
8. Bankruptcy.jmp
9. Banks.jmp
10. Baregg Daily.jmp
11. BathSoap.jmp
12. BathSoap.xls
13. bicup2006.xls
14. BostonHousing.jmp
15. CanadianWorkHours.jmp
16. Cereals.jmp
17. CharlesBookClub.jmp
18. Colleges.jmp
19. DepartmentStoreSales.jmp
20. EastWestAirlinesCluster.jmp
21. EastWestAirlinesNN.jmp
22. eBayAuctions.jmp
23. Financial Reporting 10 Companies.jmp
24. Financial Reporting Raw.jmp

Data Mining for Business Analytics: Concepts, Techniques, and Applications with JMP Pro®, First Edition.
Galit Shmueli, Peter C. Bruce, Mia L. Stephens, and Nitin R. Patel.
© 2017 John Wiley & Sons, Inc. Published 2017 by John Wiley & Sons, Inc.

25. FlightDelays.jmp
26. Flight Delays LR.jmp
27. FlightDelays NB Train.jmp
28. Fundraising.jmp
29. FutureFundraising.jmp
30. GermanCredit.jmp
31. LaptopSales.txt
32. LaptopSalesJanuary2008.jmp
33. Natural Gas Sales.jmp
34. Pharmaceuticals.jmp
35. RidingMowers.jmp
36. RidingMowerskNN.jmp
37. Sept11Travel.jmp
38. ShampooSales.jmp
39. SouvenirSales.jmp
40. SP500.jmp
41. Spambase.jmp
42. SystemAdministrators.jmp
43. Taxi-cancellation-case.xlsx
44. TaykoAllData.jmp
45. TaykoPurchasersOnly.jmp
46. Tiny Dataset.jmp
47. ToyotaCorolla.jmp
48. ToyotaCorolla1000.jmp
49. ToysRUsRevenues.jmp
50. UniversalBank.jmp
51. UniversalBankCT.jmp
52. Utilities.jmp
53. VoterPersuasion.jmp
54. WalMartStock.jmp
55. West Roxbury Housing.jmp
56. Wine.jmp

INDEX

A-B tests, 290
accident data
   discriminant analysis, 278
   naive Bayes, 181
   neural nets, 257
activation function, 248
additive seasonality, 354
adjusted-R^2, 144
affinity analysis, 15
agglomerative, 304, 311
agglomerative algorithm, 311
aggregation, 55, 65, 70, 72
AIC_c, 144
airfare data
   multiple linear regression, 152
Akaike Information Criterion, 144
algorithm, 8
all possible models, 145
ALVINN, 245
Amtrak data
   time series, 337, 346
   visualization, 53
analytics, 3

appliance shipments data
   time series, 344, 374, 394
   visualization, 79
ARIMA, 361
AR models, 361
artificial intelligence, 5, 8, 82
artificial neural networks, 245
assess variable importance, 264
association rules, 15, 17
assumptions, 235
asymmetric cost, 19, 425
asymmetric response, 425
attribute, 8
Australian wine sales data
   time series, 375
autocorrelation, 346, 356
automated procedures, 23
average linkage, 315, 319
average squared errors, 135
averaging predictions, 287

background maps, 76
back propagation, 252

Data Mining for Business Analytics: Concepts, Techniques, and Applications with JMP Pro®, First Edition.
Galit Shmueli, Peter C. Bruce, Mia L. Stephens, and Nitin R. Patel.
© 2017 John Wiley & Sons, Inc. Published 2017 by John Wiley & Sons, Inc.

backward elimination, 145, 147
bagging, 203, 288
balanced portfolios, 302
bankruptcy data, 425
bar chart, 54, 55
batch updating, 252
bath soap data, 420
Bayesian Information Criterion, 144
bias, 25, 141, 144, 276
bias-variance trade-off, 141
BIC, 144
big data, 6
binning, 71, 72
binomial distribution, 235
black-box method, 289
boosting, 203, 288
  boosted tree, 206
bootstrap, 204, 288
  forest, 204, 206
Boston housing data, 29, 82
  multiple linear regression, 150
  visualization, 52
boxplot, 51, 56, 70, 138
  side-by-side, 58
bubble plot, 64
Bureau of Transportation
    statistics, 223
business analytics, 3, 14
business intelligence, 3

C4.5, 183, 194
cab cancellations, 418
Canadian manufacturing workhours data
  time series, 344, 368
CART, 183, 194
case, 8
  deletion, 141
  updating, 252
categorical variable, 184
cell plot, 59, 60
centroid, 270, 271, 320, 321
cereals data, 90
  exploratory data analysis, 101
  hierarchical clustering, 329
CHAID, 193, 194
Charles Book Club
  data, 401
chi-square test, 194
city-block distance, 307
classification, 14, 53, 62, 71, 78,
      155, 167
  discriminant analysis, 268

logistic regression, 211
matrix, 192, 236
methods, 221
and regression trees, 100, 264
rules, 184, 196
trees, 183, 409, 413
  performance, 192
classifier, 276, 428
cleaning the data, 20
cluster analysis, 16, 17, 53, 97, 156, 301
  algorithms, 311
  average distance, 309
  average linkage, 313
  centroid distance, 309, 323
  complete linkage, 313
  dispersion, 321
  initial partition, 321
  labeling, 316
  maximum distance, 309
  minimum distance, 309
  nesting, 315
  normalizing, 305
  outliers, 320
  parallel coordinate plot, 323
  partition, 316, 322
  profile plot, 323
  single linkage, 312
  stability, 319
  summary statistics, 316
  unequal weighting, 307
  validating clusters, 316
  validity, 326
coefficients, 250
collaborative filtering, 15
collinearity, 218
color map, 59
Columns Viewer, 23
Column Switcher, 64
combining categories, 100
combining predictions, 287
comparative box plot, 58
conditional probability, 9, 167, 171
confidence interval, 8
confusion matrix, 229
consumer choice theory
  logistic regression, 212
continuous, 20
correlation, 81, 271, 307
  analysis, 87
  based similarity, 307
  matrix, 87, 97
  table, 59

correspondence analysis, 90
covariance matrix, 91, 271, 307
C_p, 144
credit card data
   neural nets, 265
credit risk score, 409
crossvalidation, 25
curse of dimensionality, 82, 164
customer segmentation, 301
cutoff, 403, 410
cutoff value, 221, 250
   classification tree, 190
   logistic regression, 213

dashboards, 3
database marketing, 402
data driven method, 245, 336, 377
data exploration, 81
data filter, 64, 67
data filtering, 16
data mining, 5
data partitioning, 192, 335
data projection, 92
data reduction, 16
data science, 7
data visualization, 16
decision-making, 212, 265
deep learning, 263
deep learning networks, 263
delayed flight data
   classification tree, 208
   logistic regression, 223
dendrogram, 301, 314
department store sales data
   time series, 344, 371, 392
dependent variable, 8
de-seasonalizing, 380
de-trending, 380
dimensionality, 81
dimension reduction, 16, 17, 81, 82,
     183, 211
discriminant analysis, 5, 156, 202,
     268
   assumptions, 275
   classification performance, 275
   confusion matrix, 276
   correlation matrix, 276
   cutoff, 273
   distance, 270
   more than two classes, 278
   multivariate normal distribution, 275
   outliers, 275

prior/future probability of membership, 276
   prior probabilities, 276
   validation set, 276
discriminant functions, 198, 272, 279
discriminators, 271
distance, 304, 319
   between clusters, 309
   matrix, 310, 311, 319
   between observations, 309
   between records, 156
distribution plots, 51, 56
domain-dependent, 306
domain knowledge, 22, 23, 82, 141, 311, 322
double exponential smoothing, 387
double (Brown) exponential smoothing, 387
dummy variables, 21, 87

East-West Airlines data
   cluster analysis, 330
   neural nets, 266
eBay auctions data
   classification tree, 207
   logistic regression, 243
effect coding, 137
efficient, 218
eigenvalues, 94
ensemble, 203, 285
   forecast, 337
entropy impurity measure, 206, 207
epoch, 252
error
   AAE, 37
   average, 138
   back propagation, 252
   MAD, 37
   MAE, 37
   RASE, 37
   rate, 195
   RSquare, 37
estimation, 8, 9
Euclidean distance, 156, 270, 305, 319
evaluating performance, 201
exhaustive search, 142
explained variability, 144
explanatory modeling, 134
exploratory analysis, 275
exponential smoothing, 377, 382
exponential trend, 68, 350
extrapolation, 264

factor analysis, 164
factor selection, 82

Fast Ward, 314
feature, 8, 9
    extraction, 82
field, 9
filtering, 52, 65, 67
finance, 302
financial applications, 268
Financial Condition of Banks data
    logistic regression, 241
first principal component, 92
fitting the best model to the data, 134
forecasting, 53, 54, 72, 335
forward selection, 145
fraudulent financial reporting data, 168
fraudulent transactions, 19
From Distances to Propensities and
        Classifications, 272
function
    nonlinear, 213
fundraising data, 423, 425

Generalized Regression, 149
German credit case, 409
Gini index, 206
global pattern, 65, 66, 78, 341
Gower, 308
graph builder, 54
graph guilder, 55, 57–59, 62, 63, 66, 68, 74,
        76, 89, 386
graphical exploration, 52

heatmap, 59, 84
hidden layer, 245
hierarchical, 304
    clustering, 301, 311
    methods, 315
hierarchies, 65
histogram, 51, 56, 72, 138, 235
holdout data, 26
holdout sample, 8
holdout set, 135
Holt-Winter's exponential smoothing, 388
home value data, 18
homoskedasticity, 135

impurity, 193
    measures, 206
imputation, 141
impute missing, 330
independent variable, 9
indicator variables, 21, 87, 137

industry analysis, 302
Informative Missing, 24
input variable, 8
integer programming, 320
interactions, 62, 246
    term, 222
interactive visualization, 51, 71
iterative search, 145

Jaquard's coefficient, 308
jittering, 70
joining tables, 177

kfold crossvalidation, 254
kitchen-sink approach, 140
k-means, 304, 320
    clustering, 301, 320, 421
k-nearest neighbor, 6, 155, 403, 405
    algorithm, 155

labels, 68
laptop sales data
    visualization, 79
large dataset, 5, 70
lazy learning, 164
learning rate, 252, 262, 265
least squares, 250
level, 335, 337
lift, 229, 415
    curve, 120–123, 222, 229, 236, 425
likelihood, 187, 207
linear, 250
    classification rule, 269
    combination, 90, 91
    regression, 5, 6, 17, 29, 100, 155, 234, 336
        models, 246
    relationship, 87, 133, 135, 141
line chart, 54
line graph, 54
linked plots, 73
loadings plot, 95
Local Data Filter, 67
local optimum, 265
local pattern, 66, 78, 341
log, 138
logistic regression, 5, 100, 211, 246, 250, 268,
        275, 403, 408, 409, 413
    classification, 217
        performance, 221
    confidence intervals, 218
    maximum likelihood, 218

model interpretation, 227
model performance, 229
negative coefficients, 219
nominal classes, 239, 240
ordinal classes, 238
parameter estimates, 218
positive coefficients, 219
prediction, 217
profiling, 235
p-value, 236
training data, 236
variable selection, 222, 230
logistic response function, 213
logit, 213, 214, 221
-LogLikelihood, 236
log-scale, 269
log transform, 251

MA, 378
machine learning, 5, 8, 20
Mahalanobis distance, 271, 307
majority class, 157
majority decision rule, 156
majority vote, 161
Make Indicator Columns, 21
Mallow's C_p, 144
Manhattan distance, 307
marketing, 192, 194, 403
market segmentation, 301
market structure analysis, 301
matching coefficient, 308
maximum coordinate distance, 307
maximum likelihood, 218, 250,
    253
Max Validation RSquare, 145
measuring impurity, 206
minority class, 264
misclassification error, 156
missing data, 17, 23
    pattern, 59
missing value coding, 24
    neural nets, 262
missing values, 23, 59, 141, 183, 331
    utility, 24, 61
mixed stepwise, 147
    regression, 145
model, 8, 19, 21, 27
    logistic regression, 213
model comparison, 264
    platform, 264
model complexity, 222
modeling type, 20

model validity, 134
mosaic plot, 55
moving average, 377, 378
    centered moving average, 378
    trailing moving average, 378, 379
multicollinearity, 87, 96, 141
multilayer feedforward networks, 246
multi-level forecaster, 337
multiple linear regression, 34, 133, 212, 222,
    236, 280, 414
    model, 136
multiple panels, 52
multiple R^2, 236
multiplicative seasonality, 356
Multivariate Methods, 84
multivariate normal imputation, 331

naive Bayes, 167
naive forecast, 341
naive model, 236
naive rule, 157
natural gas sales data
    time series, 396
natural hierarchy, 304
nearest neighbor, 156
    almost, 163
negative loglikelihood, 253
neural nets, 25, 198, 245, 409
    activation function, 248
    architecture, 246, 258
    assess variable importance, 264
    bias, 247, 248
        values, 254
    black box, 263
    boosting, 262
    categorical profiler, 264
    classification, 245, 250, 256
        matrix, 254
    cutoff, 256
    engineering applications, 245
    financial applications, 245
    hidden layer, 246, 248
    informative missing, 262
    input layer, 246
    iteratively, 251
    learning rate, 262
    logistic, 248
    maximum likelihood, 253
    neurons, 246
    nodes, 246
    output layer, 246, 249
    overfitting, 259

neural nets (*Continued*)
oversampling, 264
prediction, 245, 256
preprocessing, 251
save fast formulas, 266
sigmoidal function, 248
theta$_j$, 247
tours, 261
transfer function, 248
updating the weights, 252
user input, 260
variable selection, 264
weighted average, 249
weighted sum, 248
weights, 247
$w_{i,j}$, 247
noise, 335, 337, 340
noisy data, 264
Nominal, 20
nonhierarchical, 304
algorithms, 311
clustering, 320
nonparametric method, 155
normal distribution, 135, 235
normalize, 24, 96, 305
numerical response, 161, 198

observation, 8
odds, 213, 217
OLAP, 14
ordering of, 229
ordinal, 20
ordinary least squares, 135
orthogonally, 92
outcome variable, 9
outliers, 17, 22, 23, 32, 307
utility, 23
output variable, 9
overall fit, 236
overfitting, 5, 25, 26, 27, 81, 144, 157, 183,
193, 245
oversampling, 19, 415
oversmoothing, 157
overweight, 19

pairwise correlations, 87
Pandora, 162
parallel coordinates plot, 71
parallel plot, 318
parametric assumptions, 163
parsimony, 21, 22, 141, 232, 282
partition, 26, 226

pattern, 8, 9
performance, 134
personal loan data, 164, 180
classification tree, 192
discriminant analysis, 269, 282
logistic regression, 214
persuasion models, 290
pharmaceuticals data
cluster analysis, 329
political persuasion, 415
polynomial trend, 68, 352
predicting bankruptcy case, 425
predicting new observations, 134
prediction, 9, 15, 53, 62, 77, 198, 200
predictive analytics, 5, 14, 15
predictive modeling, 5, 134
predictive performance, 138
predictor, 9
preprocessing, 17, 20, 214, 225
principal components, 95, 149
scores, 94
weights, 97
principal components analysis, 5, 32, 90, 163,
164, 264
classification and prediction, 98
normalizing the data, 96
training data, 98
validation set, 98
weights, 96
averages, 95
probabilities
logistic regression, 213
probability plot, 138, 235
profiler, 220, 264
prediction profiler, 140, 189, 190, 201, 221,
227, 230, 234, 356
profiling, 211
discriminant analysis, 268
propensity, 159, 167
logistic regression, 213
pruning, 183, 194
Public Transportation Demand case, 428
Public Transportation Demand data, 428
public utilities data
cluster analysis, 302

quadratic discriminant analysis, 276
quadratic model, 352
Query Builder, 14

R^2, 144
random forest, 203, 288

random sampling, 17
random seed, 259
random utility theory, 212
random walk, 346, 363
ranking, 171, 420
  of records, 179
recode, 89
record, 8, 9, 23
recursive partitioning, 183–185, 194
redundancy, 91
regression, 133
  time series, 346
  trees, 149, 183, 198, 414
re-scaling, 65
residuals
  histogram, 235
  series, 359
response, 9
  rate, 19
RFM segmentation, 403
riding-mower data
  CART, 185
  discriminant analysis, 268
  k-nearest neighbor, 156
  logistic regression, 243
  visualization, 79
right-skewed, 251
  distribution, 138
robust, 218, 311, 319
  distances, 307
row, 8
row labels, 69

sample, 5, 8, 18
sampling, 19, 70
scale, 97, 305
scaling, 24
scatterplot, 54, 55, 62, 70
  animated, 64
scatterplot matrix, 63
score, 9, 403
  plot, 95, 97
scoring, 18, 38
Seasonal ARIMA, 361
seasonality, 335, 337
second principal component, 92
segmentation, 301
segmenting consumers of bath soap
    case, 420
self-proximity, 305
SEMMA, 18
sensitivity analysis, 263

separating hyper-plane, 271
separating line, 271
Sept 11 travel data
  time series, 343, 365, 390
shampoo sales data
  time series, 345, 395
similarity measures, 307
simple linear regression, 216
single linkage, 314
singular value decomposition, 164
smoothing, 157, 336
  parameter, 383, 386
  time series, 377
  weight, 383
souvenir sales data
  time series, 344, 373
spam email data
  discriminant analysis, 284
specialized visualization, 73
S&P monthly closing prices, 365
SQL, 14
squared distances, 273
standardize, 24, 96, 156, 305
statistical distance, 271, 307, 319
statistics, 5
steps in data mining, 17
stepwise, 37, 145
  in JMP, 142, 143
  regression, 145
stopping rules, 145
stopping tree growth, 193
subsets, 164
subset selection, 100, 145
  in linear regression, 140
success class, 9
summary statistics, 141
sum of squared deviations, 135, 207
sum of squared errors, 236
sum of squared perpendicular distances, 92
supervised learning, 9, 16, 32, 51, 53, 63
system administrators data
  discriminant analysis, 283
  logistic regression, 242

tabulate, 85, 87
target variable, 9
taxi cancellations, 418
Tayko data, 410
  multiple linear regression, 151
Tayko software catalog
    case, 410
test data, 18

test partition, 26
test set, 8, 9
time series
  acf plot, 359
  ARIMA models, 360
  autocorrelation plot, 359
  autoregression model, 360
  forecasting, 78, 335
  indicators, 353
  lag, 358
  lagged series, 356
  pacf plot, 359
  partial autocorrelation plot, 359
  partitioning, 341
  residuals, 348
  window width, 382, 386
total variability, 91
Toyota Corolla data, 45, 136
  classification tree, 199, 209
  multiple linear regression, 154
    best subsets, 145
    forward selection, 147
  neural nets, 266
  principal components analysis, 102
Toys R Us revenues data
  time series, 369
training, 251
  data, 16, 18
  partition, 26
  set, 9, 135, 335
transfer function, 248
transform, 25, 251
transformations, 65, 87, 246
  of variables, 202
transpose, 271
trees, 25
  depth, 193
  search, 163
trend, 65, 68, 335, 337, 350
  lines, 68, 78, 339
trial, 252
triangle inequality, 305
two model approach, 293
two-way clustering, 318

unbiased, 135, 144
underfitting, 144
Universal Bank data, 164
  classification tree, 192

discriminant analysis, 269, 282
  logistic regression, 218
university rankings data
  cluster analysis, 330
  principal components analysis, 102
unstable estimates, 228
unsupervised learning, 9, 17, 51, 53, 63,
  71, 78
UPGMC, 313
uplift modeling, 285
  UPLIFT MODELS, 290

validation data (or set), 8, 9, 18, 133, 194,
  221, 335
validation partition, 26
variables, 9
  binary dependent, 212
  categorical, 20
  continuous, 20
  nominal, 20
  numerical, 20
  ordinal, 20
  selection, 21, 133, 140
  text, 20
variation
  between-cluster, 317
  within-cluster, 317
visualization, 3
  animation, 64
  color, 62
  hue, 62
  maps, 76
  marker, 62
  multiple panels, 62, 63, 64
  networks, 73
  shape, 62, 64
  size, 62, 64
  Treemaps, 75

Walmart stock data
  time series, 369
Ward's method, 314
weight decay, 252
weighted average, 161
weighted sampling, 423
within-cluster dispersion, 323

zooming, 52, 65, 66, 72
z-score, 24, 271, 305

MPM 240817
Printed in Singapore